Foundations of
Technical Mathematics

Foundations of Technical Mathematics

Warren Donahue, Chairman
Department of Mathematics
Palomar College

Everett Robertson, Chairman
Department of Industrial Technology
Palomar College

John Wiley & Sons, Inc.

New York   London   Sydney   Toronto

Library of Congress Catalog Card Number: 75-96962
SBN 471 21774 3

Printed in the United States of America

10   9   8   7   6   5   4   3   2   1

Preface

This book is designed to introduce students with technical orientation to the basic topics in mathematics that seem to be the most important in solving practical problems. Not only is it designed to provide the mathematical tools of the beginning technician but it is also meant to serve as a foundation upon which more advanced mathematical study can be built. It is our hope that the textbook will also be useful for those engaging in self-study.

Students planning technical careers must become proficient in using arithmetic, applied geometry, fundamentals of algebra, graphing techniques, right-triangle trigonometry, and mathematical tables. We have, therefore, selected topics from these areas that technicians actually use. Emphasis has been placed upon the application of the mathematical ideas and relationships while avoiding the "cookbook" approach.

The text provides sufficient material and flexibility to allow its use in one, two, or even three semesters of study. Because of its basic nature, any section can be supplemented by the instructor if desired.

We have presented the important role that approximations play in solving technical problems and have, throughout the text, encouraged the student to make them. In a similar manner, we have attempted to present significant figures on a practical level and have given consistent attention to that topic.

In the chapter on graphing, special attention has been given to the relationship between the slope and the proportionality constant. Also, an introduction to the graphing of empirical data is presented.

Because tables are important tools of the modern technician, the student is frequently referred, both directly and indirectly, to each of the many tables contained in the text.

Frequent reference is also made to slide rule solutions, and in some sections specific slide rule procedure is given. It is assumed that the student is able to use a slide rule or is taking a slide rule course concurrently.

We wish to thank the many students who have given us considerable help in the development and testing of materials used in the text. We are also indebted to the people associated with John Wiley & Sons for their help and care in the development and production of the book.

Comments and suggestions for improvement of the book will always be appreciated.

WARREN DONAHUE and EVERETT ROBERTSON

San Marcos, California
January 1970

v

Contents

1 Review and Selected Topics

The topics discussed in this chapter are considered essential for successful work by modern technicians. Although it may seem strange, the more basic the topic, the more essential it is. A review of basic arithmetic is found in Appendix A.

1.1 POWER OF 10 FORM*

The use of the power of 10 form simplifies many calculations and approximations. When a number such as 3500 is written as 3.5×10^3, it is in the power of 10 form.* For example:

Standard Form	Power of 10 Form
782,000	7.82×10^5
0.00238	2.38×10^{-3}

Writing a Number in Power of 10 Form

To write a number in the power of 10 form:

1. Move the decimal point (to the right or left as needed) to form a number between 1 and 10.
2. Count the number of places (digits) between the original and new position of the decimal point. (The count can take place while performing Step 1.)
3. Show the number obtained in Step 1 multiplied by 10, raised to the power obtained by the count. The count obtained in Step 2 is positive $(+)$ when the decimal point is moved to the left, and it is negative $(-)$ when the decimal point is moved to the right.

examples

$$0.00045 = 4.5 \times 10^{-4} \quad \text{(Count is } -4.)$$
$$0.012 = 1.2 \times 10^{-2} \quad \text{(Count is } -2.)$$
$$8.3 = 8.3 \times 10^0 \quad \text{(Count is zero.)}$$
$$982,000 = 9.82 \times 10^5 \quad \text{(Count is } +5.)$$

There are three occasions in which we may alter the scientific form.

* The "power of 10 form" is frequently called the "scientific form."

1. To extract a square root of the number. Thus, 3.6×10^5 would be written as 36×10^4 to make the exponent of 10 divisible by the index of the root: i.e., by 2.

$$\sqrt{3.6 \times 10^5} = \sqrt{36 \times 10^4} = 6 \times 10^2$$

2. To make all exponents equal when adding or subtracting numbers written in the power of 10 form:

$$3.20 \times 10^5 + 4.80 \times 10^6 = 3.20 \times 10^5 + 48.0 \times 10^5$$
$$= 51.2 \times 10^5 = 5.12 \times 10^6$$

3. To facilitate the simplification of fractions by cancellation:

$$\frac{5.4 \times 10^3 \times 8.4 \times 10^2}{7.0 \times 10 \times 7.2 \times 10^3} = \frac{\overset{9}{\cancel{54}} \times \overset{1}{\cancel{10^2}} \times \overset{\overset{1}{\cancel{7}}}{\cancel{84}} \times \cancel{10}}{\underset{\underset{\cancel{10}}{}}{\cancel{70}} \times \underset{\cancel{6}}{\cancel{72}} \times \cancel{10^2}} = 9$$

Changing from Power of 10 Form to Standard Form

To change from the power of 10 form to standard form, the indicated multiplication is carried out.

examples

$3.5 \times 10^4 = 35,000$	(Moving decimal point four places to the right multiplies by 10,000.)
$7.27 \times 10^{-3} = 0.00727$	(Moving decimal point three places to the left divides by 1000.)
$8.8 \times 10^0 = 8.8$	(Decimal point is not moved.)

EXERCISE 1.1

Write Problems 1–10 in standard form.

1. 7.5×10^2
2. 3.85×10^3
3. 9.05×10^4
4. 7.5×10^6

5. 3.11×10^0
6. 9.1×10^{-1}
7. 3.5×10^{-3}

8. 4.21×10^{-2}
9. 1.07×10^{-3}
10. 3.925×10^0

Write Problems 11–30 in correct power of 10 form.

11. 0.0033
12. 125,000
13. 0.0123
14. 529,000
15. 0.000075
16. 0.019
17. 175,000,000

18. 7.51
19. 0.01234
20. 98,000,000,000
21. 88.6×10^5
22. 0.395×100
23. 78.6×10^3
24. 0.9×1000

25. 8.9×10
26. 72×10^{-1}
27. $58,400 \div 100$
28. 297×100
29. 31.5×10^{-3}
30. 0.075×10^3

In Problems 31–34, add the numbers.

31. $5.27 \times 10^{-4} + 3.0 \times 10^{-2}$
32. $4.15 \times 10^3 + 3.6 \times 10^2$
33. $8.00 \times 10^4 + 3.80 \times 10^5 + 9.00 \times 10^3$
34. $1.205 \times 10^{-3} + 8.7103 \times 10^{-2} + 6.09 \times 10^{-4}$

1.2 SIGNIFICANT FIGURES

Exact and Approximate Numbers

It is common practice to refer to some numbers as exact numbers and to others as approximate numbers. In fact, *all numbers are exact.* Consider the numbers $\frac{7}{9}$ and 0.778. Both numbers are exact. However, each is only approximately equal to the other; i.e., $\frac{7}{9} \approx 0.778$. The numbers $\sqrt{2}$ and 1.414 are both exact numbers, but each is only approximately equal to the other ($\sqrt{2} \approx 1.414$).

An *exact number* is a number that exactly equals the quantity being expressed. An *approximate number* is a number that approximately equals the quantity being expressed.

example A certain engine has 8 cylinders and develops 325 horsepower. The 8 is an exact number because the number of cylinders is exactly 8. However, the 325 is an approximate number because there is no way to determine the exact horsepower of the engine.

The three main reasons for using approximate numbers are:

1. Most numbers are approximate numbers because the exact quantity cannot be determined.
2. The exact numerical representation of a quantity may not be known, although it could be determined.
3. Use of approximate numbers in computations saves time and energy. It is practical whenever the results meet the restrictions dictated by the problem.

You can usually tell from the way a number is used whether it is exact or approximate. When it is impossible to determine how a number is being used, assume that the number is approximate.

EXERCISE 1.2A

For each of the numbers in the following statements, write the letter E if the number is exact or the letter A if the number is approximate.

1. A 6-cylinder engine weighs only 250 pounds (lb).
2. A hexagonal bolt head has 6 sides.

3. Under standard conditions, water boils at 100°C.
4. The thermostat opens at a temperature of 180°F.
5. All 8 spark plugs were set at the factory with a 0.035 inch (in.) gap.
6. A transformer contains 52 laminated sheets of steel which have a combined thickness of 1.25 in.
7. Seven bricks will "lay up" 1 square foot (sq ft) of wall surface.
8. The stroke of a 4-cycle engine is 3.5 in.
9. He bought a 35-piece tool set for $19.28.
10. A triangular-shaped plate has 3 sides and a perimeter of 28 in.
11. The integer 12 times the integer 13 equals 156.
12. A rectangle that measures 12 in. by 13 in. has an area of 156 sq in.
13. 28 grams (g).
14. 1 gallon (gal).
15. 1.25.
16. $3.38 is the cost of 4 tickets.
17. This is Problem 17.
18. A 1-quart (qt) can of oil contains 1 qt.
19. The typewriter has 42 keys and weighs 26 lb.
20. 3 miles per hour (mph).
21. The sum of the 3 angles of any triangle is 180°.
22. He measured the 3 angles of a triangle and found their sum to be 180°.
23. The score for the football game was 43 to 7.
24. Yards gained in rushing was 165.
25. Each team received 5 penalties.
26. It takes 3 in. to write a word with 22 letters.
27. The clock struck 9 times.
28. The clock was 6 minutes (min) fast.
29. The spark was advanced 6°.
30. The bolt has a tensile strength of 2500 lb.
31. 1 ton = 2000 lb.
32. John's foot measured 12 in.
33. $37 \times 48 = 1776$.
34. 36 in. $\times$ 36 in. = 1296 sq in.
35. 48 lb of lead.
36. 48 1-lb packages of lead.
37. 1 package of lead weighing 48 lb.
38. The solder tests 50-percent (%) tin and 50% lead.
39. The solution freezes at -32°F.
40. 75,000 people attended the game.

What are Significant Figures

Significant figures are those digits in a number that are known to be reliable, i.e., known to be correct. The last significant figure (on the right) is sometimes said to be "questionable," because it is frequently obtained by estimation or rounding off. For example, a slide rule reading of 413 has three significant figures. The 4 and the 1 are actually read from the scale, but the 3 is estimated.

The number π is an exact number which we frequently approximate as 3.14, 3.142, 3.1416, 3.14159, or 3.141593. Of these five approximations, the last one comes closest to equaling π. Each of these approximations contains a different number of significant digits. The first number has three significant digits; the second has four; the third has five; the fourth has six; and the last has seven. Consider the two approximate numbers: 120 and 3.05. Each has three digits, but the first has two significant figures and the second has three.

The following digits in a number *are* significant:

1. All nonzero digits.
2. All zeros that lie between significant digits.
3. All zeros at the end of a decimal.

The following digits in a number *are not* significant:

1. All zeros at the beginning of a decimal.
2. Zeros at the end of a whole number, unless their significance is noted by other means.*

examples
1. 29.05 is correct to four significant figures.
2. 0.005 is correct to one significant figure.
3. 1.0050 is correct to five significant figures.
4. 280,000 is correct to two significant figures.
5. 23.5000 is correct to six significant figures.
6. 1,280,0õ0 is correct to six significant figures.

Rounding Off Numbers

It is frequently necessary to round off a result because the last digits are doubtful or not needed. To round off a number, determine the position of the last significant digit to be retained. If the digits to the right of this position represent more than half a unit in the last place to be retained, add 1 to the last retained digit. If they represent less than half a unit in the last place retained, drop them. If they represent exactly half a unit in the last place retained, leave the last retained digit even by adding 1 to this digit if it is odd originally. If the last retained digit is already even, nothing is added to it. Digits dropped from a whole number are replaced with zeros. (See table of examples.)

---

* The tilde "~" may be written over the last zero known to be significant, or the number may be written in the power of 10 form retaining the zeros known to be significant; for example: 62,0õ0 = 6.200×10^4. (Both numbers show four significant figures.)

Examples of Rounding Off Numbers

Number	Significant Figures	Rounded Off to Four Figures	Rounded Off to Three Figures	Rounded Off to Two Figures
28.0078	6	28.01	28.0	28
0.89004	5	0.8900	0.890	0.89
127,600	6	127,600	128,000	130,000
44,492	5	44,490	44,500	44,000
1.2000×10^4	5	1.200×10^4	1.20×10^4	1.2×10^4
175.45	5	175.4	175	180
9.8750	5	9.875	9.88	9.9

In addition and subtraction problems, the answer is rounded off by eliminating those digits resulting from operations on broken columns on the right.

examples

Add: 28.29
3.451
76.4
9.02

117.161 rounded to 117.2   (Last unbroken column is tenths.)

Subtract: 297.312
42.85

254.462 rounded to 254.46   (Last unbroken column is hundredths.)

In multiplication and division problems, the answer is rounded off to the number of significant figures contained in the least accurate quantity involved. In any set of factors or numbers, the approximate number with the fewest significant figures is called the least accurate factor or number. Exact numbers have an unlimited number of significant figures.

examples

$$29.05 \times 3.1416 \times 0.92 = 83.96240160$$

This number is rounded off to 84 because the least accurate factor (0.92) has only two significant figures.

$$27.288 \div 12.0 = 2.274$$

This number is rounded off to 2.27 because the least accurate factor (12.0) has three significant figures.

The Last Retained Significant Digit

It is important to consider the information obtained by an analysis of the last retained significant digit of an approximate number. Suppose you have a steel rod with a length of H units. You measure the rod with a micrometer and obtain a reading of 0.8736 in. Assume that you only need the length to the closest 0.001 in. You would round off your reading to 0.874 in. and write $H = 0.874$ in. Suppose the information $H = 0.874$ in. was turned over to a technician. What would this mean to him? It should mean that the actual length H is *between* 0.8735 in. and 0.8745 in.; i.e., 0.8735 in. $< H < 0.8745$ in.

example   The diameter D of a cylinder measures 3.4987 in. Using inequality symbols, state the mathematical meaning of $D = 3.4987$ in. and then write this meaning in sentence form.

$$3.49865 \text{ in.} < D < 3.49875 \text{ in.}$$

The diameter D is between 3.49865 in. and 3.49875 in.

Significant Figures in Powers and Roots

The power or root of an approximate number has only as many significant figures as the number itself. Powers or roots of exact numbers have an unlimited number of significant figures.

examples

$$\sqrt{625} = 25.0 \qquad \text{(three figures)}$$

$$(5.21)^3 = 141.420761 = 141 \qquad \text{(three figures)}$$

EXERCISES 1.2B

Make a copy of the following table and complete as shown in problem 0.

	Approximate Number	Significant Figures	Rounded off to Four Figures	Rounded off to Three Figures	Rounded off to Two Figures	Rounded off to One Figure
0.	729.459	6	729.5	729	730	700
1.	0.44549					
2.	127,350					
3.	0.005180					
4.	1.0000					
5.	4.49					
6.	0.0056					
7.	3,485,000					
8.	9.2148					
9.	78.0273					
10.	80.045					
11.	0.003195					
12.	590.03					
13.	12,000					
14.	1,400,000					
15.	9.200×10^4					
16.	3.002×10					

In Problems 17–30, round off the given answers to the proper number of significant figures. (*Assume all numbers to be approximate.*)

17. $8.05 \times 3.13 = 25.1965$
18. $4.19 \times 0.03 = 0.1257$
19. $65.457 \times 3.15 = 206.18955$
20. $45,000 \div 1.500 = 30,000$
21. $3.1416 \times 2.08 \times 120 = 784.143360$
22. $\dfrac{51.25 \times 4.87}{0.95} = 262.7237$

23. $\dfrac{58.035 \times 4.002}{3.1416} = 73.92923$

24. $\dfrac{152 \times 3.8}{4.65 \times 2.95} = 42.1068$

25. 312.07
     48.295
     176.8705
     3.24
     92.602
     ────────
     633.0775

26.   895.203
     − 127.62
     ────────
       767.583

27. 0.00718
     0.0020
     0.01215
     0.810
     0.00035
     ────────
     0.83168

28.   26.0
     − 19.185
     ────────
       6.815

29.   140.00
     − 86.003
     ────────
       53.997

30. 523.000
     49.020
     7.810
     56.200
     149.003
     ────────
     785.033

In Problems 31–34, calculate the required answer and round off to the proper number of significant figures. (*Note that some data are exact.*)

31. What is the combined thickness of 7 shelves, each of which measure 0.78 in. in thickness?
32. Compute the area of a rectangle that measures 28.5 centimeters (cm) by 6.25 cm.
33. Calculate the volume of a rectangular solid with dimensions of 12.50 cm × 16.2 cm × 4.1 cm.
34. Compute the average of the following measurements: 5.29 in., 7.02 in., and 8.1 in.

In Problems 35–44, using inequality symbols, write each number N in the appropriate interval determined by the given approximate number. Then write this same statement in words. For example:

$$N = 5.072 \text{ cm}$$

$$5.0715 \text{ cm} < N < 5.0725 \text{ cm}$$

N is between 5.0715 cm and 5.0725 cm

35. $N = 0.0521$ g
36. $N = 4.0020°$C.
37. $N = 5.6$ in.
38. $N = 82$ ft
39. $N = 526.2$ lb
40. $N = 51.04$ milliliters (ml)
41. $N = 6.067$ in.
42. $N = 8.31$ g
43. $N = 0.125$ millimeters (mm)
44. $N = 5.6$ days

In Problems 45–48, round off the given power or root to the proper number of significant figures.

45. $\sqrt{281} = 16.76305$ 47. $(68)^3 = 314,432$
46. $\sqrt[3]{464} = 7.741753$ 48. $(169)^2 = 28,561$

Sometimes $\frac{22}{7}$ is used as an approximation for π. In Problems 49–54, if $\frac{22}{7}$ is suitable to use in the calculations involving π, write yes for the answer. If $\frac{22}{7}$ is not suitable, give the least accurate approximation for π which would be appropriate ($\pi \approx 3.1415927$), ($\frac{22}{7} \approx 3.142857$).

49. Finding the circumference C of a circle whose diameter d measures 12 in. ($C = \pi d$).
50. Finding the diameter d of a cylinder whose circumference is 28.7 in. ($d = C/\pi$).
51. Finding the cross-sectional area A of a wire whose diameter d measures 1.125 mm ($A = \frac{1}{4}\pi d^2$).
52. Finding the volume V of a cone, $V = \pi r^2 h/3$, if $r = 2.285$ in. and $h = 5.22$ in.
53. Finding the surface area S of a right circular cylinder, $S = 2\pi rh$, if $r = 5.055$ ft and $h = 10.00$ ft.
54. Using $C = \pi d$, when $d = 15.0236$ mm.

1.3   PERCENT AND INTEREST

Strange as it may seem, much of the difficulty students have with percentage problems is due to the fact that they do not know what the word *percent* means. Neither do they know the mathematical meaning of expressions such as 25%, 6%, and 0.5%.

Very simply stated, the word percent means "in each 100" or "per 100." For example, 25 percent means 25 in each 100 or 25 per 100. The statement: "a company laid off 25 percent of its employees," means that 25 out of each 100 employees (or an equivalent ratio) were laid off.

The Percent Sign

The symbol % is called a percent sign. It is used instead of the word percent. Therefore, 25 percent can be written as 25%.

examples

1. 7% means 7 in each 100 or 7 per 100.
2. 41% means 41 in each 100 or 41 per 100.
3. $3\frac{1}{2}$% means $3\frac{1}{2}$ in each 100 or $3\frac{1}{2}$ per 100.
4. 0.3% means 0.3 in each 100 or 0.3 per 100.
5. 100% means 100 in each 100 or 100 per 100.

The Word Per

The word *per* is used so frequently in mathematics that a symbol was developed to take its place. This symbol is the slanted vinculum line, /. Thus, the phrase 25¢

per pound is written 25¢/lb; 55 miles per hour is written 55 miles/hour. In a similar manner, if we replace the word per in the previous examples, we find that:

1. 7% means $7/100$.
2. 41% means $41/100$.
3. $3\frac{1}{2}\%$ means $3\frac{1}{2}/100$.
4. 0.3% means $0.3/100$.
5. 100% means $100/100$.

Observe that all of the above fractions have the same denominator. The denominator is designated by the meaning of the word percent. Percent does not mean in each (or per) 10 or in each (or per) 1000. *It means in each (or per) 100 and nothing else.*

Meaning of 100%

Most of us know that a 100% score on a test means a perfect score. But 100% also has a very important mathematical meaning: 100% means $\frac{100}{100}$ or "ONE". So 100% times some number means 1 times that number. We all know that 1 times a number equals that same number. For example:

$$7 \times 100\% \text{ means } 7 \times 1 = 7$$
$$0.72 \times 100\% \text{ means } 0.72 \times 1 = 0.72$$
$$\tfrac{3}{5} \times 100\% \text{ means } \tfrac{3}{5} \times 1 = \tfrac{3}{5}$$

Before we see why this point is so important, let us be sure that we agree with each other. We must agree that multiplying a number by 100% (or by 1) gives an answer that is equal to the original number. This leads us to a simple method of changing a fraction (decimal or common) to its equivalent as a percent. In fact, this method can be used to change *any number* to its equivalent as a percent.

Changing a Number to a Percent

You can change any whole number, mixed number, common fraction, or decimal fraction to its percentage equivalent by multiplying the number by 100%.

examples

1. Change $\tfrac{3}{8}$ to a percent:

$$\tfrac{3}{8} \times 100\% = \tfrac{300}{8}\% = 37\tfrac{1}{2}\%$$

2. Change 7.5 to a percent:

$$7.5 \times 100\% = 750\%$$

3. Change 0.352 to a percent:

$$0.352 \times 100\% = 35.2\%$$

4. Change $3\frac{5}{8}$ to a percent ($3\frac{5}{8} = \frac{29}{8}$):

$$\frac{29}{8} \times 100\% = \frac{2900}{8}\% = 362\frac{1}{2}\%$$

Changing Percents to Common or Decimal Fraction Equivalents

To change a percent to its fractional equivalent, divide the given percent by 100%.

examples

1. Change 59% to a common fraction:

$$59\% \div 100\% = \frac{59\,\%^*}{100\,\%} = \frac{59}{100}$$

2. Change 0.075% to a common fraction and then to a decimal fraction:

$$0.075\% \div 100\% = \frac{0.075\,\%^*}{100\,\%} = \frac{75}{100,000} = 0.00075$$

3. Change 525% to decimal form:

$$525\% \div 100\% = \frac{525\,\%^*}{100\,\%} = \frac{525}{100} = 5.25$$

EXERCISE 1.3A

Change the numbers in Problems 1–20 to their equivalent percents by multiplying by 100% and simplifying when necessary.

1.	0.5	8.	2.05	15.	0.002
2.	0.75	9.	$3\frac{1}{4}$	16.	1/200
3.	1.5	10.	$\frac{5}{4}$	17.	1/25
4.	0.05	11.	0.625	18.	15.25
5.	$\frac{3}{5}$	12.	$\frac{4}{5}$	19.	0.0003
6.	$\frac{7}{8}$	13.	$3\frac{3}{5}$	20.	$\frac{21}{20}$
7.	$\frac{9}{10}$	14.	7.015		

Change the percents in Problems 21–40 to their common fraction and decimal fraction equivalents by dividing each by 100%.

21.	52%	28.	1%	35.	$22\frac{1}{2}\%$
22.	3.5%	29.	0.5%	36.	$33\frac{1}{3}\%$
23.	100%	30.	2.5%	37.	$\frac{7}{8}\%$
24.	50%	31.	0.03%	38.	$\frac{4}{5}\%$
25.	10%	32.	452%	39.	$\frac{5}{12}\%$
26.	2%	33.	1000%	40.	0.05%
27.	350%	34.	$\frac{1}{2}\%$		

* The $\%$ signs are cancelled the same as any other common factor.

Finding a Specified Percent of a Given Quantity

When you are asked to find a specific percent of a number, you are really asked to find a fractional part of the number. To do this, you only need to change the given percent to a fraction (common or decimal) and multiply by the fractional equivalent.

example   Find 25% of 80.

1. Change 25% to its fractional equivalent by dividing by 100%:

$$\frac{25\%}{100\%} = \tfrac{1}{4} \quad \text{or} \quad 0.25$$

2. Multiply the 80 by $\tfrac{1}{4}$ (or 0.25):

$$80 \times \tfrac{1}{4} = 20$$

or

$$0.25 \times 80 = 20$$

example   Find 3.4% of 150.

1. Divide 3.4% by 100%:

$$\frac{3.4\%}{100\%} = 0.034$$

2. Multiply 150 by 0.034, obtaining 5.1.

example   Find 250% of 88.

1. Change 250% to its fractional equivalent by dividing by 100%:

$$\frac{250\%}{100\%} = 2.5$$

2. Multiply 88 by 2.5, obtaining 220.

EXERCISE 1.3B

Calculate the following problems, assuming the given percents to be exact. Round off answers to proper number of significant figures.

1. Find 8% of 40.0 gal.
2. Find 23% of 500.0 tons.
3. Find 6.5% of $800.
4. Find 0.3% of $280.
5. Find 0.01% of 22 lb.
6. Find 150% of 24 liters (ℓ).
7. Find 6850% of 27.5 mm.
8. Find 500% of 41.0 volts (v).
9. Find $\tfrac{1}{2}$% of 200 ohms ($\tfrac{1}{2}$% = 0.5%).
10. Find $\tfrac{3}{4}$% of 60 ohms ($\tfrac{3}{4}$% = 0.75%).
11. 12% of 800 = ?
12. 35% of 28 = ?
13. 4.5% of $50.00 = ?
14. 11.5% of 122 = ?
15. 800% of 72.0 = ?
16. Find 125% of 40.0 lb/sq in.
17. Find 0.3% of 400.0 g/sq cm.
18. Find 5.5% of 870 sq ft.
19. Find 0.02% of 450 acre-feet (acre-ft).
20. Find $\tfrac{1}{2}$% of 560 horsepower.

Expressing One Number as a Percent of Another

When asked to determine what percent one number is of another, you are actually being asked to determine what fractional part the first number is of the second number. This type of problem can be solved in two steps:

1. Write the two numbers involved as a ratio in fraction form with the first number in the numerator.
2. Multiply this fraction by 100% to change it to its percentage equivalent.

example   12 g is what percent of 48 g?

1. To compare (as a special fraction) 12 g with 48 g, write the fraction

$$\frac{12 \text{ g}}{48 \text{ g}}$$

2. Multiply by 100%:

$$\frac{12}{48} \times 100\% = \frac{1200\%}{48} = 25\%$$

example   15 ft is what percent of 150 ft?

$$15 \text{ ft is } \frac{15 \text{ ft}}{150 \text{ ft}} \text{ of } 150 \text{ ft}$$

$$\frac{15}{150} \times 100\% = 10\%$$

example   8 volts is what percent of 20 volts?

$$8 \text{ volts is } \frac{8 \text{ volts}}{20 \text{ volts}} \text{ of } 20 \text{ volts}$$

$$\frac{8}{20} \times 100\% = 40\%$$

EXERCISE 1.3C

In the following problems, compare the two numbers as a percent. (Assume that the given numbers are exact.)

1. 12 gal is what percent of 40 gal?
2. 10 ft is what percent of 500 ft?
3. 6 in. is what percent of 80 in.?
4. 0.2 yard (yd) is what percent of 40 yd? ($0.2 \text{ yd}/40 \text{ yd} = \frac{2}{400}$, change to percent).
5. 1.6 ohms is what percent of 160 ohms?
6. 10¢ is what percent of $5.00? (Be sure that the units are the same before writing the fractional part.)

7.   10 in. is what percent of 40 ft? (Round off to nearest 0.01%.)
8.   5 cm is what percent of 70 cm? (Round off to nearest 0.01%.)
9.   9 lb/sq in. is what percent of 6 lb/sq in.?
10.   15 tons is what percent of 8 tons?
11.   5 is what percent of 25?
12.   8 revolutions per minute (rpm) is what percent of 32 rpm?
13.   3° is what percent of 30°?
14.   0.05 kilometer (km) is what percent of 8 km?
15.   125 acres is what percent of 250 acres?
16.   30 in. is what percent of 5 ft?
17.   What percent of $40 is 25¢?
18.   9 seconds (sec) is what percent of 130 sec? (Round off to nearest 0.01%.)
19.   24 days is what percent of 3 days?
20.   What percent of 80 is 5? (Correct to nearest 0.1%.)
21.   12 is what percent of 12?
22.   500 is what percent of 5?

Determining a Number When a Percentage is Given

There are two key words in percentage problems which give considerable aid in solving the problem. One is the word "is" and the other is the word "of." If you replace the word "is" with an equal sign, =, and the word "of" with a multiplication sign, ×, the problem is greatly simplified.

example   5% of what number is 12? Replacing the word "of" with × and the word "is" with =, you have 5% × (what number) = 12. There is a simple method you can use to arrive at the next step. Write 7 × 8 = 56 below the previous statement:

$$.05\% \times (\text{what number}) = 12$$

$$7 \times 8 = 56$$

Now cover up the 8. To obtain this missing factor, you see that you must divide the product (56) by the given factor (7). If you apply the same reasoning to the percentage problem, and divide the product (12) by the given factor (5%), you obtain the needed factor:

$$12 \div 5\% = 12 \div 0.05 = 240$$

example   15 is 6% of what number?
1. Write

$$15 = 6\% \times (\text{what number})$$

2. Divide the product 15 by the factor 6%:

$$15 \div 0.06 = 250$$

example   40% of what number is 6?

　　1.　Write

$$40\% \times \text{(what number)} = 6$$

　　2.　Divide the product 6 by the given factor 40%:

$$6 \div 0.40 = 15$$

EXERCISE 1.3D

Assume that all given data are approximate and round off answers to proper number of significant digits.

　1.　34.2 is 5.0% of what number?
　2.　7.568 is 22.0% of what number?
　3.　0.412 is 40.0% of what number?
　4.　4.059 is 33% of what number?
　5.　0.0042 g is 12% of what amount?
　6.　93 lb is 1.5% of what amount?
　7.　4.8 tons is 1% of what amount? (Think about this one and write the answer.)
　8.　1.815 volts is 0.3% of what amount?
　9.　22.2% of what number is 198.69?
10.　4õ0% of what number is 18?
11.　8 is 2% of what number?
12.　7% of what number = 140?
13.　2.5% of what number = 0.350?
14.　8.0 gal is 0.50% of what amount?
15.　72.8 sq ft is 13% of what area?
16.　8.4 cm is $\frac{1}{5}$% of what length?
17.　12.0% of what distance equals 46.08 miles?
18.　12 acres is 40% of what area?
19.　14.4 cubic yards (cu yd) is 15% of what volume?
20.　25.2 ft/sec is $2\frac{1}{2}$% of what speed?

Percentage Increase or Decrease

　　The words increase and decrease indicate a change from some starting figure. The difference between the starting amount and the resulting or final amount is the actual change.

example   A beaker of salt solution cooled from 82.5° to 66.0° in 3 min.

Original temperature (starting temperature)	82.5°
Resulting temperature	66.0°
Change in temperature	$\overline{16.5°}$

To discover by what percent the temperature changed you would *compare the change in temperature with the starting temperature* and express this ratio as a percent.

$$\frac{\text{change in temperature}}{\text{original temperature}} \quad \text{(expressed as a percent)}$$

$$\frac{16.5°}{82.5°} = 0.20 = 20\% \quad \text{(percentage change in temperature)}$$

example   Oxidation of a bolt has decreased the original tensile strength of 18,000 lb/sq in. by 8.0%. What is its present tensile strength?

original strength − decrease in strength = present strength

18,000 lb/sq in. − (8% of 18,000) lb/sq in. = present strength

(original)   (change)   (present tensile strength)
18,000 lb/sq in. − 1440 lb/sq in. = 16,560 lb/sq in. ≈ 17,000 lb/sq in.

example   The addition of a certain chemical to the compound forming the tread material of a tire will increase the average tire mileage from 15,000 miles to 18,000 miles. By what percent is the mileage increased?
To determine percent of increase, compare the increase with the starting amount. In this problem the increase is 3000 miles and the original mileage is 15,000 miles.

$$\frac{3000 \text{ miles}}{15,000 \text{ miles}} \quad \text{expressed as a percent} = 20\% \text{ increase}$$

When working with the percent of increase or decrease, the following relationships always hold:

1. original amount + (or −) change = result or final amount
2. original amount × percent (increase or decrease) = change in amount
3. $\dfrac{\text{change in amount}}{\text{original amount}} \times 100\%$ = (percentage increase or decrease)

example   Due to inflation, the cost of producing a $12.50 item was increased by 8%. What is the new cost of the item?
One method of determining the new cost is to determine 8% of $12.50 and add it to the $12.50 (relationship 1 above):

$12.50	original cost
1.00	0.08 × $12.50
$13.50	new cost

A second method of increasing an amount by a specific percent is to multiply the original amount by (100 + specific) %. In this case, multiply the $12.50 by 108% (expressed as a decimal equivalent):

$$
\begin{array}{r}
\$12.50 \\
1.08 \\
\hline
100\ 00 \\
1250 \\
\hline
\$13.50 \qquad \text{new cost}
\end{array}
$$

EXERCISE 1.3E

Assume that the given percents are exact. Round off answers to proper number of significant digits.

1. Increase 125 miles by 10% (two methods).
2. Decrease 50Õ0 hours (hr) by 5% (two methods).
3. Find the percent of increase from 125° to 150°.
4. Find the percent of decrease from $2000 to $1000.
5. Increase $2.80 by 5%.
6. Decrease $2.80 by 5%.
7. The cubic inch displacement of an automobile engine was changed from 340 cu in. to 357 cu in. Determine the percent of increase.
8. The elapsed time for a dragster changed from 10.50 sec to 9.87 sec. By what percent was the time decreased?
9. Due to expansion, the length of a 30.00-ft railroad rail increased by 0.3%.
   (a) What was the increase in length?
   (b) What was the expanded length?
10. A bolt has a rated tensile strength of 20,000 lb/sq in. Under testing it was found to break at 23,000 lb/sq in. By what percent did the actual strength exceed its rated strength?
11. Increase 120.0 lb by 2% (two methods).
12. Decrease 120.0 lb by 2% (two methods).
13. Increase $25Õ by 10%.
14. Decrease $275.00 by 10%.
15. Find the percent of change in temperature if the water temperature of an engine changes from 7Õ° to 168° in 5.0 min.
16. Find the percent of decrease from $2.88 to $2.70.
17. The horsepower of an engine changed from 180 to 202. Find the percent of increase. (Leave answer exact.)
18. The value of a stamping machine decreased from $5722 to $4850 in 1 year's (yr) use. By what percent (nearest 0.1%) did the machine depreciate in value?
19. If the capacity of a container was increased by 10% and then decreased by 10%, would it have the original capacity? Explain. (Use 100 for the original capacity.)
20. Increase 2040 by 15% and then decrease the new amount by 15%. (Leave answer to nearest unit.)

Trade Discounts

The reduction in price from a fixed or listed price is called a discount. Discounts are usually expressed as a percent. In wholesale business transactions, two or more discounts may be allowed on the same price. Successive discounts (frequently called chain discounts) are applied one after the other.

example   A shipment of parts contained an invoice for $2460. Discounts of 25% and 10% were offered. What was the actual cost if both discounts were obtained?
1. Decrease the 2460 figure by 25%:

$$\begin{array}{r} \$2460 \\ \underline{615} \\ \$1845 \end{array} \quad (25\% \text{ of } \$2460)$$

2. Decrease the $1845 by 10%:

$$\begin{array}{r} \$1845.00 \\ \underline{184.50} \\ \$1660.50 \end{array} \quad (10\% \text{ of } \$1845)$$

The actual cost was $1660.50.

example Trade discounts of 20%, 10%, and 2% were offered on a billing for $16,500. What was the net cost if all three discounts were applied?
1. Decrease $16,500 by 20%:

$$\begin{array}{r} \$16,500 \\ \underline{3,300} \\ \$13,200 \end{array} \quad (20\% \text{ of } \$16,500)$$

2. Decrease $13,200 by 10%:

$$\begin{array}{r} \$13,200 \\ \underline{1,320} \\ \$11,880 \end{array} \quad (10\% \text{ of } \$13,200)$$

3. Decrease $11,880 by 2%:

$$\begin{array}{r} \$11,880.00 \\ \underline{237.60} \\ \$11,642.40 \end{array} \quad (2\% \text{ of } \$11,880)$$

The net price was $11,642.40.

EXERCISE 1.3F

Determine the net cost (nearest cent) for the following items after deducting the specific discounts. (Multiple percents are chain discounts.)

1. $250—22%
2. $3500—12.5%
3. $8400—35%
4. $5.65—10%
5. $275—22%
6. $3150—8.5%
7. $2400—12.35%
8. $16,800—6%
9. $2420—15%, 5%
10. $1400—18%, 4%
11. $56,500—15%, 3%, 1%
12. $825.30—20%, 5%, 2%
13. $48.70—10%, 4%, 2%
14. $72,500—18%, 7%, 3%
15. With a list price of $2400, which is the greater discount: a single discount of 25% or chain discounts of 15%, 8%, and 3%?
16. Using a list price of $250, which is the greater discount and by how much is it greater: a single discount of 22% or chain discounts of 10%, 10% and 4%?

Simple Interest

When a person obtains a loan of money or buys an article by paying for it over a period of time, he is charged interest. *Interest* is the amount of money paid for the use of money. The amount of money being used (loaned or borrowed) is called the principal. The time involved is usually expressed in years. The interest, expressed as a percent, is called the *rate*. The *amount* in an interest problem is the sum of the interest and the principal. It is the total amount of money that must be paid back.

The method of determining simple interest is contained in the formula $I = Prt$, where I = amount of interest (in dollars), P = principal, r = rate (as a percent), and t = time (in years). In most private transactions, 30 days are considered as 1 month and 360 days as 1 year.

example To calculate the interest charges on a loan of $2500 at 6% for 3 yr, substitute the given data in the formula $I = Prt$:

$$I = \$2500 \times 6\% \times 3$$

$$I = \$2500 \times 0.06 \times 3$$

$$I = \$450$$

example Calculate the interest for the following:

$$P = \$15,000 \qquad r = 8.5\% \qquad t = 6 \text{ yr } 6 \text{ mo}$$

$$I = Prt$$

$$I = \$15,000 \times 0.085 \times 6.5$$

$$I = \$8287.50$$

example Calculate the interest and the amount for a loan of $36,000 at 8% for 2 yr 6 mo 20 days

$$6 \text{ mo} = 0.5 \text{ yr}, \qquad 20 \text{ days} = \frac{20}{360} = \frac{1}{18} \text{ yr}$$

$$\text{total time} = 2\tfrac{5}{9} \text{ yr}$$

$$I = Prt$$

$$I = \$36{,}000 \times 0.08 \times 2\tfrac{5}{9}$$

$$I = \$7360$$

$$\text{amount} = P + I = \$36{,}000 + \$7360 = \$43{,}360$$

EXERCISE 1.3G

Calculate the interest and the amount for each of the following problems.

1. $2400 at 8% for 3 yr
2. $150 at 9% for 6 mo
3. $2500 at 7.5% for 4 yr 6 mo
4. $4860 at 7% for 96 days
5. $540 at 12% for 10 days

6. $8500 at 9.5% for 2 mo 12 days
7. $5200 at 6.6% for 5 yr 4 mo 9 days
8. $1250 at 8.2% for 150 days
9. $22,000 at 6.5% for 1 yr 36 days
10. $35,000 at 7.2% for 10.5 yr

Compound Interest

Interest paid on both the principal and the previously earned interest which has been added to the principal is called compound interest. Suppose that you have $100 in a savings account which earn interest at the rate of 6%. At the end of 1 yr, your account would have a balance of $100 + $6, or $106. The second year you would earn interest on the total $106 at 6%, which would amount to $6.36. Each year the interest would be added to your account and would in turn, in each succeeding year, earn additional interest.

Computing compound interest is considerably more difficult than computing simple interest. It is common practice to compute compound interest by using precalculated amounts obtained from tables. Table 14 in Appendix D lists the amounts that $1 will increase to when compounded at various rates. To determine how much a certain principal will amount to when compounded at a given rate, proceed as follows:

1. From Table 14, determine the amount $1 will become for the given rate and number of years when compounded annually. When the interest is compounded semiannually, use half the given rate of interest and twice as many periods as the given number of years.

2. Multiply this amount by the number of dollars in the principal.
3. To find compound interest, subtract the principal from the amount.

example Calculate the amount and interest earned on $1200 at the end of 10 yr at 6% compounded annually.
   1. According to Table 14, $1 will amount to $1.7908 (column 6%, line 10) when compounded at 6% for 10 yr.
   2. Multiply the $1.7908 by 1200:

$$\$1.7908 \times 1200 = \$2148.96$$

$$\text{amount} = \$2148.96$$

3. To determine interest earned, subtract the principal from the amount:

$$\$2148.96 - \$1200 = \$948.96 \text{ interest}$$

example Compute the amount and interest earned on $500 invested at 5% compounded semiannually for 15 yr.
   1. From Table 14, determine how much $1 will amount to at $2\frac{1}{2}$% for 30 periods. You find that $1 will amount to $2.0976 (column $2\frac{1}{2}$%, line 30).
   2. Multiply the $2.0976 by 500:

$$\$2.0976 \times 500 = \$1048.80 \text{ amount}$$

3. Interest earned equals amount minus principal:

$$\$1048.80 - \$500 = \$548.80 \text{ interest}$$

EXERCISE 1.3H

Using data from Table 14, Appendix D, determine the amount and interest for the following problems.

1. $2000 compounded annually at 4% for 15 yr
2. $500 compounded annually at 5% for 40 yr
3. $800 compounded annually at $3\frac{1}{2}$% for 12 yr
4. $100 compounded annually at 2% for 50 yr
5. $1500 compounded semiannually at 5% for 8 yr
6. $2200 compounded semiannually at 4% for 2 yr
7. $1600 compounded semiannually at 3% for 25 yr
8. $12,000 compounded semiannually at 4% for 16 yr
9. How much more interest is earned when $2000 is invested for 20 yr at 5% compounded semiannually rather than when compounded annually?
10. How much more interest is earned when $5000 is invested for 12 yr at 6% compounded semiannually instead of at 6% simple interest?

11. Compute the amount and interest on $4500 invested for 12 yr at 6% compounded quarterly.

12. Which will produce the greater amount: $2800 invested for 10 yr at 6% compounded annually or $3200 invested for 10 yr at 5% compounded semiannually? How much greater?

1.4   DENOMINATE NUMBERS

Numbers with specific units of measure attached are called *denominate numbers*. Numbers such as 60 gal, 4 m, and 3.25 in. are denominate numbers, whereas 60, 4, and 3.25 by themselves are called abstract numbers.

Most mistakes made in computations involving denominate numbers are due to incorrect conversion of units. When a problem involves units of measurement, be careful in setting up and using the proper conversion ratio or conversion factor.

Adding and Subtracting Denominate Numbers

Addition or subtraction of denominate numbers can only be performed with denominate numbers of like units. For example, 6 ft + 8 ft = 14 ft, or 25 cm − 13 cm = 12 cm. In contrast, attempting to add 3 gal and 4 miles does not make sense.

Denominate numbers that express the same kind of measurement may be added or subtracted after conversion to the same units of measurement. In the problem 51.2 cm + 5.0 in. both numbers express measurements of length. After converting 51.2 cm to inches (or 5.0 in. to centimeters), the resulting equivalents may be added.

$$5.0 \text{ in. } \times \frac{2.54 \text{ cm}}{1 \text{ in.}} = 12.7 \text{ cm}$$

$$51.2 \text{ cm} + 12.7 \text{ cm} = 63.9 \text{ cm}$$

Compound Denominate Numbers

Consider a board which measures 212 in. This length could also be expressed as 17 ft 8 in. or as 5 yd 2 ft 8 in. The 17 ft 8 in. number and the 5 yd 2 ft 8 in. number are called *compound denominate* numbers because they include two or more different units of measurement. An understood addition is involved in compound numbers; 5 yd 2 ft 8 in. means 5 yd + 2 ft + 8 in. This implied addition is not performed unless you wish to express the compound denominate number as a single denominate number (5 yd 2 ft 8 in. = 180 in. + 24 in. + 8 in. = 212 in.). When adding two or more compound numbers, arrange the numbers in columns of like units and add each column. The sum is usually simplified. Answers such as 4 ft 27 in. should be written as 6 ft 3 in.

example Add:

$$
\begin{array}{llll}
& 28 \text{ yd} & 2 \text{ ft} & 5 \text{ in.} \\
& 9 \text{ yd} & 1 \text{ ft} & 3 \text{ in.} \\
& 2 \text{ yd} & 0 \text{ ft} & 8 \text{ in.} \\
& 12 \text{ yd} & 2 \text{ ft} & 0 \text{ in.} \\
& 0 \text{ yd} & 2 \text{ ft} & 6 \text{ in.} \\
\hline
& 51 \text{ yd} & 7 \text{ ft} & 22 \text{ in.} \\
= & 51 \text{ yd} & 8 \text{ ft} & 10 \text{ in.} \\
= & 53 \text{ yd} & 2 \text{ ft} & 10 \text{ in.}
\end{array}
$$

(22 in. = 1 ft 10 in.)

(8 ft = 2 yd 2 ft)

simplified answer

example Subtract:

$$
\begin{array}{ll}
10 \text{ gal} & 2 \text{ qt} \\
5 \text{ gal} & 3 \text{ qt} \\
\hline
\end{array}
$$

Because 2 qt is less than 3 qt, borrow 1 gal = 4 qt from the 10 gal and rewrite the problem as:

$$
\begin{array}{ll}
9 \text{ gal} & 6 \text{ qt} \\
5 \text{ gal} & 3 \text{ qt} \\
\hline
4 \text{ gal} & 3 \text{ qt}
\end{array}
$$

You can avoid rewriting the entire problem when borrowing·

$$
\begin{array}{ll}
9 \text{ gal} & 6 \text{ qt} \\
\cancel{10 \text{ gal}} & \cancel{2 \text{ qt}} \\
5 \text{ gal} & 3 \text{ qt} \\
\hline
4 \text{ gal} & 3 \text{ qt}
\end{array}
$$

example The sum of the angles of any triangle is 180°. If two angles of a triangle measure 26° 32′ 18″ and 49° 46′ 52″, determine the magnitude of the third angle. You can determine the third angle by subtracting the sum of the two given angles from 180°.

$$
\begin{array}{l}
26° \; 32′ \; 18″ \\
49° \; 46′ \; 52″ \\
\hline
75° \; 78′ \; 70″
\end{array}
$$

becomes 76° 19′ 10″ upon simplification.

$$
\begin{array}{l}
180° \; 0′ \; 0″ \\
76° \; 19′ \; 10″ \\
\hline
\end{array}
\quad \xrightarrow{\text{borrowing } 1°} \quad
\begin{array}{l}
179° \; 60′ \; 0″ \\
76° \; 19′ \; 10″ \\
\hline
\end{array}
$$

$$
\xrightarrow{\text{borrowing } 1′} \quad
\begin{array}{l}
179° \; 59′ \; 60″ \\
76° \; 19′ \; 10″ \\
\hline
103° \; 40′ \; 50″
\end{array}
\quad \text{third angle}
$$

EXERCISE 1.4A

Simplify each of the following* in Problems 1–12.

1. 2 ft 33 in.
2. 1 hr 72 min
3. 3 ft 39 in.
4. 2 qt 13 pints (pt)

5. 11 gal 7 qt
6. 16 ft 15 in.
7. 1 lb 42 ounces (oz)
8. 15° 86′

9. 24° 75′ 83″
10. 1 yd 14 ft
11. 5 tons 2700 lb
12. 7′ 92″

Add or subtract (as indicated) the following denominate numbers and simplify.

13. Add:

          6 lb 5 oz
          8 lb 11 oz
          3 lb 7 oz
          9 lb 14 oz

14. Add:

          5 gal 2 qt 1 pt
          3 gal 1 qt 0 pt
          8 gal 2 qt 1 pt

15. Subtract:

          14 tons 800 lb
           7 tons 1200 lb

16. Subtract:

          13 miles 1265 ft
           4 miles 3871 ft

17. Add:

           3 hr 32 min 41 sec
           5 hr 29 min 38 sec
          10 hr 4 min 57 sec
          12 hr 46 min 29 sec

18. Add:

           9 yd 2 ft 8 in.
           2 yd 2 ft 5 in.
           7 yd 1 ft 10 in.
          12 yd 0 ft 6 in.
                 5 ft 4 in.

19. Subtract:

          10 hr 8 min 14 sec
           5 hr 16 min 29 sec

20. Subtract:

          14 yd 1 ft 4 in.
           8 yd 2 ft 7 in.

21. Add:

          102° 13′ 14″
           19° 46′ 5″
           36° 29′ 56″
           96° 52′ 43″

22. Subtract:

          180° 0′ 0″
          126° 47′ 33″

23. Subtract:

          6 yr 4 mo 6 days
          2 yr 5 mo 12 days

24. Add:

           5 yr 8 mo 11 days
           9 yr 7 mo 22 days
          12 yr 10 mo 19 days
           8 yr 6 mo 29 days

* Tables of weights and measures are located inside the front and back covers of the text.

Multiplication Involving Denominate Numbers

To multiply a denominate number by an abstract number, multiply each term of the denominate number by the abstract number and simplify the resulting product.

example   Multiply 2 yd 1 ft 7 in. by 5:

$$
\begin{array}{r}
2 \text{ yd } 1 \text{ ft } \ 7 \text{ in.} \\
5 \\
\hline
10 \text{ yd } 5 \text{ ft } 35 \text{ in.}
\end{array}
$$

which upon simplification, becomes:

$$12 \text{ yd } 1 \text{ ft } 11 \text{ in.}$$

EXERCISE 1.4B

Multiply and simplify the following.

1. $\begin{array}{r} 4 \text{ lb } 8 \text{ oz} \\ 6 \\ \hline \end{array}$

2. $\begin{array}{r} 8° \ 14' \ 26'' \\ 5 \\ \hline \end{array}$

3. $\begin{array}{r} 3 \text{ gal } 2 \text{ qt } 1 \text{ pt} \\ 7 \\ \hline \end{array}$

4. $\begin{array}{r} 1 \text{ hr } 17 \text{ min } 12 \text{ sec} \\ 8 \\ \hline \end{array}$

5. $\begin{array}{r} 5 \text{ yd } 2 \text{ ft } 9 \text{ in.} \\ 9 \\ \hline \end{array}$

6. $\begin{array}{r} 12° \ 51' \ 45'' \\ 6 \\ \hline \end{array}$

7. $\begin{array}{r} 3 \text{ tons } 825 \text{ lb} \\ 5 \\ \hline \end{array}$

8. $\begin{array}{r} 8 \text{ miles } 726 \text{ ft} \\ 10 \\ \hline \end{array}$

9. $\begin{array}{r} 42° \ 0' \ 24'' \\ 8 \\ \hline \end{array}$

10. $\begin{array}{r} 7 \text{ ft } 11 \text{ in.} \\ 42 \\ \hline \end{array}$

Division Involving Denominate Numbers

To divide a denominate number by an abstract number:

1. Divide the quantity with the larger unit first. (If you have a remainder, change it to the next lower unit. A remainder of 3 qt would be changed to 6 pt.)
2. Divide the next lower unit and convert the remainder as before.
3. Continue until each part of the original number has been divided.

example   Divide 8 yd 2 ft 6 in. by 6:

$$
\begin{array}{r}
1 \text{ yd} \\
6\overline{)8 \text{ yd}} \\
6 \text{ yd} \\
\hline
\text{remainder } 2 \text{ yd} = 6 \text{ ft}
\end{array}
$$

$$
\begin{array}{r}
1 \text{ ft} \\
6\overline{)8 \text{ ft}} \quad (2 \text{ ft} + 6 \text{ ft} = 8 \text{ ft}) \\
6 \text{ ft} \\
\hline
\text{remainder } 2 \text{ ft} = 24 \text{ in.}
\end{array}
$$

$$
\begin{array}{r}
5 \text{ in.} \\
6\overline{)30 \text{ in.}} \quad (6 \text{ in.} + 24 \text{ in.} = 30 \text{ in.}) \\
30 \text{ in.} \\
\hline
0
\end{array}
$$

Quotient is:

$$ 1 \text{ yd } 1 \text{ ft } 5 \text{ in.} $$

The same problem could be done as follows:

$$
\begin{array}{r}
1 \text{ yd } 1 \text{ ft } 5 \text{ in.} \quad \text{(quotient)} \\
6\overline{)8 \text{ yd } 2 \text{ ft } 6 \text{ in.}} \\
6 \text{ yd} \\
\hline
2 \text{ yd} = 6 \text{ ft} \\
6\overline{)8 \text{ ft}} \\
6 \text{ ft} \qquad 6 \text{ in.} \\
\hline
2 \text{ ft} = 24 \text{ in.} \\
6\overline{)30 \text{ in.}} \\
30 \text{ in.} \\
\hline
0
\end{array}
$$

EXERCISE 1.4C

Carry out the following indicated divisions.

1. 14 yd 2 ft 6 in. by 3
2. 5 miles 721 yd 311 ft by 2
3. 24 gal 1 qt 1 pt by 5
4. 90 lb 2 oz by 7
5. 315° 44′ 12″ by 12
6. 17 hr 35 min 30 sec by 15
7. 41 tons 592 lb by 8
8. 49° 9′ 28″ by 4
9. 109 gal 1 qt 1 pt by 7
10. 4 mo 28 days 4 hr 6 min by 3

1.5 SYSTEMS OF MEASUREMENTS

Two systems of measurements are in use today. About 80% of the countries of the world use the metric system, and about 20% still use the English system.

The story of the development of the English system parallels the history of civilized man, from Bible times with the use of the cubit* up to modern man's use of the foot. Because many units in the English system date back as far as several thousand years, it is not surprising that some units are no longer used or are no longer practical in the modern industrial world.

Because the units were developed by different people under different conditions, there was no preplanned relationship between units measuring the same dimension; e.g., the cubit and the furlong.† The main units in this system for measuring length, weight, and liquid volume are the foot, the pound, and the gallon, respectively.

In contrast to the English system, the metric system was well planned and logically structured. All units of length have a power of 10 relationship with each other. Similarly, all units of weight have a power of 10 relationship, as do all units of liquid volume. The metric system has just three basic units: (1) the *meter*—unit of length, (2) the *gram*—unit of weight, and (3) the *liter*—unit of liquid volume.

Compared to the yard, the *meter* is a little longer, measuring 39.37 in. The meter is divided into 10 equal parts called decimeters; the decimeter is divided into 10 equal parts called centimeters; the centimeter is further divided into 10 equal parts called millimeters. Larger units are formed by multiplying the meter by 10, by 100, and by 1000, forming the decameter, the hectometer, and the kilometer, respectively. The equivalents for these units are:

$$10 \text{ millimeters (mm)} = 1 \text{ centimeter (cm)}$$
$$10 \text{ cm} = 1 \text{ decimeter (dm)}$$
$$10 \text{ dm} = 1 \text{ } meter \text{ (m)}$$
$$10 \text{ m} = 1 \text{ decameter (dkm)}$$
$$10 \text{ dkm} = 1 \text{ hectometer (hkm)}$$
$$10 \text{ hkm} = 1 \text{ kilometer (km)}$$

Also,

$$1000 \text{ mm} = 100 \text{ cm} = 10 \text{ dm} = 1 \text{ m} = 0.1 \text{ dkm} = 0.01 \text{ hkm} = 0.001 \text{ km}$$

---

* The ancient cubit was the length of a man's forearm from his elbow to the tip of his middle finger.

† A furlong equals $\frac{1}{8}$ mile.

Using these same prefixes and the standard unit of weight, the *gram*, we have:

$$10 \text{ milligrams (mg)} = 1 \text{ centigram (cg)}$$
$$10 \text{ cg} = 1 \text{ decigram (dg)}$$
$$10 \text{ dg} = 1 \text{ } gram \text{ (g)}$$
$$10 \text{ g} = 1 \text{ decagram (dkg)}$$
$$10 \text{ dkg} = 1 \text{ hectogram (hkg)}$$
$$10 \text{ hkg} = 1 \text{ kilogram (kg)}$$

Also,

$$1000 \text{ mg} = 100 \text{ cg} = 10 \text{ dg} = 1 \text{ g} = 0.1 \text{ dkg} = 0.01 \text{ hkg} = 0.001 \text{ kg}$$

Similarly, for the unit of liquid volume, the *liter*, we have:

$$10 \text{ milliliters (ml)} = 1 \text{ centiliter (cl)}$$
$$10 \text{ cl} = 1 \text{ deciliter (dl)}$$
$$10 \text{ dl} = 1 \text{ } liter \text{ } (\ell)$$
$$10 \text{ } \ell = 1 \text{ decaliter (dkl)}$$
$$10 \text{ dkl} = 1 \text{ hectoliter (hkl)}$$
$$10 \text{ hkl} = 1 \text{ kiloliter (kl)}$$

Also,

$$1000 \text{ ml} = 100 \text{ cl} = 10 \text{ dl} = 1 \text{ } \ell = 0.1 \text{ dkl} = 0.01 \text{ hkl} = 0.001 \text{ kl}$$

It is important to learn the meaning of the six prefixes used in the metric system:

1. "deci" means $\frac{1}{10}$; therefore, 1 decimeter $= \frac{1}{10}$ meter.
2. "centi" means $\frac{1}{100}$; therefore, 1 centimeter $= \frac{1}{100}$ meter.
3. "milli" means $\frac{1}{1000}$; therefore, 1 millimeter $= \frac{1}{1000}$ meter.
4. "deca" means 10; therefore, 1 decameter $= 10$ meters.
5. "hecto" means 100; therefore, 1 hectometer $= 100$ meters.
6. "kilo" means 1000; therefore, 1 kilometer $= 1000$ meters.

EXERCISE 1.5

Use information found in this section or in the table inside the front or back cover to answer the following problems.

1. The meter is slightly larger than what English unit?
2. What unit in the metric system is 39.37 in. long?
3. What system of weights and measures was a planned and logically structured system?
4. Most countries of the world now use what system?

In Problems 5–30, write statements of equality involving the given units.

5. Yard and inches
6. Yard and feet
7. Foot and inches
8. Mile, feet, yards, rods, and furlongs
9. Pound and ounces
10. Ton and pounds
11. Fathom and yards
12. Square foot and square inches
13. Square mile and acres
14. Acre and square feet
15. Cubic yard and cubic feet
16. Cubic foot and cubic inches
17. Chain, yards, and feet
18. Rod, feet, and yards
19. Gallon and pints
20. Square yard and square feet
21. Day and hours
22. Day and minutes
23. Minute and seconds
24. Month and days
25. Kilometer and meters
26. Meter and centimeters
27. Hectometer and centimeters
28. Liter and milliliters
29. Kiloliter and liters
30. Deciliter and milliliters

In Problems 31–36, determine the required equivalents.

31. 552 ml equals how many liters?
32. 4.6 g equals how many milligrams?
33. 1 metric ton equals how many kilograms?
34. 1 sq km equals how many square meters?
35. 2.65 m equals how many centimeters?
36. 715 mg equals how many grams?

1.6 TABLES

A book of tables can be a valuable tool to the technician if he knows what it contains, where to find the table he needs, and how to obtain the information he seeks. When you first turn to the tables in Appendix D or inside the front and back covers, study each separate table so that you begin to know what tables are available and where they are located.

When using any particular table, the entry at the top of each column and the entry at the left of the row in which you are working should be examined for peculiarities in notation. Whenever practical, make a mental check to see that the data obtained from a table are reasonable. For example, suppose you have looked up $\sqrt[3]{38}$ (Table 7) and read by mistake $\sqrt{38} = 6.1644$. A bit of mental arithmetic gives $3^3 = 27$ and $4^3 = 64$; then $\sqrt[3]{38}$ should be between 3 and 4. Checking Table 7 again, you discover that $\sqrt[3]{38} = 3.3620$ which, compared with $\sqrt[3]{27} = 3$ and $\sqrt[3]{64} = 4$, seems reasonable. The use of a ruler or the edge of a paper as a guide to align proper row entries and column entries saves many such mistakes.

Considerable time and trouble can be saved if data from a table are checked a second time before being used in further calculations. The few seconds it takes to remove the guide from the table and look up the entry again is time well spent.

EXERCISE 1.6

1. Make a list of the tables in Appendix D. For each table, 1–19, write a sentence expressing the type of information that can be obtained from it.

In Problems 2–30, use the tables of equivalents (inside the front and back covers) to write the proper statements of equality for the specified units. Then write two unit ratios using the information in the equality.

Example: Express the relationship between inch and centimeters (cm).

$$1 \text{ in.} = 2.54 \text{ cm and unit ratios are}$$

$$\frac{1 \text{ in.}}{2.54 \text{ cm}} = 1 \qquad \frac{2.54 \text{ cm}}{1 \text{ in.}} = 1$$

2. Ounce and grams
3. Cubic feet and gallons
4. Kilogram and pounds
5. Kilogram and grams
6. Gram and milligrams
7. Liter and quarts
8. Liter and milliliters
9. Statute mile (S.M.) and feet
10. Mile (S.M.) and kilometers
11. Nautical mile (N.M.) and miles (S.M.)
12. Square mile and acres
13. Cubic foot and liters
14. Cubic foot and cubic inches
15. Hectometer and meters
16. Gallon and cubic inches
17. Gallon and liters
18. Square mile and square kilometers
19. Bushel and cubic feet
20. Fathom and yards
21. Gram and grains
22. Chain and feet
23. Statute mile and chains
24. Pound and grams
25. Kilometer and chains
26. Square yard and square feet
27. Cubic yard and gallons
28. Square inch and square centimeters
29. Chain and rods
30. Square kilometer and acres

In Problems 31–43, use the given table to determine the specified quantity. (All tables are in Appendix D.)

31. Table 2: The mathematical meaning of the following prefixes:
    (a) Kilo
    (b) Micro
    (c) Deca
    (d) Millimicro

32. Table 13: Conversion factor to convert:
    (a) B.t.u. per minute to horsepower
    (b) Atmospheres to inches of mercury
    (c) Centimeters per second to feet per minute
    (d) Square miles to acres

33. Table 1: The Greek letters:
    (a) Alpha
    (b) Pi
    (c) Phi
    (d) Beta

34. Table 12:
    (a) B.t.u. available from 1 lb of acetylene
    (b) Approximate B.t.u. available from 1 cu ft of natural gas
    (c) B.t.u./gallon petroleum (specific gravity = 0.785)
    (d) B.t.u. range/pound coke
    (e) Substance with highest B.t.u. rating/pound
35. Table 4:
    (a) Density of water
    (b) Mean polar radius of the earth
    (c) Acceleration of gravity
    (d) Speed of light
36. Table 3: Symbol for:
    (a) Less than
    (b) Perpendicular
    (c) Angle
    (d) Degree (angle)
    (e) Square root
    (f) Greater than or equal to
    (g) A sub one
37. Table 11: Weight of 1 cu ft of:
    (a) Air
    (b) Pure copper
    (c) Ice
    (d) Gold
    (e) Gravel
    (f) Loam (soil)
    (g) Wet compact snow
38. Table 7:
    (a) $\sqrt[3]{42}$
    (b) Circumference of a circle whose diameter $N = 91$ ft
    (c) $1000 \times \dfrac{1}{N}$ for $N = 89$
    (d) Area of a circle $(A = \pi N^2/4)$ for diameter $N = 68$ in.
    (e) 87^2
    (f) Reciprocal of 73
39. Table 9:
    (a) Coefficient of expansion for copper
    (b) Melting point of mercury
    (c) Electric conductivity of platinum
    (d) Tensile strength of aluminum
    (e) Weight of gold/cubic foot
40. Table 10:
    (a) Resistance in ohms/1000 ft, Number 8 gage copper wire
    (b) Weight of 1000 ft, Number 0 gage copper wire

(c)   Current capacity in amperes of rubber insulated copper wire with cross-sectional area of 500,000 circular mils

(d)   All entries for Number 000 gage copper wire

41.   Table 5: Decimal equivalent for:

   (a) $\frac{5}{64}$

   (b) $\frac{9}{16}$

   (c) $\frac{7}{8}$

   (d) $\frac{29}{64}$

   (e) $\frac{5}{32}$

42.   Table 8: Coefficient of heat transfer (K) for:

   (a)   12-in. thick brick wall with interior plaster

   (b)   Same as (a) with north exposure

   (c)   6-in. thick concrete floor on ground with terrazzo covering

   (d)   Wood shingle roof with rafters exposed

   (e)   Single windows and sky lights

43.   Table 6:

   (a)   Decimal equivalent for $\frac{1}{6}$

   (b)   Percentage equivalent for 0.625

   (c)   Common fractional equivalent for $81\frac{1}{4}\%$

   (d)   Decimal equivalent for $68\frac{3}{4}\%$

1.7   CONVERSION OF UNITS

There is little need for the usual difficulty students have with the conversion of units of measure. Instead of trying to remember when to multiply by 12, or 10, or 100, or 36, or $16\frac{1}{2}$, or when to divide by 3, or 4, or 10, there is a simple way to make conversions whether or not you are familiar with the units.

The Ratio Method

When two equal quantities are written in fraction form, they form a ratio equal to 1. For example, 36 in./1 yd = 1 because 36 in. and 1 yd are exactly the same length. Thus

$$\frac{12 \text{ in.}}{1 \text{ ft}} = 1 \qquad \frac{5280 \text{ ft}}{1 \text{ mile}} = 1 \qquad \frac{4 \text{ qt}}{1 \text{ gal}} = 1$$

because the two terms of each fraction are equivalent quantities. Let us see how these unit ratios simplify conversions.

example   Change 450 in. to yards.

   1.   From the tables of weights and measures, you find that

$$36 \text{ in.} = 1 \text{ yd}$$

   2.   Using these equivalent quantities, write unit ratios of 36 in./1 yd and 1 yd/36 in.

3. Which ratio should you use? The answer is very easy to deduce. To change *from inches* to some other unit (in this case, yards), use the ratio with inches in the denominator:

$$\frac{1 \text{ yd}}{36 \text{ in.}}$$

4. Now write

$$\frac{450 \text{ in.}}{1} \times \frac{1 \text{ yd}}{36 \text{ in.}}$$

5. Cancel the inch units by dividing the numerator and denominator by 1 in.:

$$\frac{450 \cancel{\text{ in.}} \times 1 \text{ yd}}{36 \cancel{\text{ in.}}}$$

Then multiply and simplify, obtaining 12.5 yd.

example Change 0.05 ton to ounces.
  1. From the table of equivalents, find ·

$$1 \text{ ton} = 2000 \text{ lb}$$

$$1 \text{ lb} = 16 \text{ oz}$$

2. Set up the unit ratios

$$\frac{1 \text{ ton}}{2000 \text{ lb}} \qquad \frac{2000 \text{ lb}}{1 \text{ ton}} \qquad \frac{1 \text{ lb}}{16 \text{ oz}} \qquad \frac{16 \text{ oz}}{1 \text{ lb}}$$

3. Which unit ratios should you use? Select the pound-ton ratio that has ton in the denominator because you are *changing from tons*. If you set up the problem to this point, it will help you answer the second part of the question.

$$0.05 \cancel{\text{ ton}} \times \frac{2000 \text{ lb}}{1 \cancel{\text{ ton}}}$$

To change from pounds, select the ounce-pound ratio (16 oz/1 lb) because pound is in the denominator. Putting everything together, the problem looks like this:

$$\frac{0.05 \cancel{\text{ ton}}}{1} \times \frac{2000 \cancel{\text{ lb}}}{1 \cancel{\text{ ton}}} \times \frac{16 \text{ oz}}{1 \cancel{\text{ lb}}} = 1600 \text{ oz}$$

example Change 4.5 liters (ℓ) to milliliters (ml).

1. From the tables of equivalents you see that

$$1 \ell = 1000 \text{ ml}$$

2. Write the unit ratios:

$$\frac{1 \ell}{1000 \text{ ml}} \qquad \frac{1000 \text{ ml}}{1 \ell}$$

   (Select the unit ratio having liter in the denominator.)

3. Multiply:

$$\frac{4.5 \ell}{1} \times \frac{1000 \text{ ml}}{1 \ell}$$

4. Cancel liter units and multiply:

$$4.5 \cancel{\ell} \times \frac{1000 \text{ ml}}{1 \cancel{\ell}} = 4500 \text{ ml}$$

Note: No place along the line did you stop to decide which of the two units involved in the unit ratios was the larger or smaller unit. Neither did you have to answer the question: "Do I multiply by 27 or 36 or 1728, or do I divide?"

example Change 60 miles per hour (mph) to feet/minute. This problem involves a double conversion: a conversion from miles to feet and a conversion from hours to minutes. According to the tables 1 mile = 5280 ft. Because you are changing from miles, use the unit ratio 5280 ft/1 mile. Also, because 1 hr = 60 min, use 1 hr/60 min to change from per hour to per minute:*

$$\frac{60 \cancel{\text{miles}}}{\cancel{\text{hr}}} \times \frac{5280 \text{ ft}}{1 \cancel{\text{mile}}} \times \frac{1 \cancel{\text{hr}}}{60 \text{ min}} = 5280 \text{ ft/min}$$

If you want to find feet/second, you would multiply by the ratio 1 min/60 sec:

$$\frac{5280 \text{ ft}}{\cancel{\text{min}}} \times \frac{1 \cancel{\text{min}}}{60 \text{ sec}} = 88 \text{ ft/sec}$$

$$60 \text{ mph} = 5280 \text{ ft/min} = 88 \text{ ft/sec}$$

---

* In this example the time unit (hour) is in the denominator. Therefore, select the unit ratio with hour in the numerator to allow conversion from per hour to per minute.

EXERCISE 1.7A

Using the tables of equivalents, write unit ratios and multiply to make the following conversions. Not only does the ratio method simplify the conversion, but problems set up in ratio form are easily solved by slide rule computations. (Assume given data are exact.)

1. 80 in. to feet
2. 0.02 mile to yards
3. 2 miles to rods
4. 5 gal to liters
5. 1.5 lb to ounces
6. 14.25 gal to quarts
7. 0.08 ton to pounds
8. 2.5 lb to grains
9. 2 cu yd to cubic feet
10. 20,736 cu in. to cubic feet
11. 6.2 acres to square feet
12. 2.25 acre-ft to cubic feet (1 acre-ft = 43,560 cu ft)
13. 2 days to minutes
14. 2 lb to grams
15. 206 cu ft to gallons
16. 50 yd to meters
17. 2.6 km to centimeters
18. 45 mph to feet/second
19. 2.5 acre-ft to gallons (see Problem 12)
20. 10,500 cu m to cubic yards (1 cu m = 1.308 cu yd)

Conversion Factors

The technician should be able to make conversions by using conversion factors as well as by using unit ratios from tables of equivalents. Table 13, Appendix D, is a condensed table of conversion factors used to make specified conversions. For handy reference, the first few entries in Table 13 are shown below.

Conversion factors

Multiply	By	To obtain
Abamperes	10	Amperes
Acres	43,560	Square feet
Acres	1.562×10^{-3}	Square miles
Acres	4840	Square yards
Acre-feet	43,560	Cubic feet
Acre-feet	3.259×10^{5}	Gallons

A quantity in units shown in Column 1 is multiplied by the conversion factor shown in Column 2; the conversion is to the units shown in Column 3.

example Change 1.5 acres to square yards:

$$1.5 \times 4840 = 7260 \text{ sq yd}$$

example Change 2500 acres to square miles:

$$2500 \times 1.562 \times 10^{-3} = 3.905 \text{ sq miles}$$

EXERCISE 1.7B

Use conversion factors from Table 13, Appendix D, to make the following conversions. (Assume that the given quantities are exact.)

1. 650 amperes (amp) to abamperes
2. 1.6 acre-ft to gallons
3. 10.0 bars to atmospheres
4. 1000 B.t.u./min to horsepower
5. 580 cm to inches
6. 100 cm of mercury to atmospheres
7. 250,000 circular mils to square centimeters
8. 5200 cu cm to cubic inches
9. 125 cu ft to gallons
10. 56.10 cu in. to cubic centimeters
11. 7.5 cu yd to gallons
12. 10.0 drams (dr) to grams
13. 40.0 ft of water to inches of mercury
14. 72,000 ft-lb to B.t.u.
15. 450 gal/min to cubic feet/second
16. 5.20 horsepower to foot-pounds/second
17. 5.60 kg to pounds
18. 250.0 km/hr to miles/hour
19. 58.0 links (surveyor's) to inches
20. 1.25 ohms to microhms
21. 8.0 perches to cubic feet
22. 5.5 radians to degrees
23. 0.0002 sq cm to circular mils
24. 522,000 statcoulombs to coulombs
25. 50,000 watts (w) to B.t.u./minute

Conversion Ratios Obtained from Data in Problems

Numerous problems in the technical fields require conversions using relationships not listed in tables of equivalents or tables of conversion factors. The following examples show how such ratios are established and used.

example   Find the cost of excavating 60,000 cu yd of earth and rock at a cost of $1.10/cu yd. If you compare $1.10 with the cost value of 1 cu yd, you have the ratio $1.10/1 cu yd. Multiplying the 60,000 cu yd by this ratio gives the cost equivalent of 60,000 cu yd:

$$60{,}000 \text{ cu yd} \times \frac{\$1.10}{1 \text{ cu yd}} = \$66{,}000 \text{ cost}$$

example   If 200.0 ml of a solution contain 2.20 grams of iodine, how much iodine is contained in 1230 ml of this solution? You can set up the ratio

$$\frac{200.0 \text{ ml solution}}{2.20 \text{ g iodine}} \quad \text{or} \quad \frac{2.20 \text{ g iodine}}{200.0 \text{ ml solution}}$$

Use the ratio with milliliters in the denominator so the milliliter units will cancel:

$$1230 \text{ ml} \times \frac{2.20 \text{ g}}{200.0 \text{ ml}} = 13.5 \text{ g iodine}$$

example A 0.250-lb sample of a certain fuel has a heat content of 5600 B.t.u. What is the heat content of 2000 lb of this fuel? First, set up the ratios

$$\frac{0.250 \text{ lb}}{5600 \text{ B.t.u.}} \quad \text{and} \quad \frac{5600 \text{ B.t.u.}}{0.250 \text{ lb}}$$

Using the ratio having pounds in its denominator, you have:

$$2000 \text{ lb} \times \frac{5600 \text{ B.t.u.}}{0.250 \text{ lb}} = 44,800,000 \text{ B.t.u.}$$

or 4.48×10^7 B.t.u.

example If 1240 ft of wire have a resistance of 7.20 ohms, what is the resistance of 3080 ft of this wire? First, set up the ratios

$$\frac{1240 \text{ ft}}{7.20 \text{ ohms}} \quad \text{and} \quad \frac{7.20 \text{ ohms}}{1240 \text{ ft}}$$

Multiplying by the ratio having feet in its denominator:

$$3080 \text{ ft} \times \frac{7.20 \text{ ohms}}{1240 \text{ ft}} = 17.9 \text{ ohms resistance}$$

Students often hesitate to set up ratios that involve quite different units. Remember that whatever relationship exists between two quantities (expressed with different units) as stated in the problem, the same relationship holds for any larger or smaller amounts of these same quantities. For example, if there is 0.00025 g of a certain poison in 1 oz of fly spray, then the relationship

$$\frac{\text{total amount of poison in grams}}{\text{total amount of spray in ounces}} = \frac{0.00025 \text{ g}}{1 \text{ oz}}$$

is valid.

example If 12.6 g of silver are deposited by electrolysis in 2.40 min, how many pounds of silver would be deposited in 1.51 hr? In this problem, you need to use three ratios: one involving pounds and grams, one involving hours and minutes, and one involving grams and minutes. From the given data, establish the ratios

$$\frac{12.6 \text{ g}}{2.40 \text{ min}} \quad \text{and} \quad \frac{2.40 \text{ min}}{12.6 \text{ g}}$$

From the tables of equivalents, you can write unit ratios of

$$\frac{60 \text{ min}}{1 \text{ hr}} \quad \text{and} \quad \frac{1 \text{ hr}}{60 \text{ min}}$$

and

$$\frac{1 \text{ lb}}{453.59 \text{ g}} \quad \text{and} \quad \frac{453.59 \text{ g}}{1 \text{ lb}}$$

To decide on the procedure and which ratios to use, your thoughts could go something like the following. Restate the question as : In 1.51 hr, how many pounds of silver will be deposited? If you start with 1.51 hr, your first conversion would be *from* hours *to* minutes. Your problem so far has become

$$\frac{1.51 \text{ hr}}{1} \times \frac{60 \text{ min}}{1 \text{ hr}}$$

Next, you need to change *from* minutes *to* grams of silver deposited *in this number of minutes*. *It is important to note that you are not converting minutes to grams*. Two of the six ratios have minutes in the denominator. However, you do not want to convert back to hours, so you only have a choice of the ratio 12.6 g/2.40 min. Next, indicate multiplication by this ratio, obtaining

$$\frac{1.51 \text{ hr}}{1} \times \frac{60 \text{ min}}{1 \text{ hr}} \times \frac{12.6 \text{ g}}{2.40 \text{ min}}$$

The ratio 1 lb/453.59 g lets you change from grams and end up with pounds. The entire problem now becomes

$$\frac{1.51 \text{ hr}}{1} \times \frac{60 \text{ min}}{1 \text{ hr}} \times \frac{12.6 \text{ g}}{2.40 \text{ min}} \times \frac{1 \text{ lb}}{453.59 \text{ g}}$$

$$= 1.05 \text{ lb}$$

EXERCISE 1.7C

Use conversion ratios established from given data in solving the following problems. (Watch significant figures.)

1. If 128 ft of copper wire has a resistance of 0.025 ohm, what would be the resistance of 2480 ft of this wire?
2. If 16.4 ft of a certain wire has a resistance of 1.02 ohms, what is the resistance of 402 ft of the wire?
3. An automatic oiler uses 3.50 gal of oil in 1.25 hr. How many gallons should a supply reservoir hold to supply the oiler for 48 hr?
4. If 2.80 gal of water flow through a cooling jacket in 0.75 min, how many gallons of water will flow through the jacket in 60.0 min?
5. A stamping machine will stamp out 52 copper caps in 0.25 min. How many caps will the machine stamp out in 1 hr 32 min?

6. A still will produce 0.125 gal of distilled water in 2.5 min. How much distilled water will it produce in 24 hr (24 hr = 1440 min)?

7. An automatic stamp-canceling machine will cancel 122 envelopes in 9.0 sec. How many envelopes will the machine cancel in 1 hr (1 hr = 3600 sec)?

8. A data-processing machine will alphabetize 80$\tilde{0}$ cards in 1.2 min. How many cards will the machine handle in 24 hr?

9. Experimental tests on a new automobile tire gave the following results: 1650 miles driven with 0.025 in. of tread wear. What mileage can be expected from the tire if the original tread depth is 0.575 in.?

10. In Problem 9, what tread depth would be needed if the manufacturer wished to guarantee the tire for 4$\tilde{0}$,000 miles?

11. If 8 sheets of a certain grade paper have a combined thickness of 0.016 in., what is the thickness of 1 ream of this paper (1 ream = 500 sheets)?

12. If 12 laminated sheets used in a condenser have a combined thickness of 0.0125 in., what is the combined thickness of 72$\tilde{0}$ such sheets?

13. A 2.22-g force will stretch a certain spring 3.12 cm. What force will be required to stretch the same spring 16.5 cm?

14. A change in temperature of 1.60°F. causes a metal rod in a thermostat to increase in length by 0.00250 in. What change in temperature would be needed to produce a change in length of 0.0440 in.?

15. A water company finds that 1.58 lb of chlorine must be added to the water during each 36.0 min of pumping time from a certain reservoir. How much chlorine must be stored to provide an emergency reserve for ten days of continuous pumping from the same reservoir?

16. A timing device gains 3.5 sec in 1.2 hr. How many seconds will the device gain in 84 hr?

17. An alert manager of a department store discovered that for an increase of 10$\tilde{0}$ in the student-enrollment figure in the local school district, he could expect his sales of children's winter clothes to increase by $750. If the increase in 1 yr was 452 students, what would be the expected increase in his sales?

18. A modern printing machine uses 12.5 lb of ink for each 500$\tilde{0}$ newspapers printed. How much ink is used to print 72,250 papers?

19. Tests on an automobile spring show that for each 0.22 mm added to the width of the spring, the carrying capacity of the spring is increased by 4.8 kg. If the manufacturer needs to increase the carrying capacity of the spring by 5$\tilde{0}$ kg, what increase in width is required?

20. For each 2.5°C. increase in water temperature above 20.0°C., a certain thermostatically controlled valve opens 0.035 in. Find the valve opening at:
    (a) 62°C.
    (b) 98°C.
    (c) What is the water temperature when the valve opening is 0.84 in.?

Challenge Problems   (*Use tables of equivalents when necessary.*)

21. If milk testing 3.60% butterfat (3.60 g butterfat/100 g milk) is passed through a modern cream separator at the rate of 1.25 lb of milk each 3.80 sec, how many kilograms of butterfat would be produced (separated) in 1 hr? (Assume 100% separation.)

Hint: You will need the following ratios:

$$\frac{1.25 \text{ lb milk}}{3.80 \text{ sec}} \qquad \frac{453.59 \text{ g}}{1 \text{ lb}} \qquad \frac{3600 \text{ sec}}{1 \text{ hr}} \qquad \frac{1 \text{ kg}}{1000 \text{ g}} \qquad \frac{3.60 \text{ g butterfat}}{100 \text{ g milk}}$$

22. During the summer months, a certain reservoir loses water through evaporation at the rate of 1.50 acre-ft/mo (30 days). If the reservoir has a surface area of 42.0 acres, find the loss in gallons/day/acre of surface.

23. How many tons of cement are required to produce 10,000 concrete pipe sections if 1 cu yd of concrete will produce 12.5 pipe sections? Each cubic yard of concrete requires 5.50 cu ft of cement weighing 94.0 lb/cu ft.

24. Water flowing from a river into the water system of a town contains 22.5 g of suspended solids/liter. If 98.1% of the suspended solids are removed during the purification and treatment of the water, how many tons of suspended solids are removed from each acre-foot of water?

1.8   SQUARE ROOTS

The square root ($\sqrt{}$) of a number is a factor which, when multiplied by itself, equals the original number. For example:

$$\sqrt{25} = 5 \qquad \text{(because } 5 \times 5 = 25\text{)}$$
$$\sqrt{49} = 7 \qquad \text{(because } 7 \times 7 = 49\text{)}$$

Square roots are commonly determined by using the slide rule,* tables, logarithms, or a process called Newton's method.† We assume that the reader is already familiar with the use of the slide rule to determine square roots, so we shall consider here the use of tables and Newton's method. Finding square roots by the use of logarithms is covered in Appendix C.

Tables

Table 7, Appendix D, lists the square roots of integers from 1 to 100. To determine the square root of an integer such as 67, locate the number (67) in the left-hand column headed N. On the line with the 67 under the column headed $\sqrt{N}$, find 8.1854 for $\sqrt{67}$.

Newton's Method

To understand the principle of this method, start with a perfect square, such as 64, with a known root of 8. You know that $8 \times 8 = 64$. In general, you could

---

* With hairline set over the number on A or B scale, the square root of the number is found under the hairline on D or C, respectively.

† Developed by Sir Isaac Newton.

write

$$(\text{factor}) \times (\text{factor}) = \text{number}$$

When the factors are equal, the factor is called the square root of the number. Now suppose that you increase the first factor from 8 to 16. Then the second factor must be decreased to 4: $(16) \times (4) = 64$. This procedure tells you that if the first factor is larger than the square root, the other factor must be smaller than the square root. The square root is between the first factor and the second factor.

You can use this fact in the following way to obtain an approximation for the square root of a number.

1. Guess at the value of the square root of the given number.
2. Divide the given number by the guess.
3. The square root lies between the divisor and the quotient. If the guess is smaller than the actual root, the quotient is larger than the root.
4. Average the guess with the quotient and use this value as the next divisor. Repeat this process until the desired accuracy is obtained.

One big advantage of this method is that it is self-correcting.

example $\sqrt{210.25}$. Suppose that you guess the root to be 12:

$$\frac{17.52 \approx 18}{12\overline{)210.25}}$$

Average of 12 and 18 = 15. Divide 210.25 by 15:

$$\frac{14.01 \approx 14}{15\overline{)210.25}}$$

Average of 14 and 15 = 14.5. Divide 210.25 by 14.5:

$$\frac{14.5}{14.5\overline{)210.25}}$$

The root is 14.5.

EXERCISE 1.8

Use Newton's method to determine the following indicated square roots.

1. $\sqrt{144}$ 5. $\sqrt{10.89}$ 8. $\sqrt{64.3204}$

2. $\sqrt{225}$ 6. $\sqrt{27.04}$ 9. $\sqrt{26.2144}$

3. $\sqrt{289}$ 7. $\sqrt{144.2401}$ 10. $\sqrt{533.61}$

4. $\sqrt{529}$

Use Table 7, Appendix D, to determine answers to Problems 11-30.

11. $\sqrt[3]{86}$ 15. $35 \times \pi$ 18. $\dfrac{1000}{86}$

12. 29^3 16. 99π 19. $\sqrt{9604}$

13. $\sqrt{2601}$ 17. $\dfrac{1000}{45}$ 20. $\sqrt[3]{658,503}$

14. $\sqrt[3]{571,787}$

21. The area of a circle can be computed by the formulas $A = \pi r^2$ and $A = \pi d^2/4$ where r = radius of the circle and d = diameter of the circle. Using Table 7, Column headed $\pi N^2/4$, determine the area of a circle whose diameter is 47 units.
22. Same as Problem 21 with diameter equal to 89 units.
23. If the area of a circle is 1134.11 sq in., what is its diameter? Circumference? (*Hint:* Circumference $= \pi d$.)
24. Determine the diameter and circumference of a circle whose area is 5153 sq cm.
25. What is the approximate square root of 2250?
26. What is the approximate cube root of 16,000?
27. Determine the reciprocal of 82. ($1000/N$ is 1000 times $1/N$.)
28. Express $\frac{1}{47}$ as a decimal.
29. Express 0.0344828 as a common fraction. (*Hint:* First multiply by 1000; then use $1000/N$ column.)
30. Express 0.0112360 as a common fraction.

1.9 APPROXIMATION PROCEDURES

One important step in solving most problems is making an estimate or approximation of the answer. There are four good reasons for making approximations.

1. The approximation can be used to determine the correct location of decimal points when calculations are performed on the slide rule or on some desk calculators.

2. The approximation serves as a check to determine the reasonableness of a calculated result.

3. The mental steps involved in making an approximation give a better feel and understanding for the overall problem.

4. The information gained may save unnecessary calculation if the approximation reveals an impractical result.

For these reasons, it is essential to give attention to basic approximation procedures.

method I *Rounding each factor (or term) to nearest 1, 10, 100, 1000, etc.*

example Estimate the product of

$$295 \times 38.0 \times 4.12$$

If you round off the factors to $300 \times 40 \times 4$, you can do the arithmetic in your head and come up with 48,000 for your approximation. (The calculated answer is 46,200).

example Estimate the result for

$$\frac{2.95 \times 3.05 \times 21.4}{42.6 \times \pi \times 18.2}$$

If you round off the factors as follows:

$$\frac{3 \times 3 \times 20}{40 \times 3 \times 20}$$

you can cancel common factors, leaving

$$\frac{3 \times 3 \times 20}{40 \times 3 \times 20} \quad \text{or} \quad \frac{3}{40} \approx \frac{1}{13} \approx 0.08$$

The estimated result is 0.08. (The calculated result is 0.0790.)

method II *Using power of 10 form.*

By writing each factor in the power of 10 form and rounding the coefficient to the nearest 1, you can make estimates of answers for problems involving numerous operations.

example   Estimate the answer for

$$295 \times 36.0 \times 8.29 \times 0.0130$$

1. Write each factor in the power of 10 form to obtain $(2.95 \times 10^2) \times (3.60 \times 10^1) \times (8.29 \times 10^0) \times (1.30 \times 10^{-2})$.
2. Round off each coefficient to the nearest one. You now have $(3 \times 10^2) \times (4 \times 10^1) \times (8 \times 10^0) \times (1 \times 10^{-2})$.
3. Multiply the coefficients and add the exponents of 10 to obtain

$$3 \times 4 \times 8 \times 1 \times 10^{2+1+0-2} = 96 \times 10^1$$

or 960 for your estimation. (A desk calculator gives 11445174 for the digits of this product.) Using the estimate of 960, locate the decimal point, obtaining 1144.5174 which, rounded to three significant figures, is 1140.

example   Estimate the quotient for $\frac{7895}{36.4}$. By writing the factors in scientific form, you obtain

$$\frac{7.895 \times 10^3}{3.64 \times 10^1}$$

Rounding the coefficients, you have

$$\frac{8 \times 10^3}{4 \times 10^1} = 2 \times 10^{3-1} = 2 \times 10^2 = 200$$

(The exponent of 10 in the denominator is subtracted from the exponent of 10 in the numerator.)

The desk calculator gives 21689 for the first five digits in the quotient. Using the estimate of 200, locate the decimal point, obtaining 216.89. Rounding this off, you have 217 for the quotient.

example   Estimate the answer for the following problem. Locate the decimal point in the given calculator reading and round off to the proper number of significant figures:

$$\frac{528 \times 3.07 \times 912}{6.62 \times 88.7 \times 96.4} = \text{``2611611''}$$

(The `` '' means that the decimal point has not been located.) In scientific form, this becomes:

$$\frac{5.28 \times 10^2 \times 3.07 \times 10^0 \times 9.12 \times 10^2}{6.62 \times 10^0 \times 8.87 \times 10^1 \times 9.64 \times 10^1}$$

By rounding coefficients to the nearest 1 and adding exponents on the 10 factors in the numerator and denominator separately, you have

$$\frac{(5 \times 3 \times 9) \times 10^4}{(7 \times 9 \times 10) \times 10^2} = \frac{(5 \times 3 \times 9) \times 10^4}{(7 \times 9) \times 10^3} = \frac{15}{7} \times 10^1$$

(The exponent 3 is subtracted from the exponent 4.)

$$\frac{15}{7} \times 10 \approx 2 \times 10 \qquad \text{or} \qquad 20$$

Using the given reading and the estimate, locate the decimal point, obtaining 26.11611 which, rounded to three places, is 26.1.

example   Estimate the quotient, locate the decimal point in the given reading, and round off to the proper number of significant figures for the following:

$$\frac{62.4}{0.018} = \text{``346666''}$$

In scientific form this becomes

$$\frac{6.24 \times 10^1}{1.8 \times 10^{-2}}$$

Rounding coefficients, you have

$$\frac{6 \times 10^1}{2 \times 10^{-2}} = 3 \times \frac{10^1}{10^{-2}} = 3 \times 10^3 \qquad \text{or} \qquad 3000$$

The negative exponent ($^{-2}$) in the denominator is subtracted from the exponent (^{1}) in the numerator. (The way to subtract a negative number is to change its sign and add.)

Using the estimate of 3000, locate the decimal point in the given calculator reading, obtaining 3466.66. Rounding to two significant figures, you obtain 3500.

EXERCISE 1.9

In Problems 1–10, use Method I to estimate the answers and use the estimate to locate the decimal point in the given calculator reading. Then round off to the proper number of significant figures.

1. $256 \times 3.14 \times 82.7 \times 0.95 =$ "631536"
2. $4.8 \times 0.51 =$ "2448"
3. $5.12 \times 8.07 =$ "41318"
4. $612 \times 3.141 \times 8.09 =$ "155513"

5. $\dfrac{295}{38} =$ "77631"

6. $\dfrac{15.31}{6.6} =$ "23196"

7. $\dfrac{0.828}{0.21} =$ "39428"   (*Hint:* Multiply numerator and denominator by 10.)

8. $\dfrac{85.3 \times 7.26}{12.5} =$ "49542"

9. $\dfrac{51 \times 38 \times 76}{12 \times 27 \times 82} =$ "55438"

10. $\dfrac{12.65 \times 182.1}{0.9128 \times 51.03} =$ "494537"

In Problems 11–20, use Method II to estimate answers and locate the decimal point in the given readings. Then round off to the proper number of significant figures.

11. $286.4 \times 388 =$ "111123"
12. $4.59 \times 0.9281 =$ "425997"

13. $\dfrac{68.08}{2.02} =$ "337029"

14. $\dfrac{125,000}{3.1416}$ = "397886"

15. $\dfrac{72.14}{0.023}$ = "313652"

16. $\dfrac{82.6 \times 3.141}{53.9}$ = "481348"

17. $\dfrac{1.072 \times 3.56}{47.5}$ = "803435"

18. $\dfrac{56 \times 3.9 \times 42}{9.2 \times 18 \times 5.7}$ = "971777"

19. $\dfrac{2.035 \times 9.219 \times 0.0895}{14.02 \times 39.71 \times 8.625}$ = "349675"

20. $\dfrac{5.12 \times 3.1416 \times 0.9201}{8.08 \times 62.4 \times 12.75}$ = "230223"

2 Geometric Properties and Constructions

Much of the geometry in use today was developed by the early Greeks and Egyptians. Early geometry was the science of land measurement. From the measurement of land areas, geometry has been extended to the study and measurement of lines and figures in any flat surface (plane geometry) and, finally, to the study and measurement of three-dimensional objects (solid geometry).

Most of the geometry needed by the modern technician can be developed from a study involving three basic concepts: the plane, the point, and the line. As you proceed to study each new geometric relationship, keep in mind the roles of these three fundamental concepts.

A set of data cards is a helpful tool when learning the many basic geometric facts. Number a set of 3 in. × 5 in. unlined index cards from 1 to 82 in the upper left-hand corner on the back side. These 82 cards will serve as a "basic set." Draw a figure and ask a question on the front of each card, numbered to correspond to the 82 geometric facts listed in Appendix B. On the back of the card, write the corresponding fact or answer. Most of the 82 facts are covered in this chapter.

Figure 2-1 shows the front and back sides of five different cards. Avoid putting too much information on any one card. Use your imagination in making the cards and do not copy exactly the statements in Appendix B.

Half the value of a set of cards is obtained from the thinking involved in making them; the other half comes from using the cards for study and review purposes. Adding tabs to cards numbered 1, 4, 9, 16, 19, 49, 50, 56, 63, 76, and 79 will allow you to pull out the set of cards on a particular group of facts. For example, cards 19 through 49 cover facts about triangles.

2.1 PLANES, POINTS, AND LINES

To grasp the concept of a _plane,_ think about the flat surface of a table top or of still water. If you think of this surface being extended in length and width without limit, you approach the concept of a geometric plane. Because a plane is without thickness, you are limited to the use of only two dimensions. For example, a rectangle in a plane can have length and width but no thickness. It is helpful to think of a plane as a surface which a precision straightedge will touch at all points, regardless of the direction in which the surface is tested. In other words, a plane

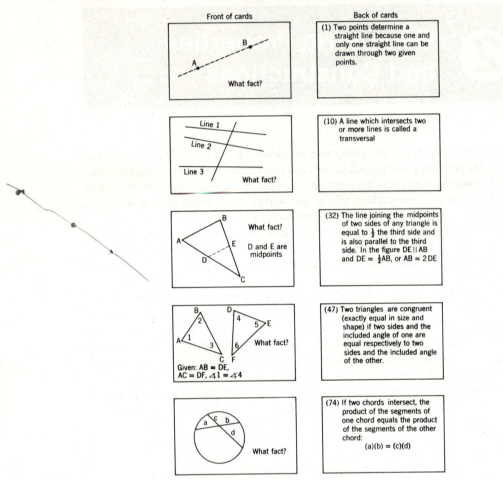

Figure 2-1   Sample Data Cards

is a surface on which any number of straight lines may be drawn (through any point on the surface).

A *point* is the location where two lines intersect. Its location may be represented by two intersecting lines with either a number or letter placed near it or by a dot with a small circle drawn around it. (See Figure 2-2.) A point has neither length nor width; it indicates position only.

A *straight line* could be thought of as the path of a point which moves in a specified direction. It may also be thought of as containing an infinite number of points. Any two given points in a plane determine a specific line, because only one straight line can be drawn through the two points. A line is usually named by lettering or numbering two of its points. Although a line has infinite length, you usually will work with only a part of a line called a _segment_. That portion between two points is a segment. A _half line_, or ray, is a line extending in only one direction from a specific point. _The shortest path between any two points is the line segment joining them._

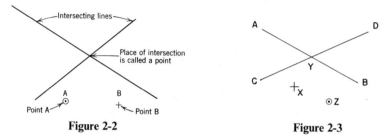

| Figure 2-2 | Figure 2-3 |

Figure 2-3 shows points X, Z, and the intersection of two line segments at point Y. Y is the point common to both lines because both lines pass through point Y. Any desired number of other lines could be drawn through point Y. Either or both of the line segments AB and CD may be extended in either direction. Extending these lines beyond points A, B, C, or D does not alter the position of these points, nor does it change the length of the line segments AB and CD. Figure 2-4 shows the lines AB and CD extended.

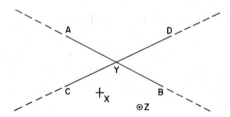

Figure 2-4   Lines AB and CD Extended

A _curved or bent line_ is one that continually changes direction. A curved line may represent the arc of a circle or an irregular curve. See Figure 2-5.

A _broken line_ might be considered as being made up of any number of intersecting straight line segments (Figure 2-6).

If a plumb bob was tied to a string and allowed to come to rest, the line represented by the position of the string would be a _vertical_ or _plumb_ line. A vertical or plumb line is a line which is straight "up and down" and, if extended, would pass through the center of the earth.

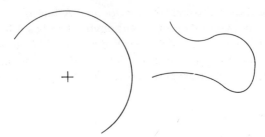

Figure 2-5   Curved Lines

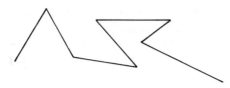

Figure 2-6   Broken Line

A *horizontal line* is a straight line which is perpendicular to a vertical line. A horizontal line is also defined as a line which is tangent to (touching but not cutting) the surface of the earth. A vertical and horizontal line lying in the same plane are perpendicular to each other. For many practical purposes (such as in plane surveying), all horizontal lines are considered perpendicular to all vertical lines whether they intersect or not.

In Figure 2-7, you are given line segment *AB* and arc *AB* denoted by the symbol $\overset{\frown}{AB}$. To *bisect the segment or arc*, use points *A* and *B* as swing points and a compass radius (distance between the steel point and pencil point) greater than half of *AB*; strike arcs intersecting at points *C* and *D*. Connect points *C* and *D* with a line to locate points *E* and *F*. Point *E* is the bisector of $\overset{\frown}{AB}$ and point *F* is the bisector of line segment *AB*.

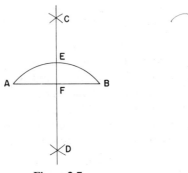

Figure 2-7

EXERCISE 2.1

1. How long is a straight line?
2. What is a point?
3. How may a straight line be named?
4. How may a particular point be located (i.e., shown)?
5. What are two common methods of naming a point?
6. What is a line extended in only one direction from a specific point called?
7. What is the only property possessed by a point?
8. How many straight lines can be drawn through two given points?
9. How many straight lines can be drawn through a particular point?
10. What determines the shortest distance between any two points?
11. How are measurements on a line always made?
12. An extended vertical or plumb line would pass through what point?
13. How are vertical lines related to horizontal lines for many practical purposes?
14. What two dimensions does a plane surface have?
15. Draw a line segment, AB, $3\frac{7}{16}$ in. long. Locate point P $1\frac{5}{8}$ in. from one end. Through this point draw a second line segment CD extending $4\frac{5}{8}$ in. on one side of AB and $2\frac{3}{16}$ in. on the other side.
16. Draw a horizontal line segment, AB, 5.70 in. long (across paper). Use your compass and ruler to bisect this line segment.
17. Draw a 5.0-in. line segment and bisect it.
18. Bisect each half of the 5.0-in. segment in Problem 17.

2.2 PERPENDICULAR LINES

When two lines intersect and form equal angles, they are said to be *perpendicular* (Figure 2-8). This fact is used in the operation of practically all machine tools such as lathes, milling machines, drill presses, grinders, and shapers. The perpendicular relationship of two legs of a triangle is basic in trigonometry with many practical uses in plane surveying, graphing, and electrical circuit problems.

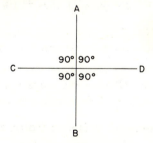

Figure 2-8 Perpendicular Lines Form Right Angles

Constructing a Perpendicular to a Line at a Given Point on the Line

To construct a perpendicular to line AB (Figure 2-9) at point A, first extend line AB. Then using any convenient radius with point A as a center, swing arcs intersecting AB at C and D. Using points C and D as center points and using a radius greater than half CD, swing arcs intersecting at E. Line EA is the required perpendicular.

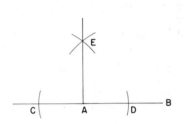

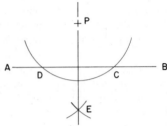

Figure 2-9 AE Constructed
Perpendicular to AB

Figure 2-10 PE Constructed
Perpendicular to AB Through
Point P

Constructing a Line Perpendicular to a Given Line from a Given Point Not on the Line

In Figure 2-10, you are given line AB and point P. With P as a center and any convenient radius, strike an arc intersecting AB at C and D. With C and D as centers and a radius greater than half CD, strike arcs intersecting at E. PE is the required perpendicular to AB.

2.3 PARALLEL LINES

Two or more straight lines in a plane are parallel if the distance between them remains constant. They may be extended indefinitely without intersecting. In Figure 2-11, AB, CD, and EF are parallel lines, but line XY, called a _transversal_, intersects all three.

Constructing a Line Parallel to a Given Line a Given Distance Away

Let AB be the given line and CD the given distance (Figure 2-12). With CD as a radius and any two points on AB as swing points, such as X and Y, strike arcs on the same side of the line. The line EF drawn tangent to both arcs is parallel to AB.

Constructing a Line Through a Given Point Parallel to a Given Line

You are given line AB and point P in Figure 2-13. With P as a center and any convenient radius, strike an arc intersecting AB at C. With C as a center and the

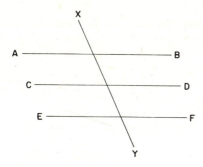

Figure 2-11 A Transversal *xy* Intersecting Three Parallel Lines

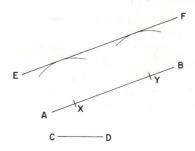

Figure 2-12 Line *EF* Constructed Parallel to *AB* a Distance *CD* from *AB*

same radius, strike an arc through *P* intersecting *AB* at *D*. With a radius equal to *PD* and *C* as a center, strike an arc intersecting the first arc at *E*. A line drawn through *P* and *E* is the required line.

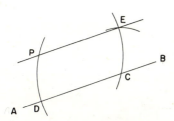

Figure 2-13 Line *PE* Constructed Parallel to *AB* Through Point *P*

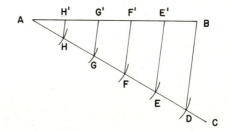

Figure 2-14 *AB* Divided into Five Equal Parts

Dividing a Line into any Number of Equal Parts

Divide *AB* (Figure 2-14) into five equal parts. Draw a construction line *AC* at any convenient angle. With a divider, compass, or scale, set off five equal divisions to locate point *D*. Connect *D* to *B*. Draw lines parallel to *DB* through points *E*, *F*, *G*, and *H* that intersect *AB*. Then $AH' = H'G' = G'F' = F'E' = E'B$.

EXERCISE 2.3

1. Draw a vertical line segment, *CD*, $6\frac{7}{16}$ in. long. Use a compass and straightedge to construct a perpendicular bisector to line *CD*.
2. Swing an arc about 4 in. long and then bisect it.

3. Draw a horizontal line segment, AB, $5\frac{7}{16}$ in. long. Measure $2\frac{1}{4}$ in. from the left end and construct a perpendicular through this point which extends $3\frac{5}{16}$ in. on each side of AB. Connect the ends of the two lines, forming a kite-shaped or diamond-shaped figure. Measure the distance around the outer edge of the figure.

4. Draw a vertical line segment, RS, 5.32 in. long. Locate point P about $2\frac{1}{2}$ in. to the right of RS. From P, construct a perpendicular line to RS, meeting RS at point Q.

5. Construct a line parallel to line PQ (Problem 4) which passes through point S.

6. Draw a horizontal line segment, AB, 5.0 in. long and divide it into three equal parts.

2.4 ANGLES

Angles are formed by the intersection of line segments. The point of intersection is called the vertex of the angles. In Figure 2-15(a), lines AB and CP intersect at point P, forming angles 1 and 2. Point P is the vertex of both angles. CP and BP are called the sides of angle 1; AP and CP are the sides of angle 2. Figure 2-15(b) shows an angle formed by two half lines (or rays). Figure 2-15(c) shows two intersecting lines forming four angles, each having its vertex at point O.

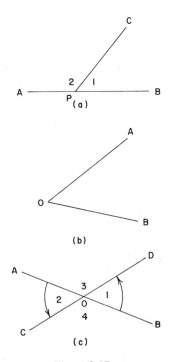

Figure 2-15

Angles 1 and 2 form a pair of angles called *vertical angles*, and angles 3 and 4 also form a pair of vertical angles. However, angles 1 and 3 are not vertical angles. If you think of line *AB* turning (in the direction of the arrows) about point *O* until it reaches line *CD*, you can see that the turning required to generate angle 1 is exactly the turning required to generate angle 2. This indicates that angle 1 equals angle 2; similarly, angle 3 equals angle 4. In general, vertical angles are equal.

Angle Notation

We shall use the symbol $\measuredangle$ to mean angle. It is customary to name a specific angle by the use of three letters. For example, you could write $\measuredangle AOC$ to designate $\measuredangle 2$ in Figure 2-15(c). When the three-letter notation is used, the center letter denotes the vertex point of the angle. When a number is used to name an angle, the number is written between the sides of the angle and near the vertex point. Sometimes an angle is named by using only a single letter at its vertex. Although you should become familiar with the three-letter notation, it is much easier to follow a discussion when angles are referred to by number. We are at liberty to number any angle in a figure to fit our own needs and to replace the corresponding three-letter notation with this number. Whenever practical, we shall use either the number or single-letter notations.

When two straight lines intersect forming four equal angles, the lines are said to be perpendicular to each other and the angles are called *right angles*. (Symbols for perpendicular are $\perp$ or $+\!\!-$). In Figure 2-16, *AB* and *CD* are $\perp$ to each other.

Figure 2-16 Line *CD* is Perpendicular to *AB*, Forming Two Right Angles

If an angle is less than a right angle (RT $\measuredangle$), it is called *acute* (Figure 2-17(a)). If an angle is greater than a right angle and less than two right angles, it is called *obtuse* (Figure 2-17(b)). A straight line is frequently called a *straight angle*, i.e., a 180° angle (Figure 2-17(c)).

Two angles having the same vertex and a common side between them are called *adjacent angles.* The sum of two adjacent angles may be less than or greater

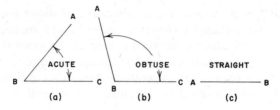

Figure 2-17   Three Common Angles

than 90°. In Figure 2-18, *B* is the vertex of the adjacent angles 1 and 2 and *DB* is the common side. *AB* and *BC* are called the exterior sides of the adjacent angles. Because the exterior sides of two adjacent 90° angles form a straight line, it is frequently called a straight angle.

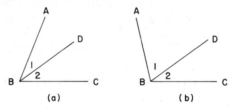

Figure 2-18   Angles 1 and 2 Are Adjacent
Angles

Two angles are called *complementary* if their sum is equal to 90°. In Figure 2-19, $\angle 3 + \angle 4 = 90°$; they are called complementary angles. Complements of the same angle or of equal angles are equal.

Two angles are *supplementary* if their sum is equal to 180°. In Figure 2-20, *AB* is a straight line intersected by *CP* at point *P*. Angles 1 and 2 are supplementary because their sum equals 180°. Supplements of the same angle or of equal angles are equal.

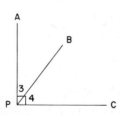

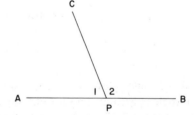

Figure 2-19   Angles 3
and 4 Are Comple-
mentary Angles

Figure 2-20   Angles 1 and 2 Are
Supplementary; $\angle APB$ Is a
Straight Angle

Bisecting an Angle

In Figure 2-21, you are given $\angle ACB$ to be bisected. With any convenient radius and C as a swing point, strike an arc intersecting CA at point X and CB at point Y. With a radius greater than half XY and using points X and Y as swing points, strike arcs intersecting at D. Draw line CD, making $\angle ACD = \angle DCB$. Line CD is called the bisector of $\angle ACB$.

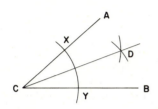

Figure 2-21 CD Constructed to Bisect $\angle ACB$

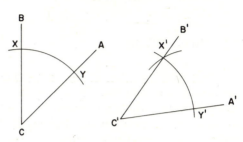

Figure 2-22

Transferring an Angle

In Figure 2-22, $\angle ACB$ is to be transferred to the new position $A'C'$. Using any convenient radius, strike arcs from centers C and C' intersecting CB at X, CA at Y, and $C'A'$ at Y'. With radius distance XY and Y' as a center, intersect the other arc at X'. Draw $C'X'$ to complete the transfer.

EXERCISE 2.4

1. How many and what kind of angles are formed by intersecting perpendicular lines?
2. What are the symbols for perpendicular?
3. If $\angle A + \angle B = 90°$, what are the two angles called?
4. If $\angle A < 90°$, it is called what kind of an angle?
5. What is the abbreviation for a right angle?
6. If $90° < \angle C < 180°$, $\angle C$ is called what kind of angle?
7. How do the opposite or vertical angles formed by two intersecting lines compare?
8. What term describes two lines that intersect to form right angles?
9. What are adjacent angles?
10. Refer to Figure 2-15(c) and use three-letter notation to write equivalent statements for:
    (a) $\angle 1 = \angle 2$.
    (b) $\angle 3 = \angle 4$.
11. Construct an angle of approximately 30°. Then construct an adjacent angle so that the two angles are complementary. (Use compass and ruler only.)
12. Construct an angle of approximately 90° and then bisect it.
13. Transfer one of the two equal angles formed in Problem 12 to a new position on the same page.

14. Construct two perpendicular lines. Bisect all four right angles. Using the common vertex point as a swing point, swing an arc cutting equal segments on all four lines (eight rays). Join adjacent end points to form a regular polygon.
15. Draw any triangle (sides about 3 to 5 in. long) and bisect all three angles. Do the bisectors meet at a common point?
16. Draw any triangle (sides about 3 to 5 in. long) and bisect all three sides. Do the bisectors meet at a common point?

2.5 POLYGONS

The word polygon means "many sides." A polygon is that portion of any plane surface having three or more sides. If all the sides enclosing the polygon are equal in length (*equilateral*) and the angles are equal in size (*equiangular*), it is called a *regular* polygon. See Figures 2-23(a), (b), and (c). A regular polygon is both equilateral and equiangular. Figures 2-23(d), (e), and (f) show three irregular polygons. The *perimeter* of a polygon is the distance around the outside edge of the polygon.

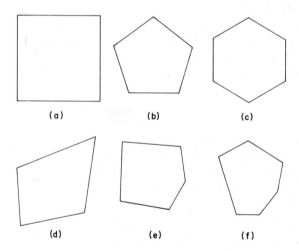

Figure 2-23 Regular and Irregular Polygons

Triangles

A triangle is a polygon formed by three straight line segments (not in the same straight line). The symbol Δ represents a triangle.

The *sides* of a triangle may be identified by using letters showing the end points of the line segments such as *AB*, *AC*, and *BC* (Figure 2-24). The side opposite an angle may be identified by the same letter (lower case) as that used for the angle

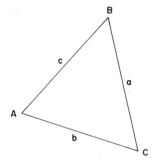

Figure 2-24

notation as *a*, *b*, and *c* in Figure 2-24. If the three sides are equal, the triangle is equilateral and equiangular (Figure 2-25(a)). If two sides of a triangle are equal, it is an *isosceles triangle* and the angles opposite the equal sides are equal (Figure 2-25(b)). If all three sides of a triangle are unequal and each angle is less than 90°, it is an *acute triangle* (Figure 2-25(c)). If one angle of a triangle is equal to 90°, it is a *right triangle*. The two sides which form the right angle are called the *legs* of the triangle and the third and longest side is called the *hypotenuse* (Figure 2-25(d)). If a triangle contains an obtuse angle (an angle greater than 90°), the triangle is called an *obtuse triangle* (Figure 2-25(e)).

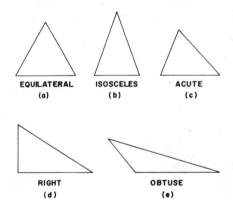

EQUILATERAL　　ISOSCELES　　ACUTE
(a)　　　　　　　(b)　　　　　(c)

RIGHT　　　　　OBTUSE
(d)　　　　　　(e)

Figure 2-25　Triangles

Constructing a triangle with three sides given. Given sides *a*, *b*, and *c* in Figure 2-26, draw one side such as *b* in the desired position. With one end point of *b* as a center and a radius equal to side *a*, strike an arc. With the other end point of side *b* as a center and a radius equal to side *c*, strike an arc intersecting the first arc. Connect this point with the end points of *b* for the required triangle.

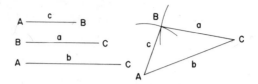

Figure 2-26   Constructing a Triangle with Specified Sides

Constructing an equilateral triangle.   With side *AB* given (Figure 2-27), and using *AB* as a radius and *A* and *B* as centers, strike arcs to intersect at *C*. Draw *AC* and *BC* to complete the triangle.

The two sides of a triangle having a common vertex are called *adjacent sides*; the angle formed between these two sides is the *included angle*. *AC* and *AB* in Figure 2-27 have a common vertex; therefore, ∠*A* (or ∠*CAB*) is the included angle.

The *base* of a triangle is usually identified as the side upon which the triangle rests. Any side may be used as the base. In Figure 2-27, *AB* is shown as the base. If a line is drawn from the vertex of the angle opposite the base, perpendicular to the base, it is called the *altitude* of the triangle. In Figure 2-28, *CM* is the altitude of △*ABC*. Although a triangle has three altitudes (one from each vertex perpendicular to the opposite side), you usually work with only one at a time.

The *perimeter* is the distance around the triangle. If the lengths of the line segments making the triangle are added together, their sum is the perimeter.

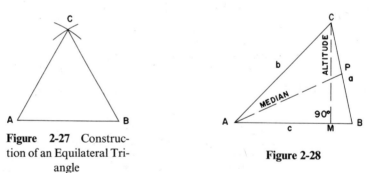

Figure   2-27   Construction of an Equilateral Triangle

Figure 2-28

A *median* of a triangle is a line drawn from any vertex to the midpoint of the opposite side. In Figure 2-28, *AP* is a *median* of △*ABC*. A triangle has three medians which intersect at a common point (Figure 2-29). The common point of intersection divides each median, forming a long segment which is twice the length of the short segment. Therefore, $AO = 2OP$ or $OP = \frac{1}{2}AO$. Additional ratios are $AO = \frac{2}{3}AP$ and $OP = \frac{1}{3}AP$.

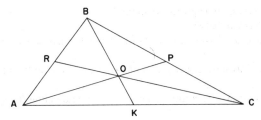

Figure 2-29 The Medians of a Triangle Intersect at
a Common Point

The *bisector* of an angle of a triangle is a line that divides the angle into two
equal angles. The bisectors of the three angles of a triangle intersect at a point
which is equidistant from the sides of the triangle. In Figure 2-30 the angle bisectors
AO, BO, and *CO* intersect at point *O*, equidistant from the three sides, making
the altitudes (of the three small triangles) $OL = ON = OM$. Point *O* would be
the center of an inscribed circle of radius *OL*.

Constructing a right triangle with the hypotenuse and one leg given. Starting
with the hypotenuse *AB* and leg *AC* (Figure 2-31), draw *AB* in the desired position.
Find the midpoint *X* of *AB*. Using *X* as a center and *XA* as a radius, draw a semicircle. With *A* as a center and a radius equal to *AC*, scribe an arc intersecting the
semicircle at point *C*. Draw *AC* and *CB* to complete the triangle.

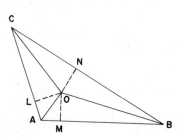

Figure 2-30 The Three Angle
Bisectors Intersect at a Common
Point

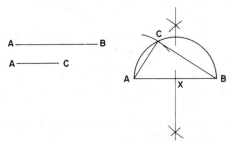

Figure 2-31 Construction of a Right Triangle with Two Sides Given

EXERCISE 2.5A

In problems not requiring constructions, answer with the appropriate word or words.

1. What is that portion of a plane bounded by three sides called?
2. What triangle has two equal sides?
3. What triangle has three equal sides?

4. What triangle has an angle greater than 90°?
5. What triangle has two legs?
6. What is the line drawn from any vertex of a triangle perpendicular to the opposite side called?
7. What is the line drawn from any vertex of a triangle to the midpoint of the opposite side called?
8. What is the line drawn from any vertex of a triangle dividing the angle into two equal parts called?
9. What is the side upon which the triangle rests usually called?
10. What is the angle opposite the base called?
11. What angles does a right triangle have?
12. What do all right triangles have in common?
13. What kinds of angles are found in an acute triangle?
14. What is the distance around a triangle called?
15. Referring to Figure 2-29, set up several ratios, using median segments *RO*, *OC*, and *RC*. Write their numeric values.
16. What is the abbreviation for "right angle"?
17. What is the symbol for "right triangle"?
18. What angle is formed by the two legs of a right triangle?
19. What relationship do the legs of a right triangle have to each other?
20. What is the longest side of a right triangle called? Where is this side located?
21. Draw any right triangle and label the vertex points *A*, *B*, and *C* so that $\angle C$ is the right angle. Then label the three sides with the proper lower case letters.
22. Describe $\angle A$ and $\angle B$ in $\triangle ABC$ (Problem 21).
23. In $\triangle ABC$ (Problem 21), $\angle A$ is opposite what side? What is side *AB* called and where is it located?
24. In $\triangle ABC$ (Problem 21), $\angle A$ could be written as $\angle BAC$. In a similar manner, how could $\angle B$ and $\angle C$ be written?
25. What is a median of a triangle? Show a sketch.
26. What do the three medians of a triangle have in common?
27. Draw a triangle having sides of approximately 3, 4, and 5 in. Then construct an inscribed circle.
28. Construct a right triangle having an hypotenuse of 2.5 in. and a leg of 2 in. Then measure the other leg.

Challenge Problems

29. Construct a triangle having one median 90 mm long and a second median 120 mm long.
30. Draw a circle with a diameter of 2.0 in. Draw in two diameters (not forming right angles). Extend the two diameters in one direction and construct a triangle so that the two extended diameters are medians of the triangle.
31. Construct a circle with a radius of 30 mm. Then construct an isosceles triangle, with a base of 100 mm, which circumscribes the circle.
32. Draw any triangle (unequal sides) and construct a circumscribed circle. (See geometric fact 35, Appendix B.)

33. Place three dots at random on a page (from 2 to 5 in. apart). Construct a circle through these three points.
34. Using information in Problems 32 and 33, draw a circle and locate its center by proper construction.

Quadrilaterals

A quadrilateral is any four-sided polygon. If all four sides are equal in length and the opposite sides are parallel, it is called a *rhombus* (Figure 2-32(e)). A rhombus with four equal angles is called a *square* (Figure 2-32(a)). A quadrilateral with

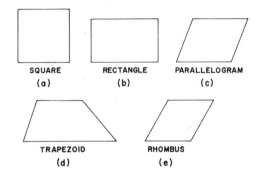

Figure 2-32   Special Quadrilaterals

opposite sides parallel and equal is called a *parallelogram* (Figure 2-32(c)). A parallelogram with four equal angles is called a *rectangle* (Figure 2-32(b)). A quadrilateral with one pair of parallel sides is called a *trapezoid* (Figure 2-32(d)). See Figure 2-33.

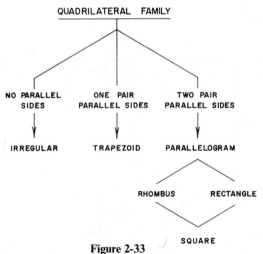

Figure 2-33

Constructing a square.   Given side *AB* in Figure 2-34, erect a perpendicular at point *A*. With *A* as a center and *AB* as a radius, draw an arc to intersect the perpendicular at *C*. With *B* and *C* as centers and *AB* as a radius, strike arcs to intersect at *D*. Draw the sides *CD* and *BD* to complete the square.

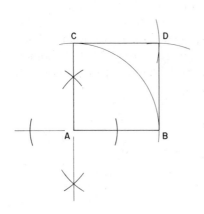

Figure 2-34 Construction of a Square with Sides Equal to *AB*

Figure 2-35 Construction of a Rectangle with Length *AB* and Width *BC*

Constructing a rectangle.   To construct a rectangle with two given sides (Figure 2-35), draw the given side *AB* in the desired position. At point *B*, construct a perpendicular. Using a compass setting of *BC* and point *B* as a center, swing an arc cutting the perpendicular at *C*. Using the same setting and point *A* as a center, swing an arc cutting an imaginary perpendicular at *D*. With *AB* as a radius and point *C* as a center, swing an arc cutting the arc at *D*. Draw sides *CD* and *DA* for the required rectangle.

Constructing a right triangle with two given legs.   To construct this triangle, follow some of the procedure for the construction of a rectangle. Referring to Figure 2-35, you need not locate point *D* but simply connect points *A* and *C*, forming the desired right triangle with legs *AB* and *BC*.

Other Polygons

Although the number of polygons is infinite, only five others are encountered frequently.

1. The *pentagon* is a polygon of five sides.
2. The *hexagon* is a polygon of six sides.
3. The *octagon* is a polygon of eight sides.
4. The *decagon* is a polygon of ten sides.
5. The *dodecagon* is a polygon of twelve sides.

Constructing a regular pentagon. Draw a circle and construct two diameters perpendicular to each other (Figure 2-36). Bisect a radius such as OB to locate point E. With a radius equal to EC and point E as center, strike an arc intersecting AB at F. With C as a center and CF as a radius, strike an arc to intersect the circle at H. Draw the chord CH; then set off (chord) distances CH around the circumference of the circle and draw the other four sides through these points.

Constructing a regular hexagon. To construct a regular hexagon with a specified side of length S, use the fact that each side of a regular hexagon is equal to the radius of the circumscribed circle. First draw a circle whose radius is equal to S (Figure 2-37). Then, using a compass setting equal to the radius of the circle and starting at any desired point on the circle, set off six chord distances around the circumference. Draw the six equal chords to form the hexagon.

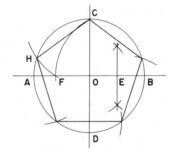

Figure 2-36 Construction of a Regular Pentagon

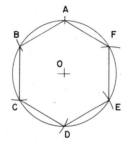

Figure **2-37** Construction of a Regular Hexagon

Constructing a regular octagon. The distance between parallel faces of a regular polygon is called the distance across the flats. To construct a regular octagon with a specified distance across the flats, first draw a square with sides equal to the required distance (Figure 2-38). Draw diagonals AD and BC. With

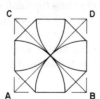

Figure 2-38 Construction of a Regular Octagon

the corners of the square as centers and with half the diagonal as a radius, scribe arcs cutting the sides of the square. Draw the sides of the octagon connecting these points.

EXERCISE 2.5B

1. What is a polygon having six equal sides and six equal angles called?
2. Construct a regular decagon (ten sides) by first constructing a regular pentagon and bisecting the five equal arcs of the circumscribed circle.
3. Construct a regular hexagon with sides equal to 3 in. From the six vertices, draw line segments to the center of the circumscribed circle. What comments can be made about the six triangles?
4. Construct a regular octagon with the distance across the flats equal to 4 in.
5. Construct a regular octagon with a 3-in. diagonal by drawing a circle whose diameter equals the diagonal. (Construct perpendicular diameters and bisect the four arcs.)
6. Construct an equilateral triangle whose sides equal 1 in. Then construct a second equilateral triangle adjacent to the first (sharing a common side). Continue in this manner to complete a regular hexagon.
7. Construct a regular dodecagon (polygon with twelve sides) by bisecting the six arcs obtained in constructing a hexagon.
8. Construct a trapezoid with parallel sides of $3\frac{1}{8}$ in. and $5\frac{3}{8}$ in. which are $2\frac{3}{4}$ in. apart. Is this trapezoid unique? That is, could you draw any other trapezoid meeting these same specifications?
9. Construct a rectangle which is twice as long as it is wide.
10. How can you locate the center of any regular polygon with an even number of sides? With an odd number of sides? Can you devise a method which will work for any regular polygon?
11. Using a pencil and ruler, construct a rhombus in less than 10 sec.
12. What figure is formed by the intersection of two streets of equal width which:
    (a) Meet at right angles?
    (b) Do not meet at right angles?
13. Construct a rhombus and a circle which passes through its vertices. (Think.)
14. Construct a circle whose radius equals 3.00 in. on 5 squares/in. engineering work paper. Construct a regular hexagon ($S = 3.00$ in.). Estimate the area of the hexagon as a percent of the area of the circle. Then by counting the small squares and fractional parts (1 small square = $\frac{1}{25}$ sq in.), compute the area of the hexagon as a percent of the area of the circumscribed circle. The area of a circle = (3.1416) × radius × radius).
15. Using the results of Problem 14, compute the weight of a hexagonal bronze plate 4.00 in. thick with a diagonal measurement of 20.0 in. if bronze weighs 555 lb/cu ft.

2.6 THE CIRCLE

A circle (Figure 2-39) is a closed curve lying in a plane, all points of which are equidistant from a fixed point called the *center*. The *diameter* of a circle is a straight-line segment drawn through the center and terminated at both ends by the circle. Any diameter divides the circle into two equal parts called semicircles. The *radius* of a circle is a straight line drawn from its center to any point on the circle. In Figure 2-39, OC is a radius and equals one-half the diameter AB. The *circumference* of a circle is the distance around the circle (sometimes called the perimeter). The circumference is an arc of 360°.

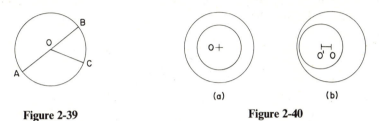

Figure 2-39   **Figure 2-40**

Concentric Circles

Circles are said to be concentric if they have the same center but unequal radii (Figure 2-40(a)). If the circles are not concentric, their centers may be joined by a straight line segment called the *line of centers* (Figure 2-40(b)). OO' is the line of centers.

Chords

A chord is any straight line segment with both end points on the circle. In Figure 2-41, segments AB, CD, and EF are all chords. The *diameter* is a special chord which passes through the center of the circle. Any chord, other than a diameter, subtends a minor and a major arc. Unless otherwise specified, the minor arc is referred to as the subtended arc.

Chords of equal lengths (in the same circle or in equal circles) subtend equal arcs. Such chords are also equidistant from the center.

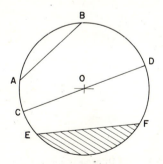

Figure 2-41

The area bounded by a chord and its subtended arc is called a *segment* of the circle (shaded area in Figure 2-41).

The perpendicular bisectors of two or more chords intersect at the center of the circle.

Secants

A secant is any straight line that intersects a circle at two distinct points. See Figure 2-42, line *CD*.

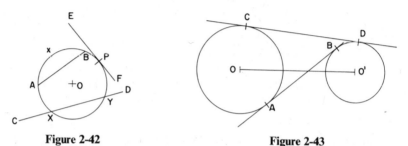

Figure 2-42   Figure 2-43

Tangents

A tangent is a line that touches the circle at only one point. In Figure 2-42, line *EF* is tangent to the circle at point *P*. The circle is also said to be tangent to the line.

A line which is tangent to two circles is called a *common* tangent. In Figure 2-43, line *OO'* is the line of centers. When the line of centers is cut by a tangent (line *AB*), the tangent is called an *internal tangent*. If the common tangent line (*CD*) does not cut the line of centers, it is an *external tangent*.

Tangent circles are circles that are tangent to the same line at the same point. Two circles are tangent *internally* if one lies within the other (Figure 2-44(a)).

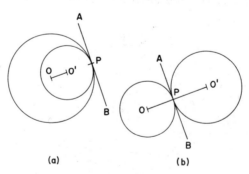

(a)   (b)

Figure 2-44   (a) Internally Tangent Circles   (b) Externally Tangent Circles

They are tangent *externally* if they lie on opposite sides of the common tangent (Figure 2-44(b)).

Central Angle

An angle with its vertex at the center of a circle is called a central angle. In Figure 2-45, $\angle AOB$ is a central angle. A central angle is measured by its intercepted arc. Thus, a central angle of 35° intercepts an arc of 35°, and an arc of 35° subtends a central angle of 35°.

Figure 2-45

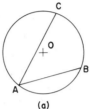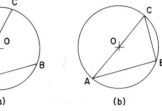

(a)   (b)

Figure 2-46

Inscribed Angle

An inscribed angle is an angle whose vertex is on the circle and whose sides are chords of the circle. In Figure 2-46(a), $\angle BAC$ is an inscribed angle with vertex point A on the circle. An inscribed angle is measured by one-half its intercepted arc. In Figure 2-46(b), AC is a diameter; $\angle ABC$ is an inscribed angle. What kind of triangle is $\triangle ABC$?

EXERCISE 2.6

For those problems not requiring construction, answer with the proper words.

1. What is a line segment joining two points on the circumference of a circle called?
2. How is the diameter of a circle related to the radius?
3. What is the distance from the center of a circle to the circle called?
4. What is another name for the perimeter of a circle?
5. What are circles having the same center called?
6. Draw three circles having a common line of centers.
7. Draw three mutually tangent circles having a common line of centers.
8. What name is given to the longest chord that can be drawn in a circle?
9. A chord equal in length to the radius of a circle would form the side of what regular inscribed polygon?
10. What is the measure (in degrees) of an arc formed by a chord equal in length to the radius of the circle?

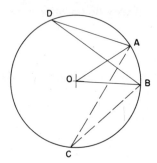

Figure 2-47

11. Construct a circle and a chord which subtends an arc of 45°. (Use compass and ruler only.)
12. Construct five circles having one common tangent line.
13. Draw a line segment AB and then construct a circle tangent to AB.
14. Draw a line segment AB approximately $2\frac{1}{2}$ in. long. Then construct a circle with a radius of 2 in. passing through points A and B.
15. If arc $\overset{\frown}{AB}$ in Figure 2-47 is $\frac{1}{12}$ of the circumference, what is the measurement of $\angle AOB$? $\angle ACB$? $\angle ADB$?
16. Explain why the largest angle that can be inscribed in a semicircle is a right angle.
17. Describe and illustrate a simple method of constructing a right triangle with a given hypotenuse.
18. Construct three equal inscribed angles in the same circle.
19. Tangents drawn to a circle from an external point are equal in length and make equal angles with the line joining the point to the center of the circle. Select any point outside a circle and draw a figure to illustrate this statement.

Challenge Problems

20. An inscribed angle intercepts an arc of 36.0° on a circle whose radius is 5.00 in.
    (a) What is the length of the intercepted arc?
    (b) What is the magnitude of the central angle formed by the arc in part (a).
21. What is the magnitude of an inscribed angle that intercepts an arc which measures $2.40 \times 10\,\text{cm}$ on a circle whose radius is $1.20 \times 10\,\text{cm}$?

2.7 THE THEOREM OF PYTHAGORAS

The famous Greek mathematician Pythagoras (580–501 B.C.) was the first to give a proof of the relationship that the square on the hypotenuse of a right triangle equals the sum of the squares on the other two sides. When applied to a right triangle with legs a and b and hypotenuse c, this theorem (which bears his name), results in the algebraic statement $a^2 + b^2 = c^2$ or $c^2 = a^2 + b^2$. Figure

2-48 illustrates this theorem with a special right triangle known as the 3-4-5 right triangle.

The relationship expressed in the theorem is one of the basic foundations upon which the subject of trigonometry was developed. The theorem allows you to solve

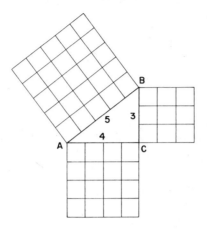

Figure 2-48   The Pythagorean Theorem

for an unknown side of a right triangle when any two of its sides are known. For example, a triangle with legs of 8 in. and 6 in. would have a hypotenuse side c such that

$$c^2 = (8 \text{ in.})^2 + (6 \text{ in.})^2$$
$$c = \sqrt{64 \text{ in.}^2 + 36 \text{ in.}^2}$$
$$c = \sqrt{100 \text{ in.}^2} = 10 \text{ in.}$$

You can check this answer by constructing a right triangle with 6-in. and 8-in. legs and measuring the hypotenuse.

The algebraic formulas developed from this theorem are:

$$1. \ c = \sqrt{a^2 + b^2}$$
$$2. \ a = \sqrt{c^2 - b^2}$$
$$3. \ b = \sqrt{c^2 - a^2}$$

where c equals the length of the hypotenuse and a and b are the lengths of the two legs.

example Solve for the unknown leg b of a right triangle whose hypotenuse $h = 17$ in. and leg $a = 15$ in. By formula 3 above, $b = \sqrt{h^2 - a^2}$.

$$b = \sqrt{(17 \text{ in.})^2 - (15 \text{ in.})^2}$$

$$b = \sqrt{289 \text{ in.}^2 - 225 \text{ in.}^2}$$

$$b = \sqrt{64 \text{ in.}^2}$$

$$b = 8.0 \text{ in.}$$

EXERCISE 2.7

In the following problems, a and b designate the legs of a right triangle and c designates the hypotenuse. Solve these problems by constructing the triangles to scale and measure the unknown side. Check with the Pythagorean theorem.

1. If $a = 10$ in. and $b = 24$ in., solve for c.
2. If $c = 25$ mm and $b = 24$ mm, solve for a.
3. If $c = 13$ ft and $a = 5.0$ ft, solve for b.
4. If $a = 8.5$ in. and $b = 7.1$ in., solve for c.
5. If $c = 11.5$ cm and $a = 3.25$ cm, solve for b.
6. A surveyor runs a traverse having three courses forming a right triangle. If the two courses forming the legs of the right triangle are 324 ft and 512 ft, determine the course distance along the hypotenuse side of the triangle.

2.8 PARALLEL LINES AND TRANSVERSALS

A line intersecting two or more lines is called a *transversal*. Figure 2-49(a) and (c) shows two parallel lines cut by a transversal; Figure 2-49(b) shows two nonparallel lines cut by a transversal.

A transversal forms eight angles when intersecting two lines. The four angles formed between the two lines are *interior angles*. The four angles formed outside the two lines are *exterior* angles. In Figure 2-49, $\angle$'s 1, 5, 8, and 4 are exterior angles and $\angle$'s 6, 2, 3, and 7 are interior angles. Two nonadjacent interior angles which lie on opposite sides of the transversal are called *alternate interior angles*. Two nonadjacent exterior angles which lie on opposite sides of the transversal are called *alternate exterior angles*. In Figure 2-49, $\angle 2$ and $\angle 3$ are alternate interior angles and $\angle 6$ and $\angle 7$ are also alternate interior angles. But $\angle 1$ and $\angle 4$ are alternate exterior angles and $\angle 5$ and $\angle 8$ are also alternate exterior angles.

Two angles having the same relative positions with respect to the pair of lines and the transversal are called *corresponding angles*. In Figure 2-49, $\angle 5$ and $\angle 7$, $\angle 2$ and $\angle 4$, $\angle 1$ and $\angle 3$, and $\angle 6$ and $\angle 8$ are pairs of corresponding angles.

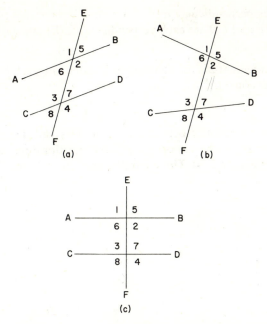

Figure 2-49

When parallel lines are cut by a transversal, several important relationships result.

1. If two parallel lines are cut by a transversal, the alternate interior angles are equal. You can see from this statement that $\angle 6 = \angle 7$ and $\angle 2 = \angle 3$ in Figure 2-49(a).

2. If two lines are cut by a transversal so that a pair of alternate interior angles are equal, the lines are parallel. If you know that $\angle 2 = \angle 3$ or that $\angle 6 = \angle 7$ in Figure 2-49(a), then you would also know that lines AB and CD are parallel.

3. If two parallel lines are cut by a transversal, the interior angles on the same side of the transversal are supplementary. If the lines in Figure 2-49(a) are parallel then $\angle 2 + \angle 7 = 180°$ and $\angle 3 + \angle 6 = 180°$.

4. If two lines are cut by a transversal so that two interior angles on the same side of the transversal are supplementary, the lines are parallel. If in Figure 2-49(a) $\angle 2$ and $\angle 7$ or $\angle 3$ and $\angle 6$ are known to be supplementary, then lines AB and CD are parallel.

5. If a line is perpendicular to one of two parallel lines, it is perpendicular to the other also. In Figure 2-49(c) if EF is perpendicular to AB, it also is perpendicular to CD. If two parallel lines are cut by a perpendicular transversal, all eight angles formed are right angles.

6. Two lines perpendicular to the same line are parallel to each other. If in Figure 2-49(c) *AB* and *CD* are each perpendicular to *EF*, they are parallel to each other.

Note that statement 2 is the converse of 1, statement 4 is the converse of 3, and statement 6 is the converse of 5.

Proportions

A statement of equality between two ratios is called a *proportion*. Many proportions arise from the properties of, and relationships among, geometric figures. Consider Figure 2-50. Three parallel lines L_1, L_2, and L_3 cut transversal

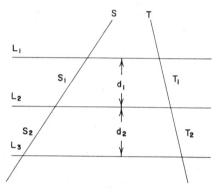

Figure 2-50   Parallel Lines Cut All Transversals Proportionally

S into segments S_1 and S_2 and cut transversal T into segments T_1 and T_2. Important relationships exist among these segments. If the distance d_1 equals the distance d_2, then segment $S_1 = S_2$ and $T_1 = T_2$. If distance d_1 is twice distance d_2, then $S_1 = 2S_2$ and $T_1 = 2T_2$. If $d_1 = \frac{1}{2}d_2$, then $S_1 = \frac{1}{2}S_2$ and $T_1 = \frac{1}{2}T_2$. By expressing the ratio of the two distances d_1 and d_2 as d_1/d_2, you can write equivalent ratios of S_1/S_2 and T_1/T_2; that is,

$$\frac{d_1}{d_2} = \frac{S_1}{S_2} = \frac{T_1}{T_2}$$

This can be summarized as follows: Parallel lines cut transversals into segments which are proportional to the distances between the parallel lines.

In addition to the relationships previously expressed, you can also write:

$$\frac{S_1}{T_1} = \frac{S_2}{T_2} \qquad \frac{S_1}{d_1} = \frac{S_2}{d_2}$$

$$\frac{d_1}{T_1} = \frac{d_2}{T_2} \qquad \frac{S_1}{S_1 + S_2} = \frac{d_1}{d_1 + d_2}$$

Proportions may be solved by using the slide rule or by algebraic operations. Algebraic solutions are discussed in Chapters 3 and 4. For reference and review purposes, we include here a brief discussion of the solution of simple proportions with the slide rule.

Slide Rule Solution of Proportions

To start, consider the ratio $\frac{1}{2}$. If you place 1 on the C scale over 2 on the D scale, you can observe the following equal ratios:

$$\frac{2\,(\text{on } C)}{4\,(\text{on } D)}, \qquad \frac{3\,(\text{on } C)}{6\,(\text{on } D)}, \qquad \frac{6\,(\text{on } CF)}{12\,(\text{on } DF)}, \qquad \frac{7\,(\text{on } CF)}{14\,(\text{on } DF)}$$

and many others.

example Solve the following proportion for N:

$$\frac{1}{2} = \frac{7.5}{N}$$

Set up the ratio $\frac{1}{2}$ as before, move H (hairline) to 7.5 on CF, and read $N = 15$ on DF. If you use A and B scales instead of C and D, you need not change scales to place H over any desired setting.

example Solve the following proportion for K:

$$\frac{3.75}{8.22} = \frac{K}{12.3}$$

1. Approximate K as being a little less than 6 or about 5.5.
2. Move the slide to place 3.75 on C over 8.22 on D.
3. Move H to 123 on DF.
4. Read "561*" under H on CF.
5. Use the approximation in Step 1 to obtain 5.61 for K.

EXERCISE 2.8

For Problems 1–10, refer to Figure 2-49(a).

1. If $AB\|CD$, what do you know about $\angle 1$ and $\angle 3$? What angles equal $\angle 8$?
2. List the pairs of equal corresponding angles if $AB\|CD$.
3. If AB is not parallel to CD, list the pairs of equal angles.

* Quotation marks indicate the digits of a number without regard to the location of the decimal point.

4. If $AB\|CD$, what two angles equal $\angle 5$ plus $\angle 6$?
5. If $\angle 4 = \angle 2$, what do you know about lines AB and CD?
6. If $\angle 8 = \angle 7$, what do you know about lines AB and CD?
7. If $AB\|CD$, list the pairs of alternate interior angles.
8. If $AB\|CD$, list the pairs of alternate exterior angles.
9. If $AB\|CD$, list the angles that are supplements of $\angle 7$.
10. If $AB\|CD$, list all angles equal to $\angle 7$.

For Problems 11–14, refer to Figure 2-49(b).

11. If AB is not parallel to CD, what do you know about $\angle 1$ and $\angle 3$?
12. If $\angle 1 = \angle 8$, what would you know about lines AB and CD?
13. List all angles equal to $\angle 1$.
14. If $\angle 2 = \angle 7$, would $\angle 5 = \angle 4$? Why?

For Problems 15–20, refer to Figure 2-49(c).

15. If $\angle 1 = \angle 3$, what would you know about $\angle 5$ and $\angle 7$? Why?
16. If $AB\|CD$, list all angles that are supplements of $\angle 3$.
17. List all angles equal to $\angle 8$ if $AB\|CD$.
18. If $\angle 5 = \angle 8$, what would you know about lines AB and CD?
19. If $\angle 5 \neq \angle 7$, what would you know about lines AB and CD?
20. If $AB \perp EF$ and $\angle 4 = 1$ RT $\angle$, what do you know about:
    (a) Lines CD and EF?
    (b) Lines CD and AB?
    (c) Angles 1, 2, 5, and 6?

In Problems 21–26, refer to Figure 2-50. Assume that lines L_1, L_2, and L_3 are parallel.

21. What segment is needed to complete the proportion

$$\frac{S_1}{T_1} = \frac{?}{T_2}$$

22. Write two other ratios equal to S_2/S_1.
23. Write a proportion involving $S_2/(S_1 + S_2)$.
24. If $d_1 = \frac{5}{7}d_2$, how does S_1 compare with S_2? How does d_2 compare with d_1?
25. If $S_1 = 12$ ft, $S_2 = 15$ ft, $d_1 = 8.0$ ft, and $T_2 = 12.5$ ft, determine the lengths of d_2 and T_1.
26. If $L_1\|L_2$ and $L_3\|L_2$, what do you know about L_1 and L_3?

2.9 CONGRUENT AND SIMILAR TRIANGLES

Two triangles having the same shape and size are called *congruent*. To be more specific, two triangles are congruent when their corresponding angles are equal *and* their corresponding sides are equal. To determine the congruency of two triangles, it is necessary to know the equality of only three corresponding parts (one of which must be a side). Two triangles are congruent (exactly the same) if:

1. All three sides of one are equal to all three sides of the other.

OR

2. Two sides and the included angle of one are equal to two sides and the included angle of the other.

OR

3. Two angles and their included side (common side) of one are equal to two angles and the included side of the other.

Additional conditions of congruency are listed in Appendix B.

Two triangles are called *similar* when the three angles of one equal the three angles of the other. For example, if a specific triangle has angles of 20°, 75°, and 85°, any other triangles with angles of 20°, 75°, and 85° would be similar to the first triangle. In fact, all triangles with these angles would be similar.

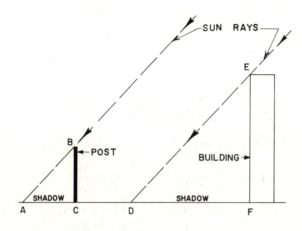

Figure 2-51

Figure 2-51 shows two similar triangles, $\triangle ABC$ and $\triangle DEF$. The rays of the sun are considered parallel so $\angle A$ and $\angle D$ are equal. (Corresponding angles formed by a transversal cutting parallel lines are equal.) Also, $\angle C$ and $\angle F$ are both right angles and $\angle B$ must equal $\angle E$ because $\angle A + \angle C + \angle B = 180° = \angle D + \angle F + \angle E$. Therefore, from the definition, the two triangles are similar.

If you start with any $\triangle ABC$ (Figure 2-52(a) or 2-52(b)), extend two sides (such as AB and AC), and construct $B'C'$ parallel to BC, then $\triangle ABC$ is similar to $\triangle AB'C'$. To establish the equality of corresponding angles, recall that corresponding angles formed by parallel lines and a transversal are equal. Now visualize (or draw) a line through point A parallel to BC and $B'C'$. Then from the discussion (Section 2.8) regarding the proportions which exist among segments of transversals

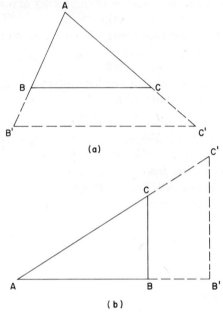

(a)

(b)

Figure 2-52

cut by three parallel lines, you are able to write the following:

$$1.\ \frac{AB}{AC} = \frac{AB'}{AC'} \qquad 2.\ \frac{AC}{AB} = \frac{AC'}{AB'}$$

$$3.\ \frac{AB}{AB'} = \frac{AC}{AC'} \qquad 4.\ \frac{AB}{BC} = \frac{AB'}{B'C'}$$

There are many more proportions in addition to these four. Note that in number 4, BC and $B'C'$ are not segments of transversals but are corresponding

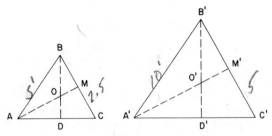

Figure 2-53   Corresponding Parts of Similar Triangles
Are Proportional

sides of the two triangles. Without attempting to prove number 4, we shall simply state that *corresponding parts* (altitudes, sides, medians, perimeters) *of similar triangles are proportional.* Another way of stating this fact is: Ratios formed by corresponding parts of similar triangles are equal. To illustrate this fact further, consider the similar $\triangle$'s *ABC* and *A'B'C'* in Figure 2-53. If each side in $\triangle A'B'C'$ is twice the length of the corresponding side of $\triangle ABC$, then:

$$\text{altitude } B'D' = 2 \times BD$$

$$\text{perimeter of } A'B'C' = 2 \times \text{perimeter of } ABC$$

$$\text{median } A'M' = 2 \times AM$$

$$\frac{\text{altitude}}{\text{median}}(\triangle ABC) = \frac{\text{altitude}}{\text{median}}(\triangle A'B'C')$$

$$\frac{BD}{AC} = \frac{B'D'}{A'C'}$$

$$\frac{BD}{B'D'} = \frac{AC}{A'C'}$$

and many others.

EXERCISE 2.9

1. What are two triangles having the same size and shape called?
2. Explain why congruent triangles are also similar triangles.
3. Explain why similar triangles may be congruent but are usually not congruent.
4. True or false: Only the sides of similar triangles are proportional. Give reasons for your answer.
5. Refer to Figure 2-53. If $\angle A' = \angle A$, $A'C' = 2AC$, and $A'B' = 2AB$, what do you know about the triangles?
6. Refer to Figure 2-53. If $AB = A'B'$, $\angle A = \angle A'$, and $AC = A'C'$, what do you know about the two triangles?
7. If two triangles are similar, what do you know about their corresponding altitudes?
8. If two triangles are congruent, what do you know about their respective areas?
9. Construct any $\triangle ABC$ and then construct a smaller similar $\triangle A'B'C'$ such that $A'B' = \frac{1}{2}AB$.
10. Using only six different line segments, draw four similar triangles. (*Hint:* Draw four parallel lines first.)
11. Construct any $\triangle ABC$ and locate the midpoints of two of its sides. Connect these two midpoints with a line segment.
    (a) Give reasons why this line segment is parallel to the third side.
    (b) Give reasons why this segment is equal to one-half the length of the third side.

12. Referring to the similar triangles in Figure 2-53 with medians AM and $A'M'$ and altitudes BD and $B'D'$, what number, segment, or ratio is needed to complete each of the following proportions?

   (a) $\dfrac{AM}{MC} = \dfrac{A'M'}{?}$ (e) $\dfrac{BM}{MC} = \dfrac{?}{1}$

   (b) $\dfrac{BC}{AC} = \dfrac{?}{A'C'}$ (f) $\dfrac{AB}{AB + BC} = \dfrac{A'B'}{?}$

   (c) $\dfrac{AO}{OM} = \dfrac{?}{O'M'}$ (g) $\dfrac{AD}{DC} = \dfrac{?}{?}$

   (d) $\dfrac{AO}{AM} = \dfrac{A'O'}{?}$ (h) $\dfrac{AO}{BO} = \dfrac{?}{?}$

13. Refer to the similar triangles in Figure 2-53 for the following: If $AB = 6.0$ ft, $A'B' = 12$ ft, $MC = 3.0$ ft, how long is $B'C'$? (AM and $A'M'$ are medians.)

14. Referring to Figure 2-54 (showing parallel lines L_1, L_2, and L_3), explain why $\triangle AOB$ and $\triangle COD$ are similar.

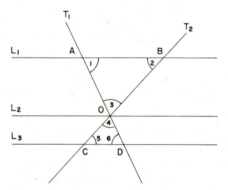

Figure 2-54

2.10   APPLICATIONS

Technicians in many fields encounter problems which require knowledge of geometry for their solution. This section contains a large variety of such problems. Most of them can be solved by using the relationships dealing with similar or congruent triangles, properties of circles, parallel lines and transversals, and the Pythagorean theorem.

Some suggestions for solving these problems follow.

1. Construct a scale drawing of the figure when appropriate. This drawing will give perspective to the relationships among the data and will also help you

to approximate the solution. Frequently the desired solution can be measured directly from the scale drawing.

    2. If two or more circles are involved, it may prove helpful to draw their line of centers.

    3. It is usually a good idea to draw a radius from the center of a circle to any point of tangency. Remember that such a radius is perpendicular to the tangent line.

    4. Watch for proportional segments of lines formed by parallel lines and for proportions resulting from the recognition of similar triangles.

    5. In some problems it may be helpful to construct a needed line segment such as an altitude, a bisector, or a third parallel line.

EXERCISE 2.10

1.   Determine the magnitudes of $\angle$'s A, B, C, D, and E in Figure 2-55.
2.   An exterior angle of a triangle is 96°. One opposite interior angle is 72°. Find the other two interior angles. [30]*
3.   A conveyor runs from point A to point B in Figure 2-56. Determine $\angle \theta$. [8]
4.   A surveyor ran a closed traverse shown in Figure 2-57. Determine angle x. [56]

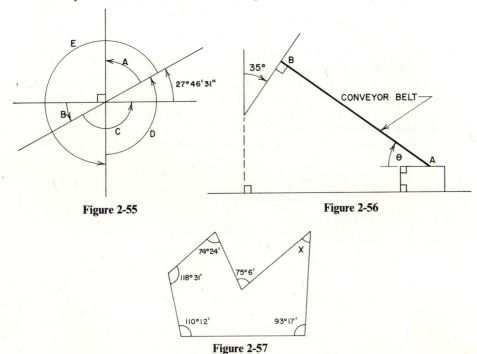

      Figure 2-55   **Figure 2-56**

                     Figure 2-57

    * Numbers in brackets refer to geometric facts with corresponding numbers in Appendix B.

5. Three parallel pipes (Figure 2-58) P_1, P_2, and P_3 are supported by two braces which intersect at O. Determine ∡'s A, B and C. [4] [11] [12] [19] [30]

6. Determine ∡A and side c in Figure 2-59. [18] [28]

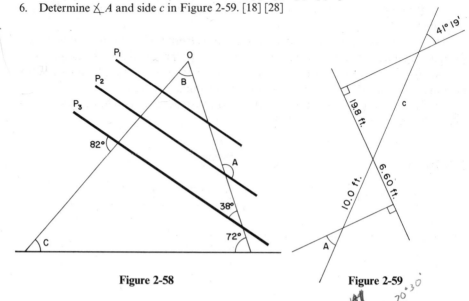

Figure 2-58

Figure 2-59

7. A streetlight is supported as shown in Figure 2-60. Determine ∡A and ∡B. [23]

8. In Figure 2-61, find the lengths of segments a and b. [28] [33]

9. Figure 2-62 shows the cross-sectional view of a flow gage orifice with a right triangular-shaped bar placed inside. The bar has legs of 2.500 mm and 1.625 mm and touches at points A, B, and C. Does AB pass through the center of the circle? If it does, determine the inside diameter of the orifice. [42] [65] [66] [67]

10. Use the parallel line method to divide a $5\frac{3}{4}$-in. line segment into three equal segments.

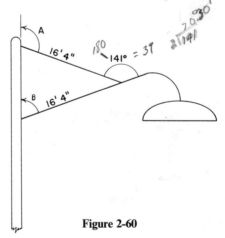

Figure 2-60

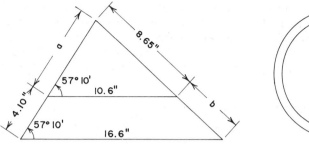

Figure 2-61

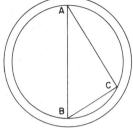

Figure 2-62

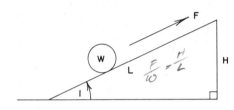

Figure 2-63

Figure 2-63 shows a weight W being held on an inclined plane by a force F. L is the length of the inclined plane and H is the height (frequently called rise). Neglecting friction, the ratios F/W and H/L are equal; i.e., F/W = H/L. If this proportion is solved for F, you obtain the formula F = (H/L) × W.

11. If $W = 526$ lb, $H = 12.5$ ft and $L = 25.0$ ft, what force would be needed to hold the weight on the ramp?

12. What is the magnitude of angle 1 (Problem 11)? (This angle is called the angle of inclination.) [26]

13. Neglecting friction, what force is required to move a 284-lb weight up a ramp inclined at 30° with the horizontal? [26]

14. A winch which will exert a maximum force of 840 lb is to be used to pull a loaded ore car weighing 6200 lb up an inclined ramp to a height of 19 ft. What must be the minimum length of the ramp?

15. A conveyor belt loaded with crushed oil shale has an average loaded weight of 18.5 lb per running foot. Neglecting friction, what force must be applied to move the belt if the conveyor is 528 ft long and must elevate the shale to a height of 88 ft?

16. Two angles of a triangular-shaped die are 48° 24′ and 83° 12′. The included side is 12.4 cm. What is the length of the shorter missing side? [24] [25]

17. The sides of a triangle are 160 mm, 140 mm, and 120 mm long. A new triangle is formed by line segments joining the midpoints of the sides of the original triangle. What is the perimeter of the new triangle? [32]

18. Four pipes (Figure 2-64) are to be encased in plastic within a square conduit. If the outside diameter (O.D.) of each pipe is 12.0 in., what is the cross-sectional area of the plastic? (*Hint:* Draw two perpendicular common tangent lines.) (Area $\odot = \pi r^2$.)

19. In Figure 2-65, determine $\angle 1$ if $\angle 2 = 34°$, $\angle 3 = 80°$, and points A, B, and C lie on a circle which is tangent to PO at point C. (*Hint:* Construct the circle through points A, B, and C.) [65] [73]

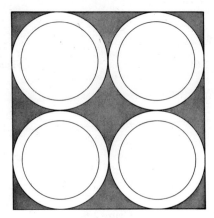

Figure 2-64

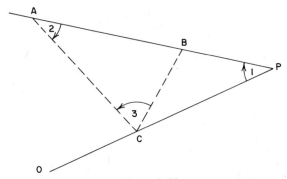

Figure 2-65

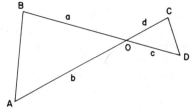

Figure 2-66

20. In Figure 2-66, points A, B, C, and D lie on the same circle. If $c = \frac{1}{2}b$ and $AB = 54$ units, determine the length of chord CD. [28] [29] [71]

21. Three holes are to be drilled in a rectangular plate as shown in Figure 2-67. Determine distances x, y, z, and w. (Slide rule accuracy is acceptable.) [13]

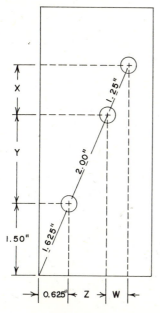

Figure 2-67   **Figure 2-68**

22. Figure 2-68 shows a triangle with segments joining the midpoints D, E, and F. The following statements are all true. Give the number of a geometric fact in Appendix B that supports the statement. For statements or reasons not listed in Appendix B, write the reason in your own words.

   (a) $EF \| AB$
   (b) $EF = \frac{1}{2}AB$
   (c) $DB = \frac{1}{2}AB$
   (d) $EF = DB$
   (e) $\angle 8 = \angle 11$

   (f) $DE \| FC$
   (g) $\angle 7 = \angle 12$
   (h) $\angle 9 = \angle 10$
   (i) $\triangle 3$ is congruent to $\triangle 4$

Challenge Problems

23. Referring to Figure 2-68, state sufficient facts to show that $\triangle 1$ is congruent to $\triangle 2$. (Points F, D, and E are midpoints).

24. In Figure 2-69, lines L_1, L_2, and L_3 are parallel. Determine distances x, y, and z. [37] [13] [28]

25. Figure 2-70 shows two belt-driven wheels. For efficient operation the arc of contact should not be less than $130°$. What is the arc of contact for the driver shown? [11] [18] [50]

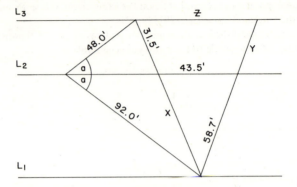

Figure 2-69

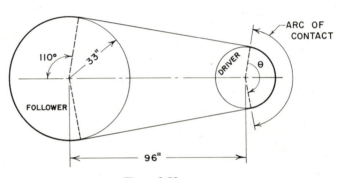

Figure 2-70

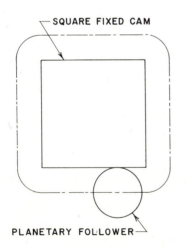

Figure 2-71

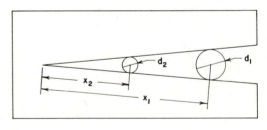

Figure 2-72

26. Figure 2-71 shows the path traced by the center point of a planetary cam follower as it moves around a square cam. Find the perimeter of the path in terms of e and r where e = edge of the square and r = radius of the follower.

27. Figure 2-72 shows a wire gage. If $d_1 = 0.385$ in. when $x_1 = 2.64$ in., find d_2 when $x_2 = 1.48$ in. [28]

28. Towns A and B are to be supplied with water from a pumping station at point O on the river (Figure 2-73). When $\angle 1 = \angle 2$, the total length of pipe is a minimum. Find the minimum length of pipe required. (Assume three figure accuracy for data.) [28] [42]

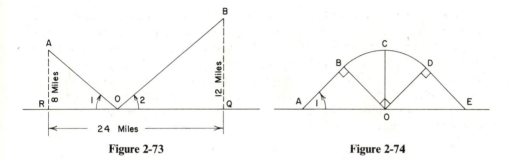

Figure 2-73   **Figure 2-74**

29. Two circles with 10-in. diameters have their centers 8 in. apart. Find the length of their common chord. [42] [70]

30. In Figure 2-74, the circular arc BCD has a radius of 12.0 in. and $OC \perp AE$. Determine $\angle 1$ and the perimeter around $ABCDEOA$. [8]

31. Determine $\angle 1$, the diameter of the circle, and the angular measure of $\widehat{MN}$ in Figure 2-75. [26] [34] [73]

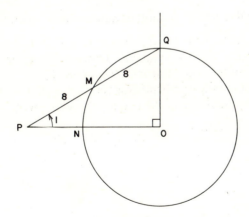

Figure 2-75

3 Fundamentals of Algebra

"Is" means equals. "Of"

In this chapter we shall study the basic operations of addition, subtraction, multiplication, and division as they apply to algebraic expressions. We shall start by using algebra in its very basic form—by using a symbol, such as a letter of the alphabet, to stand for a number. Actually, we have already been doing this. For example, we have used the letters a, b, and c to represent the lengths of line segments, and the letters A, B, C, and θ to represent the number of degrees in an angle.

To start this study, let the letter y stand for the number of years of your present age. For example, if you are 19 years old, then $y = 19$, not 19 years.

1. How old are you? Answer: y years.
2. How old were you last year? Answer: $y - 1$ years.
3. How old will you be next year? Answer: $y + 1$ years.
4. How would you express twice your present age? Answer: $2y$ years.
5. How would you express one-half your present age? Answer:

$$\frac{1}{2}y \text{ years} \qquad \text{or} \qquad \frac{y}{2} \text{ years}$$

If R = one number and Z = another number, how could you express:

1. The sum of the two numbers? Answer: $R + Z$
2. The difference of the two numbers? Answer: $R - Z$, or $Z - R$
3. The product of the two numbers? Answer: $R \cdot Z$
4. Twice the sum of the numbers? Answer: $2(R + Z)$
5. One-third the difference of the two numbers? Answer:

$$\frac{R - Z}{3} \qquad \text{or} \qquad \frac{1}{3}(R - Z)$$

EXERCISE 3.0

1. Using the letter y to stand for the number of years of your present age, write algebraic expressions for each of the following:
   (a) Your present age.
   (b) Twice your present age.
   (c) Half your present age.

(d) Your age next year.

(e) Your age last year.

2. Let the letter *x* stand for the number of years in your little sister's age (assume you have a younger sister) and write algebraic expressions for the following:

(a) Your age plus your sister's age.

(b) Twice your age plus twice your sister's age.

(c) Twice the sum of your ages.

(d) Three times your age minus twice your sister's age.

(e) Half the sum of your ages.

(f) One-half your age plus one-half your sister's age.

3. Using *x* and *y* as in Problems 1 and 2, write the following statement in words:

(a) $(x + y)$ years

(b) $2y$ years

(c) $3x$ years

(d) $(2x + 3y)$ years

(e) $\dfrac{x}{2}$ years

(f) $\dfrac{y}{3}$ years

(g) $(x + 1)$ years

(h) $(y - 2)$ years

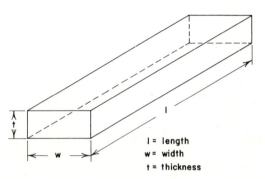

I = length
w = width
t = thickness

Figure 3-1

4. Figure 3-1 shows a rectangular-shaped steel bar. Using the letters *l* for the length, *w* for the width, and *t* for the thickness (or depth), write equivalent algebraic expressions for the following relationships:

(a) The length plus the width.

(b) Twice the length.

(c) Twice the width.

(d) The length plus twice the thickness.

(e) The perimeter *P* of the top face equals twice the length plus twice the width.

(f) The area *A* of the base equals the product of the length times the width.

(g) The volume *V* equals the product of the length times the width times the thickness.

(h) The length plus the width equals 48 in.

(i) The length minus the width equals 16 in.

(j) The diagonal *d* of the base equals the square root of the sum of the length squared* plus the width squared.

* The square of a number is the product of the number times itself; i.e., $N^2 = N \cdot N$.

5. Referring again to Figure 3-1, write equivalent word statements for the following algebraic expressions:

(a) $2l$

(b) $3w$

(c) $2l + 2w = P$ (of top or bottom face)

(d) $\frac{1}{2}P = l + w$

(e) $w = \frac{1}{2}l$

(f) $l = 2w$

(g) $l + 2w = 64$ in.

(h) $w + 3t = 32$ in.

(i) $V = l \cdot w \cdot t$

(j) d (diagonal of base) $= \sqrt{l^2 + w^2}$

(k) $d^2 = l^2 + w^2$

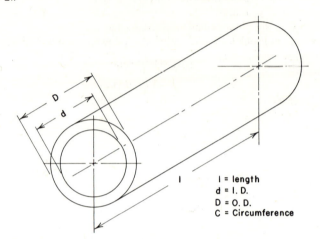

I = length
d = I. D.
D = O. D.
C = Circumference

Figure 3-2

6. Figure 3-2 shows a pipe whose length is l units, inside diameter (I.D.) is d units, outside diameter (O.D.) is D units, and circumference is C units. Using the appropriate letters, write equivalent algebraic statements for the following:

(a) Twice the length. $2\,l$

(b) Three times the inside diameter. $3\,d$

(c) Half the outside diameter. $\frac{D}{2}$

(d) The circumference equals π times the outside diameter. $C = \pi\,D$

(e) The inside radius r_i equals one-half the inside diameter. $r_i = \frac{d}{2}$

(f) The inside diameter equals twice the inside radius r_i. $d = 2\,ri$

(g) The difference in the diameters. $diff = D - d$

(h) Wall thickness W equals one-half the difference in the diameters. $W = \frac{D-d}{2}$

7. Referring again to Figure 3-2, write equivalent word statements for the following algebraic expressions:

(a) $2d + l$

(b) $\frac{1}{2}l + D$

(c) $D - d$

(d) $l - D$

(e) d^2

(f) $r_i = \dfrac{d}{2}$

(g) $C = \pi D$

(h) $W = \dfrac{D - d}{2}$

(i) $D = d + 2W$

8. Referring to Problem 7, calculate values for the expressions in parts (a) through (e) and for the indicated dimensions in parts (f) through (i) if:

(a) $d = 10$ in., $l = 54$ in.   (f) r_i if $d = 7.4$ cm
(b) $l = 14$ ft 8 in., $D = 6.0$ in.   (g) C if $D = 212$ mm
(c) $D = 152$ mm, $d = 12.7$ cm   (h) W if $D = 15.2$ cm, $d = 127$ mm
(d) $l = 32$ ft 3 in., $D = 18$ in.   (i) D if $d = \frac{1}{2}$ in., $W = \frac{1}{16}$ in.
(e) $d = 4.500$ in.

3.1   OPERATIONS WITH SIGNED NUMBERS

You are already familiar with positive and negative numbers as they are used in everyday life. Weather reports show temperature readings such as $+15°$ or $-12°$, meaning 15° above the zero point or 12° below the zero point. Such numbers written with a "plus" $(+)$ sign or with a "minus" $(-)$ sign are called *signed* numbers.

We use signed numbers for two purposes: (1) to show a numerical amount (or position) above or below a zero position, and (2) to show an increase or decrease in some quantity. Plus and minus signs also are sometimes used to indicate directions such as up $(+)$ and down $(-)$. The temperature readings mentioned above indicate actual temperature readings compared with the zero position on the thermometer. In contrast, $+15°$ or $-12°$ might be used to indicate that the temperature has increased by 15° or decreased by 12°. You can usually tell which meaning is indicated from the way a signed number is used in a particular situation. A number written without a sign is understood to be positive.

The *absolute value* (or numerical value) of a signed number is simply the number divorced from its sign. The numerical value of -19 is 19. The *absolute value* of $+19$ is also 19. In other words, it is the "how many" part of the number. The absolute value of a number may be indicated by writing the number between vertical bars ($\|$); thus $|-4| = 4$, and $|+4| = 4$.

Addition of Signed Numbers

To add numbers having like signs, add their absolute values and prefix their common sign.

examples   Add:

$$
\begin{array}{rr}
+4 & -14 \\
+2 & -6 \\
\hline
+6 & -20
\end{array}
$$

To add two numbers with opposite signs, subtract the smaller absolute value from the larger absolute value and prefix the sign of the larger.

examples   Add:

$$-18$$
$$+4$$
$$\overline{-14}$$

(Subtract 4 from 18 and prefix the sign of the 18.)

$$-18$$
$$+24$$
$$\overline{+6}$$

(Subtract 18 from 24 and prefix the sign of the 24.)

When adding several numbers having different signs, you may add all the positive numbers together first and then add all the negative numbers together and then add the two partial sums. Or you may add in the order the numbers are presented, keeping a "running tally" as you go.

example   Add: $+18$, -7, -22, $+9$, -3, $+5$, and -1.

$$
\begin{array}{ccc}
+18 & -7 & +32 \\
 & -22 & -33 \\
+9 & -3 & \overline{-1} \quad \text{sum} \\
+5 & -1 & \\
\overline{+32} & \overline{-33} & \\
\end{array}
$$

or:

$$
\begin{array}{l}
\left.\begin{array}{l}
+18 \\
-7
\end{array}\right\} = 11 \\
-22 \\
+9 \\
-3 \\
+5 \\
-1 \\
\overline{-1}
\end{array}
$$

$\left.\right\} = -11$ $\left.\right\} = -2$ $\left.\right\} = -5$ $\left.\right\} = 0$ $\left.\right\} = -1$

You will frequently have occasion to add numbers such as $-18y$, $+4y$, and $-12y$ or $+4x$, $-3x$, and $+8x$. You can add these algebraic expressions when the literal parts (the letter parts) of the numbers are exactly the same by adding their coefficients. In the algebraic expression $-12xy$, the -12 is the coefficient of the number and the xy is the literal part of the number.

example   To add $8Y$, $+4Y$, $-2Y$, $-6Y$, and $+12Y$, add the coefficients the same as if you were adding 8 ft, $+4$ ft, -2 ft, -6 ft, and $+12$ ft. That is, add the coefficients, obtaining $16Y$ in the first case and 16 ft in the second.

Collecting terms is an expression commonly used to mean the addition of like terms. (C.T. is the abbreviation for collecting terms.) For example, by collecting terms in the expression $12y + 3x - 4y + 5x$, you obtain the simplified equivalent expression of $8y + 8x$.

EXERCISE 3.1A

Add the groups of signed numbers in Problems 1–6.

1. $+ 18, - 7, + 8, + 5, - 7, - 3, - 2$
2. $+ 26, + 46, - 18, - 5, + 3$
3. $425, - 762, - 395, - 861$
4. $+ 7, + 3, - 8, - 2, - 5, + 6$
5. $- 18, - 25, - 0, + 7, + 3$
6. $469, - 318, - 425, - 218, + 45$
7. If the temperature was $52°$ at 8 p.m., and the recorded hourly temperature changes (starting at 9 p.m.) were $- 2°, - 5°, - 8°, + 1°, + 3°, - 2°, + 6°, + 7°, - 3°, - 6°, + 8°,$ and $+ 2°,$ what was the temperature at:
   (a) Midnight?
   (b) 3 a.m.?
   (c) 6 a.m.?
   (d) 8 a.m.?
   (e) What was the lowest temperature reading?
8. Stock market changes over a 10-day period for a certain stock valued at $10.85 were as follows: $+ 0.05, - 0.25, - 0.13, - 0.14, + 0.07, - 0.11, + 0.44, + 0.26, - 0.03,$ and $- 0.01.$ (Changes are in dollar units.) What was the value of the stock on the sixth day? On the last day?

Add the groups of signed numbers in Problems 9–20.

9. $+ 3x, - 5x, + 8x, - 4x, + 3x$
10. $4xy, - 2xy, - 9xy, + 12xy, + 16xy$
11. $8R, + 5R, - 7R, - 4R, + 16R, - 12R$
12. $13W, - 3W, - 5W, - 6W, + 22W, + 3W$
13. $4RT, + 5RT, - 3RT, + 8RT, - 6RT$
14. $5xyz, - 4xyz, + 8xyz, - 12xyz, + 9xyz$
15. $\frac{1}{2}X, + \frac{2}{3}X$
16. $\frac{3}{4}Y, - \frac{1}{5}Y$
17. $\frac{4}{5}W, + \frac{2}{3}W, - \frac{1}{6}W$
18. $\dfrac{X}{3}, + \dfrac{X}{4}$ (Hint: This is the same as $\frac{1}{3}X, + \frac{1}{4}X$.)
19. $\dfrac{Y}{6}, - \dfrac{Y}{5}$
20. $\dfrac{2P}{3}, + \dfrac{4P}{9}$

Collect terms to simplify each of the expressions in Problems 21–26.

21. $2ax + 3by - 5ax + 7by - 3ax - 4by$
22. $125x - 37y + 19x - 8y + 35x - 14y$
23. $425ab + 31ab - 27b + 36b - 402ab + 81b$
24. $13.75VR - 4.15VK + 6.38VR + 17.35VK$
25. $56.4Z - 13.8R + 22.7Z - 16.9R + 66.3R$
26. $119N + 37X - 102N - 43X + N - X$

In Problems 27–34, signed numbers occur which indicate either the relative position (or value) of some quantity or a change in the quantity. Write the word "position" or "change" to indicate the use of each signed number.

27. The low temperature reading was $-35°F$. *P*
28. The temperature increased to $-28°$. *C*
29. The stock report showed -0.25 point. *C*
30. The maximum tolerance was $+0.002$ in. *P*
31. The temperature variation was $4°C$. *C*
32. Water freezes at $+32°F$. *P*
33. The steel bar was cooled to $-10°C$. *C*
34. Only two sizes of bearings were available, standard and -0.003. *C*

Multiplication of Signed Numbers

Multiplication of two numbers with like signs yields a positive product.

examples

$$(-3)(-4)^* = +12$$

$$(+3)(+4) = +12$$

Multiplication of two numbers with opposite signs yields a negative product.

examples

$$(-3)(+4) = -12$$

$$(+3)(-4) = -12$$

To multiply three or more signed numbers, multiply the product of the first two factors by the third factor, that product by the next factor, and so on until you have multiplied by the last factor.

example

$$(-4)(+5)(-2)(-3) = -120$$

The product of the -4 and $+5$ is -20; the -20 multiplied by the $-2 = +40$; the $+40$ multiplied by the $-3 = -120$.

Division of Signed Numbers

Division of two numbers with like signs yields a positive quotient. Division of two numbers with opposite signs yields a negative quotient.

---

* Think of the minus sign as directing you to reverse the direction of count; then the double minus sign produces two reversals, giving a positive result. Or think of a car with two transmissions, one behind the other and both shifted into reverse; the car would move forward.

examples

$$(-28) \div (-4) = +7$$

$$(-28) \div (+4) = -7$$

$$(+28) \div (-4) = -7$$

$$(+28) \div (+4) = +7$$

EXERCISE 3.1B

In Problems 1–26, compute the product or quotient, writing each with its proper sign.

1. $(-21)(-5)$
2. $(-3)(-2)$
3. $(4)(-5)$
4. $(8)(+2)$
5. $(-8)(+2)$
6. $(+8)(-2)$
7. $(+8) \div (-2)$
8. $\dfrac{416}{-2}$
9. $\dfrac{-18}{-9}$
10. $\dfrac{-21}{-7}$
11. $\dfrac{-24}{+8}$
12. $\dfrac{+32}{-16}$
13. $(-3)(-4)(-5)$

14. $(4)(-2)(+2)$
15. $(-9)(-3)(-2)$
16. $(+3)(-2)(-2)$
17. $\dfrac{(-8)(-2)}{-4}$
18. $\dfrac{(-6)(-5)(-1)}{-2}$
19. $\dfrac{(+6)(-5)(-1)}{+15}$
20. $\dfrac{(-2)(-3)(-4)(-5)}{(-6)(-5)}$
21. $(-\tfrac{1}{2})(-\tfrac{2}{3})$
22. $(-\tfrac{3}{4})(-\tfrac{2}{3})(-\tfrac{4}{5})$
23. $(-\tfrac{4}{7})(+\tfrac{4}{5})$
24. $(-\tfrac{4}{5}) \div (\tfrac{8}{10})$
25. $\tfrac{17}{5} \div (-\tfrac{17}{5})$
26. $(-\tfrac{18}{3}) \div (-\tfrac{4}{3})$

Subtraction of Signed Numbers

The rule for subtraction of signed numbers is easy to apply. However, it may cause you difficulty if you get in a hurry. Let us consider an actual situation to get some insight into the subtraction process.

Figure 3-3(a) shows a thermometer with the high and low temperature readings for a 24-hr period. What was the temperature range for the 24 hr? According to the figure, the range was from 8 units below the zero position to 12 units above the zero position—a range of 20°. To arrive at the answer of 20°, you could have counted the number of degrees between the two positions; i.e., counting from the -8 position up to the $+12$ position (or from $+12$ down to -8). Or you could have added the absolute values of the two readings, obtaining $12° + 8° = 20°$.

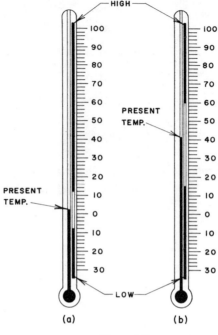

Figure 3-3

To subtract with signed numbers, change the sign of the subtrahend and proceed as in the addition of signed numbers

Much of the difficulty in this procedure can be avoided by applying the following steps *one at a time* and *in the order given*.

1. Determine the quantity being subtracted, i.e., the subtrahend.
2. Change the sign of this subtrahend.
3. Add the subtrahend and minuend together, following the rules for the addition of signed numbers.

Let us now return to the thermometer problem. Because the question asks for the range rather than for the change in temperature readings, it is asking: "How many degrees separate the two readings?" In other words, it is asking for the absolute value of the difference between the two readings. First pick the $-8°$ reading as the subtrahend. Changing the sign of this reading gives $+8°$. Adding these signed numbers gives, $+12° + 8° = +20°$. The absolute value is the "how many" part of this number, so you can write 20° for the answer. In contrast, now choose the $+12°$ as the subtrahend. Changing the sign of this subtrahend makes it $-12°$. Adding the two signed numbers gives $-12° - 8° = -20°$. So

$-20°$ also expresses a correct difference between the two readings. The absolute value of $-20° = +20°$ or simply $20°$. You could pick either of the two readings as the subtrahend.

Let us consider a second example involving temperatures. See Figure 3-3(b). First, what was the 24-hr temperature range? As was pointed out in the last example, you can subtract either reading (high or low) from the other to determine the difference between them. If you pick the $60°$ reading for the subtrahend, change its sign, and add it to the $15°$, you obtain $-45°$ for the difference. To answer the question of temperature range, you need the absolute value of the difference $(-45°)$ which is $45°$.

If the temperature had changed directly from the high of $60°$ to the low of $15°$, what would have been the temperature change? To answer this question, you must subtract the starting temperature reading from the resulting temperature. The $60°$ is the starting temperature and therefore the subtrahend. Changing its sign to $-60°$ and adding it to the $+15°$ reading, you obtain a temperature change of $-45°$. The minus sign indicates a decrease in temperature of $45°$ and not a position with respect to zero, i.e., not $45°$ below zero.

example   Subtract the second term from the first. Given:

$$- 13, \quad - 18$$

1. -18 is the subtrahend.
2. Changing its sign makes it $+18$.
3. Adding -13 and $+18$ yields $+5$ for the difference.

example   Subtract the first term from the second term. Given:

$$- 24, \quad + 43$$

1. The subtrahend is -24.
2. Changing its sign makes it $+24$.
3. Adding $+24$ and $+43$ yields $+67$ for the difference.

example   Subtract -31 from the sum of $+12$ and $+16$.
1. The sum of $+12$ and $+16 = +28$.
2. The subtrahend is -31.
3. Changing its sign makes it $+31$.
4. Adding $+31$ and $+28$ yields $+59$ for the difference.

EXERCISE 3.1C

In Problems 1–4, subtract the first term from the second term.

1. $-23, +44$
2. $+44, -12$
3. $-12, +12$
4. $+12, -12$

In Problems 5–8, subtract the second term from the first term.

5. $-61, -43$ 7. $206, -312$
6. $-44, +44$ 8. $-414, +318$

In Problems 9–12, subtract the first term from the sum of the remaining terms.

9. $-14s, +8s, +9s$ 11. $-23d, -14d, +16d$
10. $+16m, -14m, +16m$ 12. $-8j, -8j, +8j$

In Problems 13–16, subtract the sum of the first two terms from the last term.

13. $-18x, +14x, -5x$ 15. $402w, -313w, 205w$
14. $+16y, -14y, -14y$ 16. $609z, -392z, 418z$

3.2 GROUPING SYMBOLS AND THE UNDERSTOOD COEFFICIENT

It is frequently necessary to use symbols to group certain quantities together. For example, $(\frac{1}{2})(2 + 6) = 4$ means one-half the sum of $2 + 6$. However, $(\frac{1}{2})2 + 6 = 7$ means that only the 2 is to be multiplied by the one half.

To indicate that two or more terms are to be treated as a single quantity and also for the purpose of clarity, use parentheses ( ), brackets [ ], or braces { }. There is no difference in the meaning of the different symbols. Several sets of grouping symbols may be used in the same algebraic statement, and it is much easier to identify each particular group if different symbols are used. Contrast $(3x - (4y + 2(x + 3) + 2) - 5x)$ with $\{3x - [4y + 2(x + 3) + 2] - 5x\}$.

The Number 1 as an Understood Coefficient

When you say "I saw a car" or "I saw a horse," you really mean that you saw *one* car or *one* horse. When you write literal numbers such as x or y, you really mean $1x$ or $1y$. The coefficient "1" is supposed to be understood.

In the next two exercises, write in all the 1 coefficients that are missing. When you see $x + 4y$, write $1x + 4y$; for $2x - y$, write $2x - 1y$.

Coefficients of Groups of Terms

When several terms are grouped together with parentheses, brackets or braces, each group has a coefficient. It is common practice to leave off coefficients of 1. For example, with $(3x - 2y)$, the coefficient of the group is understood to be a 1; $5(x + y)$ has a coefficient of 5; $-(3x + 4y)$ has a coefficient of -1.

EXERCISE 3.2A

Write in all missing 1 coefficients.

1. $x - 4$ 6. $-[2z - s]$
2. $2x - y$ 7. $\{h + w\}$
3. $x + y$ 8. $(x - y) + (2x + y)$
4. $-(2x + y)$ 9. $(x + y + z) - (x + y - z)$
5. $(x - y)$ 10. $\{a + [(2x + y) - 3x + y] + z\}$

Removing Parentheses, Brackets, or Braces

To remove the grouping symbols from an algebraic expression, multiply each term in the group by the coefficient of that group.

example   To remove the parentheses from the group $3(2x - 3y)$, multiply each term inside the group by the coefficient 3. Thus $3(2x - 3y) = 6x - 9y$.

example   To remove the parentheses from the group $+ (2x - 3y)$, write in the missing coefficient: $+ 1(2x - 3y)$. Then remove the parentheses by multiplying each term by the $+ 1$ coefficient, obtaining $2x - 3y$.

important example   To remove parentheses from the group $- (2x - 3y)$, first write in the 1 coefficient, obtaining $- 1(2x - 3y)$. Then remove the parentheses by multiplying each term in the group by the -1 coefficient, which gives $-2x + 3y$.

You must multiply each term within a group by the coefficient of that group whether the coefficient of the group is written in or "just understood." The trouble with the "just understood" coefficient is that it is too frequently not understood.

You sometimes need to write certain terms as a group within a larger group. Consider the grouping $- [2x - (4x + 2) - 3]$. To remove the symbols of grouping in this situation, first write in the missing coefficients, which yields $- 1[2x - 1(4x + 2) - 3]$.

The simplest method is to remove the inner grouping symbols first, and then the next inner grouping symbols, working on out until all symbols are removed. In this example, you would remove the parentheses first and the brackets last. You are less apt to make a mistake if you first copy the entire expression exactly as it is down to the innermost group, and then remove the inner group symbols by multiplying each term by the coefficient of the group. Copy all remaining terms and symbols exactly as they appear in the previous step.

example

1. $-1[2x$ $- 1(4x + 2)$ $- 3]$
              same   remove   same

2. $-1[2x$ $- 4x - 2$ $- 3]$

3. Then remove [ ]:

$$- 2x + 4x + 2 + 3$$

4. Collecting terms, you obtain

$$2x + 5$$

You could just as correctly have collected terms before the brackets were removed and obtained $- 1[- 2x - 5]$, which upon removal of the brackets yields $2x + 5$, as before.

example   Remove all symbols of grouping and collect terms:

$$2\{3x - [4x - (x - 5) + 3] - 4\}$$

First write in the missing 1 coefficients:

$$2\{3x - 1[4x - 1(1x -.5) + 3] - 4\}$$

Copy the expression exactly as it is down to the ( ) and remove the ( ):

$$2\{3x - 1[4x - 1x + 5 + 3] - 4\}$$

Then removing the [ ], you have

$$2\{3x - 4x + 1x - 5 - 3 - 4\}$$

If you collect terms at this point, you have $2\{- 12\}$ which becomes $- 24$ when the $\{ \}$ are removed.

Do not yield to the temptation to remove the grouping symbols from the outside working in. Although it can be done, you will make fewer mistakes by following the procedure used in the examples.

EXERCISE 3.2B

Write in any "understood" coefficients, remove parentheses, and collect terms, i.e., add like terms.

1. $2(x - 3y)$
2. $- 3(4x + 4y)$
3. $(x - 4y)$
4. $-(3x - y)$
5. $-(x - 4) + (2x - 5)$
6. $(3z - 5) - (2z - 2)$
7. $[8x - y] - [3x - y]$

8. $(2x - 5) - 3[x + 2] - (- 6x - 8)$
9. $-(2x - 2) - (- x + 2) - (3x - 2)$
10. $2[3 + (x - 5) - x]$
11. $2\{3x - 5\} - (x - 8)$
12. $-[3 - (2x - 4) + 3x]$
13. $2\{3x - [4x + 5(x - 2) + 6] - 2\}$
14. $-\{2 - [3 - (2x - 3) - 3x] + 4\}$

3.3 MULTIPLICATION OF ALGEBRAIC EXPRESSIONS

To multiply two or more algebraic factors together, multiply their respective coefficients and multiply their respective literal parts together. For example,

$$(3x)(4y) = 12xy$$

$$(-12x)(2y)(3z) = -72xyz$$

From use of the formulas for the area of a square, $A = s^2$, and the volume of a cube, $V = e^3$, you know that the exponent 2 on the s^2 means $s \cdot s$ and that the exponent 3 on the e^3 means $e \cdot e \cdot e$. (The dot means multiplication.) Use an exponent to write a repeated factor in what could be called a shortened form. Contrast y^5 with $y \cdot y \cdot y \cdot y \cdot y$.

The Understood Exponent of 1

The factor x in the expression $2x$ has an "understood" exponent of 1, i.e., $2x^1$. The factor 2 in the expression also has an "understood" exponent of 1. If you multiply $2x$ by $3x^2$, you obtain $6x^3$. You obtain the 6 by multiplying the coefficients 2 and 3, and the exponent 3 by adding the exponents from x^1 and x^2.

Exponents are frequently referred to as powers. For example, x^3 is sometimes read "x to the third power." Exponents by themselves are meaningless. They must apply to a *base* number. In the expression $3x^5$, the 5 is the exponent or power and the x is the base.

To Multiply Quantities Expressed as Powers of the Same Base, Add Their Exponents

examples

$$10^2 \cdot 10^3 = 10^5$$

$$x^5 \cdot x^6 = x^{11}$$

$$2xy^3 \cdot 3x^2 = 6x^3y^3$$

To multiply factors with different bases, only indicate what multiplication is to take place if and when values for the literal numbers become known.

examples

$$3x^3 \cdot 4y^4 = 12x^3y^4$$

$$5x^4z^5 \cdot 6y^3z^2 = 30x^4y^3z^7$$

EXERCISE 3.3A

Make a copy of the following table, and complete as shown in Problem 1.

	Algebraic Expression	Numerical Coefficient	Literal Part	Base	Exponent
1.	$-3y^5$	-3	y^5	y	5
2.	$12Z^2$				
3.	$-R^3$				
4.	$+27z$				
5.	$+x$				
6.	$-y$				

Carry out the indicated multiplication and collect terms when possible in Problems 7–24.

7. $(3x)(4x^2)$
8. $(3x^3y)(8xy^2)$
9. $(-6x^3y^2)(-4xy)$
10. $(-3x)(-4xy)(-5yz)$
11. $(-8x^3y^2z^3)(-4xyz^5)$
12. $(-22xz^5)(2x^5z^2)$
13. $5y(2y^2 + 3y - 3)$
14. $-z(3z^3 - 5)$
15. $-4(4x - 5z)$

16. $-6w(-2w^3 - 4w^2 + 3w - 5)$
17. $-11z(-8z^4 + 4z^3 - 6z^2 + 9)$
18. $-9kr(-2kr^3 + 3k^2r^2 - 7k^3r)$
19. $-12n(9n^3 - 8n^2 + 7n + 6)$
20. $12xyz(12x^2y - 8xy^2z^3 + 11x^3z^2)$
21. $9m^2(3m - 5)$
22. $x^2y^3(4x^3 - 3y^2)$
23. $3y^2(8xy - 4xy^2 + 3x^2y^5 + 4x^3y)$
24. $8xyz(3xy - 4xz)$

Multiplication with Two or More Terms

Up to this point, our multiplication has involved single-term multipliers. But you will have occasion to multiply by factors with two and sometimes three terms. The difference in the processes involved is more a difference in how you view the problem than in the actual mechanics. Consider the multiplication $(2x)(3x + 5)$. You know that the $2x$ factor multiplies both the $3x$ and the 5, which produces $6x^2 + 10x$ for the product. Now reverse the position of the two factors, obtaining $(3x + 5)(2x)$. If you think of the two-termed factor as the multiplier, then to obtain the product $6x^2 + 10x$, the $3x$ must multiply the $2x$ and the 5 must also multiply the $2x$.

In short, every term of a multiplier must multiply every term in the multiplicand:

$$(3y - 9)(4y^2) = 12y^3 - 36y^2$$

$$(14a^2 + 3b)(2a) = 28a^3 + 6ab$$

$$(2a - 3b + 4)(5ab) = 10a^2b - 15ab^2 + 20ab$$

example Now consider the multiplication problem $(3x - y)(2x + 4y)$. To multiply each term in the multiplicand by each term in the multiplier, it is helpful to rewrite the two factors as

$$
\begin{array}{ll}
2x + 4y & \text{multiplicand} \\
3x - y & \text{multiplier} \\
\hline
6x^2 + 12xy & \\
 - 2xy - 4y^2 & \\
\hline
6x^2 + 10xy - 4y^2 &
\end{array}
$$

Then multiply each term in the multiplicand by each term in the multiplier. We usually start with the left-hand term of the multiplier and work to the right. The factors and their partial products in the order we have used them in this example are:

$$(3x)(2x) = 6x^2$$

$$(3x)(+4y) = +12xy$$

$$(-y)(2x) = -2xy$$

$$(-y)(4y) = -4y^2$$

Then add the partial products. They are usually written in an alphabetical order when several different letters are involved or in a decreasing or increasing order of exponents when only one or two letters are used. Collect all like terms by adding their coefficients.

example Multiply $2x - 4y$ by $3x + 2y - z$. Because the product of two numbers is the same regardless of the order in which they are multiplied, use either of the expressions as the multiplier and the other as the multiplicand. Therefore, write the problem as

$$
\begin{array}{lll}
2x - 4y & \quad\text{or}\quad & 3x + 2y - z \\
3x + 2y - z & & 2x - 4y \\
\hline
\end{array}
$$

Because the second method involves multiplying with only two terms in the multiplier, you might prefer that form.

The Order of Terms or Factors

Experience tells you that if you add $5 + 7 + 6$, obtaining 18, you will also obtain 18 if you change the order in which you add the terms; $5 + 6 + 7 = 18$, $7 + 5 + 6 = 18$, $6 + 7 + 5 = 18$, $6 + 5 + 7 = 18$, etc. Also, if you multiply $2 \times 5 \times 7 = 70$, you can change the order of the factors to $2 \times 7 \times 5 = 70$, $7 \times 5 \times 2 = 70$, $7 \times 2 \times 5 = 70$, etc.

Very often you will obtain algebraic expressions which have several terms. To reduce the chance of error and to facilitate ease in reading or checking these algebraic expressions, follow these common practices:

1. Add (collect) similar terms by adding their numerical coefficients.
2. Write the numerical coefficient as the first factor of each term.
3. Write the literal factors in each term in alphabetical order.
4. Write the terms in alphabetical order or in a decreasing (or increasing) order of exponents.

examples

1. $5zn$ would be written as $5nz$.
2. $3x^3 + 2x^5 - 2x^2 + 4 - x$ would be written as $2x^5 + 3x^3 - 2x^2 - x + 4$.
3. $3xy^2 - 4x^2y - 5x^3 + 2y^2 - y^3$ would be written with decreasing exponents on x and increasing exponents on y: $5x^3 - 4x^2y + 3xy^2 + 2y^2 - y^3$.

EXERCISE 3.3B

In Problems 1–8, rewrite the given expression in the order discussed in the previous examples. Collect terms when possible.

1. $RNWZ^2$
2. $x \cdot y \cdot 12 \cdot z$
3. $5x^2 - 3x^3 + 2 - x + 5x^2 + 4x$
4. $-7yr + 2yr - 5z + 2z - 8$
5. $-4x^4 - y^2 + 3x + y^3 - 8y$
6. $11s - 3s^2 + 2r - 3r^2 + 5r^3 - 8r + s^2$
7. $6b - 3a + 4b - 7a + 2c - 4 + 5c - 2$
8. $N^3W^4 + 8 + NW + N^2W^2$

In the following problems, carry out the indicated multiplication and write the products in simplified form.

9. $(3x - 2y)(5x)$
10. $(11W - 2k)(4W + 5k)$
11. $(24x - 2y)(3x + 5y)$
12. $(R - 2)(R + 4)$
13. $(W + 5)(W - 5)$
14. $(2N + 3W)(3N - 4W + 5R)$
15. $(-7k + 3n + J)(5k - 2J)$
16. $(k + w - 2)(2w - k + 3)$
17. $(R^2 - 2Z^3)(R + Z)$
18. $(x^3y - yx^2)(2xy + 3xy^2)$
19. $(z^2 - 5)(z + 2)(z - 3)$
20. $(n^2 + 2)(n - 3)(n + 1)$

3.4 DIVISION OF ALGEBRAIC EXPRESSIONS

To start our discussion, consider the inverse relationship that exists between multiplication and division operations. When you multiply 7 times 8, you get 56 for the product. When you divide 56 by 7, you obtain the factor 8; when you divide 56 by 8, you obtain the factor 7.

We shall make the same assumption for the multiplication and division of factors written as powers of the same base; i.e., when the product of two factors is divided by one of the factors, the other factor is obtained as a quotient. Consider $(3x^2)(5x^5) = 15x^7$. Dividing the product $15x^7$ by $3x^2$, you would (by our assumption) obtain $5x^5$ as a quotient. Also, $15x^7/5x^5$ would equal $3x^2$.

Let us consider a division problem in another way:

$$\frac{35x^5}{5x^3} \quad \text{really means} \quad \frac{(35)(x)(x)(x)(x)(x)}{(5)(x)(x)(x)}$$

Reducing this fraction by dividing the numerator and denominator first by 5, then by the factor x, again by x, and a third time by x, you obtain

$$\frac{\overset{7}{\cancel{35}}(x)(x)(\cancel{x})(\cancel{x})(\cancel{x})}{\cancel{5}(\cancel{x})(\cancel{x})(\cancel{x})} = 7x^2$$

If you do the problem again by subtracting the exponent 3 from the exponent 5, you obtain x^2 for the literal part of the quotient. The $35 \div 5$ yields the coefficient 7 as before.

To divide powers of the same base, subtract the exponent of the divisor from the exponent of the dividend. This difference becomes the exponent of the quotient (same base). The coefficients of the two expressions are divided in the usual manner.

EXERCISE 3.4A

Carry out the following divisions.

1. $45x^5 \div 5x^3$
2. $210y^8 \div 35y^5$
3. $54d^4 \div 6d^3$
4. $18r^5y^6 \div 3r^3y^2$
5. $27s^6t^3 \div s^4t^2$
6. $48n^5w^3 \div 3nw$
7. $132b^5f^7 \div 11b^3f^3$
8. $444u^5p^9 \div 3u^4p^7$
9. $\dfrac{512V^{15}}{32V^7}$

10. $1024D^{66} \div 64D^{37}$
11. $720K^{46} \div 18K^{23}$
12. $\dfrac{24n^5j^7k^9}{4n^3j^3k^3}$
13. $121p^{74} \div 11p^{49}$
14. $\dfrac{108g^{54}q^{33}}{9g^4q^7}$
15. $48s^3 \div 16s$

16. $540h^{27}g^{43} \div 18h^6g^{10}$
17. $\dfrac{36xy^2z^3}{6yz^2}$
18. $\dfrac{55r^3y^5n^8}{5rn^3}$
19. $40x^4 \div 40x^4$
20. $x^3 \div x^3$

The Zero Exponent

In Problem 20 in the last exercise, you might have correctly obtained x^0 for the answer. What is the meaning of the zero exponent? A nonzero number with a zero exponent is defined as equal to 1. This seems reasonable when you consider

$$\frac{x^3}{x^3} = 1 \qquad \text{and} \qquad \frac{x^3}{x^3} = x^{3-3} = x^0$$

To maintain consistency in the rules of exponents, the definition $N^0 = 1$ is very practical ($N \neq 0$).

examples

$$5^0 = 1 \qquad 3x^0 = 3 \cdot 1 = 3 \qquad (3x)^0 = 1$$
$$(256y^3 + 4x^2)^0 = 1 \qquad (\sqrt{38})^0 = 1$$

Divisors and Dividends with Two or More Terms

Consider the division problem $(12x^3 - 4x^2) \div 2x$. Both terms in the dividend are to be divided by the $2x$. Another way to indicate this same division would be to write it in fraction form:

$$\frac{12x^3 - 4x^2}{2x} \qquad \text{or} \qquad \frac{12x^3}{2x} - \frac{4x^2}{2x}$$

In both cases, each term of the dividend is divided by the divisor $2x$, obtaining the quotient $6x^2 - 2x$. Divisors with only one term cause us relatively little trouble. Division by a two-termed divisor is a little more involved.

example Divide $10x^2 + x - 24$ by $2x - 3$. Write the problem in the form shown below, writing the terms of the dividend and divisor in the same order:

$$2x - 3)\overline{10x^2 + x - 24}$$

1. Start by dividing $10x^2$ by $2x$, obtaining $5x$:

$$
\begin{array}{r}
5x \\
2x - 3)\overline{10x^2 + x - 24} \\
10x^2 - 15x \\
\hline
0 + 16x
\end{array}
$$

2. Multiply $2x - 3$ by $5x$, writing the terms of the product under the similar terms of the dividend, and subtract.
3. Bring down the next term of the dividend, forming the new dividend,

and divide $16x$ by $2x$, obtaining 8:

$$
\begin{array}{r}
5x \;+\; 8 \\
2x - 3\overline{)10x^2 + x - 24} \\
\underline{10x^2 - 16x } \\
+ 16x - 24 \\
\underline{+ 16x - 24} \\
0
\end{array}
$$

4. Multiply $2x - 3$ by 8, obtaining $16x - 24$; subtract $16x - 24$ from the dividend, obtaining a remainder of zero.

To check your work, multiply the quotient by the divisor and add any remainder:

$$
\begin{array}{r}
5x \;+\; 8 \\
2x \;-\; 3 \\
\hline
10x^2 + 16x \\
- 15x - 24 \\
\hline
10x^2 + 1x - 24 \quad \checkmark
\end{array}
$$

Suppose that the -24 in the original dividend had been a -30; you end up with a remainder of -6. Then in the checking you would add the -6 to the product:

$$
\begin{array}{r}
10x^2 + 1x - 24 \\
- 6 \\
\hline
10x^2 + 1x - 30 \quad \checkmark
\end{array}
$$

example

$$(12x^3 + 10x^2 - 4) \div (2x - 1)$$

Note that there is no first power of x term in the dividend. To avoid making a mistake in locating the partial products, place a zero in the position of the missing term:

$$
\begin{array}{r}
6x^2 + 8x \;+ 4 \\
2x - 1\overline{)12x^3 + 10x^2 + 0 - 4} \\
\underline{12x^3 - 6x^2 } \\
16x^2 + 0 \\
\underline{16x^2 - 8x } \\
+ 8x - 4 \\
\underline{+ 8x - 4} \\
0
\end{array}
$$

Check:

$$(6x^2 + 8x + 4)(2x - 1) = 12x^3 + 10x^2 - 4 \qquad \checkmark$$

EXERCISE 3.4B

Carry out the indicated divisions.

1. $(12a^2 + 14a) \div 2a$

2. $\dfrac{24x^2 y^3 + 8xy^2}{2xy}$

3. $(36x^3 - 12y^2 + 18z^3) \div 3$
4. $(27n^4 w^3 - 24n^2 w^2 + 15n^3 w^4) \div 3n^2 w^2$

5. $\dfrac{64r^3 z^2 - 48r^2 z^3}{16rz}$

6. $(3a^3 b^2 - 5a^2 b^2 - 9ab) \div 7ab$
7. $(6x^2 + 5x - 6) \div (2x + 3)$

8. $\dfrac{10y^2 - 3y - 4}{2y + 1}$

9. $\dfrac{5k^2 - 8k + 3}{5k - 3}$

10. $(11r^2 + 8rw - 3w^2) \div (r + w)$
11. $(n^2 - w^2) \div (n + w)$
12. $(x^2 - y^2) \div (x - y)$
13. $(a^2 + 2ab + b^2) \div (a + b)$
14. $(x^2 - 2xy + y^2) \div (x - y)$
15. $(x^3 - y^3) \div (x - y)$ (Watch missing terms in the dividend.)
16. $(r^3 + n^3) \div (r + n)$
17. $(9a^2 - 81) \div (3a - 9)$
18. $\dfrac{12x^2 - 10x - 8}{4x + 2}$

3.5 EVALUATION OF ALGEBRAIC EXPRESSIONS

The algebraic expression resulting from various operations is usually not the desired form for the answer. The technician frequently needs to know the numerical value for an expression. For example, the answer to a gas volume problem may turn out to be $P_1 V_1 / P_2$. When this is evaluated with $P_1 = 760$ cm, $V_1 = 42\,\ell$, and $P_2 = 76$ cm, the answer turns out to be $420\,\ell$.

There are some simple steps to follow which will aid us in evaluating algebraic expressions when values for the literal numbers are known.

1. Replace any "understood" coefficients by writing the necessary 1.
2. Remember that there is an understood multiplication to be performed between adjacent factors; $5xy$ means 5 times x times y. We can use a dot or parentheses to emphasize this; $5xy$ means

$$5 \cdot x \cdot y \qquad \text{or} \qquad (5)(x)(y)$$

Parentheses are very helpful when the factors are negative.

3. An exponent applies only to the number written to its lower left. In $5x^2$, the exponent 2 applies only to the x, not to the 5. To make it apply to the 5 also, write 5^2x^2 or $(5x)^2$.

4. Replace each literal factor with its proper value and carry out the indicated operations.

examples   Evaluate $3xy^2$, when $x = -2$, $y = -3$:

$$3xy^2 = 3(-2)(-3)^2 = (-6)(+9) = -54$$

Evaluate $-x^2$, when $x = -2$:
$$-x^2 = (-1)(x)^2 = (-1)(-2)^2 = (-1)(+4) = -4$$

Evaluate $-y^3$, when $y = -3$:
$$-y^3 = (-1)(y)^3 = (-1)(-3)^3 = (-1)(-27) = +27$$

Evaluate $6ax^2$, when $a = -2$, $x = -3$:
$$6ax^2 = (6)(-2)(-3)^2 = (-12)(+9) = -108$$

Evaluate $27(n - w)$, when $n = 285$, $w = 119$:
$$27(n - w) = (27)(1 \cdot n - 1 \cdot w) = 27(1 \cdot 285 - 1 \cdot 119)$$
$$= 27(285 - 119) = (27)(166) = 4482$$

EXERCISE 3.5

Evaluate each expression in Problems 1–10 for the given values of the literal numbers.

1. $25x^2$,   for $x = -3$
2. $-3z^2$,   for $z = -2$
3. $-N^3K^2$,   for $N = -3, K = 0$
4. $k^2 - n^2$,   for $k = 0, n = -2$
5. $T^3 - t^2$,   for $T = 5, t = 2$
6. $(T - t)^3$,   for $T = 5, t = -2$
7. $5kr - 2kr + 3kr + k - r$, for $k = 2, r = -2$
8. $3yz^3 - 2y + z$,   for $y = 5, z = 1$
9. $4s - w^2$,   for $s = 5, w = 2$
10. $\dfrac{2w^3 - 3r^2}{w + r}$,   for $w = 3, r = -2$

In Problems 11–14, perform the indicated operations and evaluate the resulting expression for the indicated values of the literal numbers. Check by substituting the given values into the original expressions and then perform the indicated operations.

11. $(x + 2)(x - 2)$, for $x = -5$

12. $\dfrac{R^2 - 2R + 1}{R - 1}$, for $R = 3$

13. $(2z - 3)(z + 4)$, for $z = -3$

14. $(A^2 + A - 12) \div (A - 3)$, for $A = 4$

In Problems 15–16, evaluate the given expressions.

15. $|-x| + |y|$, for $x = +5, y = -10$

16. $|(x - y)(x + y)|$, for $x = +8, y = -7$

3.6 KEEPING TRACK OF DIMENSIONAL UNITS

Proper dimensions are of such vital importance to the work of the technician that they deserve adequate study.

example The formula $s = at^2/2$ expresses the relationship between the distance s that a freely falling object will fall (starting at rest) in t sec of time. Find s when the acceleration of gravity $a = 980\ \text{cm/sec}^2$ and $t = 5$ sec.

Watch the units.

$$s = \frac{(980\ \text{cm})(5\ \text{sec})^2}{2\ \text{sec}^2} = \frac{(980\ \text{cm})(25)(\cancel{\text{sec}^2})}{2\ \cancel{\text{sec}^2}}$$

$$s = 12{,}250\ \text{cm}$$

It is important to note that the resulting unit (centimeter) is a proper unit for expressing distance.

example The length l of a rectangular solid can be calculated from the formula $l = V/wd$, where V = volume, w = width, and d = depth of the solid. Compute the length of the rectangular solid whose volume is 17,100 cu. in. (in.^{3}), whose width is 38 in., and whose depth is 1.0×10 in.

$$l = \frac{17{,}100\ \text{in.}^3}{(38\ \text{in.})(10\ \text{in.})} = \frac{17{,}100\ \text{in.}^3}{380\ \text{in.}^2}$$

$$l = 45\ \text{in.}^{3-2} = 45\ \text{in.}$$

Note that the rules of exponents apply equally well to dimensional units.

example   The length of the hypotenuse h of a right triangle can be computed from $h = \sqrt{a^2 + b^2}$, where a and b are the lengths of the two legs of the triangle. Compute the length of the hypotenuse of the triangle whose legs are $a = 5.0$ ft and $b = 12$ ft. Substituting for a and b, we have:

$$h = \sqrt{(5.0 \text{ ft})^2 + (12 \text{ ft})^2} = \sqrt{25 \text{ ft}^2 + 144 \text{ ft}^2}$$

$$h = \sqrt{169 \text{ ft}^2} = 13 \text{ ft}$$

example   Sometimes a formula requires only the numerical parts of each dimension, and the units must then be dropped. Consider the formula known as Ohms law, $I = E/R$. In this formula, only the number of volts is substituted for E and only the number of ohms of resistance is substituted for R. This ratio E/R is equal to the number of amperes of current I flowing in the circuit. Compute I where $E = 880$ volts and $R = 22$ ohms:

$$I \text{ (in amperes)} = \frac{880}{22} = 40 \text{ amp}$$

Whenever you start to use a formula, particularly the first few times, study the given statements about each literal number involved. Some formulas have been developed which use specific dimensions and may require only the numerical part of the dimensions. Others are more general in nature and may require that all appropriate dimensional units be used.

Contrast the following two statements for the literal number P in a formula involving a pressure P of 500 lb/sq in.

1. P is the pressure expressed in p.s.i. (pounds/square inch).
2. P is the number of p.s.i.

Using the first statement, we would substitute the factor (500 lb/1 sq in.) for P in the formula. If we used the second statement, we would substitute the factor 500 for P in the formula.

EXERCISE 3.6

The formulas used in these problems are taken from various technical fields. Solve each one for the indicated quantity, using the values given for the literal numbers. Pay close attention to the use of dimensional units and to the significant figures in the answers.

1. Another form for Ohms law is $E = IR$, where E is the number of volts, I is the number of amperes of current, and R is the number of ohms resistance. Solve for E if $I = 2.80$ amp and $R = 400$ ohms.
2. For a uniformly accelerating object, its velocity V is given by $V = at$. Find V when $a = 32$ ft/sec², and $t = 3.0$ sec.

3. The distance s that a freely falling object will fall, starting at rest, in a period of time t (in seconds) is given by the formula $s = at^2/2$. Compute s when $a = 32$ ft/sec² and $t = 5$ sec.

4. The volume of a gas V_2 is given by $V_2 = P_1V_1/P_2$. Calculate the volume V_2 of a gas when $P_1 = 78\bar{0}$ cm, $V_1 = 80.0$ cu ft, and $P_2 = 60.0$ cm.

5. The velocity V of water flowing through an orifice is given by the formula $V = 0.6\sqrt{2gh}$, where $g =$ the constant 980 cm/sec² and $h =$ the height of the water (in centimeters) above the orifice. Compute the velocity V where:
   (a) $h = 2.00 \times 10^2$ cm.
   (b) $h = 5.00 \times 10$ cm.
   (c) $h = 18.0$ cm.
   Do the decreasing answers seem reasonable? Explain.

6. According to A.S.A. standards, the bursting tensile strength S in pounds/square inch for cast iron pipe is obtained by the formula $S = Pd/2t$, where P is the internal pressure (pounds/square inch) at bursting, d is the average inside diameter (inches) of the pipe, and t is the minimum average thickness (inches) of the walls of the pipe along the principal line of break. Compute S when $P = 1900$ lb/sq in., $d = 11.10$ in., and $t = 0.48$ in.

7. Using the formula in Problem 6, compute S when $P = 1800$ lb/sq in., $d = 16.32$ in., and $t = 0.54$ in.

8. The time t (called the period) for one complete swing of a simple pendulum is given by the formula $t = 2\pi\sqrt{L/g}$, where t is the time in seconds, L is the length of the pendulum in feet, and g is the acceleration constant 32.2 ft/sec². Compute t for a pendulum whose length is 18.0 in.

9. Using the formula in Problem 8, compute the period of a pendulum whose length is 48.0 in.

10. If the formula in Problem 8 is solved for L, it becomes $L = gt^2/4\pi^2$. Solve for the length of a pendulum with a period $t = 2.00$ sec.

11. From $L = gt^2/4\pi^2$, find the length (in centimeters) of a pendulum whose period $t = 1.00$ sec.

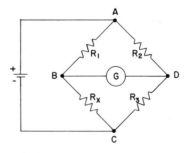

Figure 3-4

12. The fundamental equation of the Wheatstone bridge (Figure 3-4) is

$$\frac{R_x}{R_1} = \frac{R_3}{R_2}$$

If this equation is solved for R_x, we obtain $R_x = R_1 R_3 / R_2$; $R_x =$ the resistance in ohms of the unknown resistor; R_1, R_2, and R_3 are the known resistances in ohms for the respective resistors. Compute R_x when $R_1 = 15$ ohms, $R_2 = 48$ ohms, and $R_3 = 88$ ohms.

13. Using the formula in Problem 12, compute R_x when $R_1 = 250\,\Omega$ (ohms), $R_2 = 45\,\Omega$, and $R_3 = 75\,\Omega$.

14. A modification of the Wheatstone bridge, called a Murray loop, is used to determine the point at which a telephone line is grounded. The formula derived from the Murray loop is

$$x = \left(\frac{R_2}{R_1 + R_2} \right) L$$

where x is the unknown distance to point of ground, L is the known length of the loop, and R_1 and R_2 are known resistances. Compute the distance x to the point of ground when $R_1 = 382\,\Omega$, $R_2 = 412\,\Omega$, and $L = 52.0$ miles.

15. Using the Murray loop formula from Problem 14, determine the distance to the point of ground when $R_1 = 542\,\Omega$, $R_2 = 126\,\Omega$, and $L = 32.0$ km.

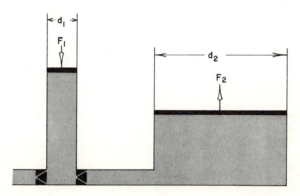

Figure 3-5

16. Figure 3-5 shows a diagram of a hydraulic press. The force F_2 that can be exerted by the press can be computed from

$$F_2 = \frac{(d_2^2)(F_1)}{d_1^2}$$

where F_1 is the force applied to the pump, d_1 is the diameter of the pump, and d_2 is the diameter of the press. (d_1 and d_2 are expressed in the same linear units.) Compute the force F_2 that can be exerted if $F_1 = 52$ lb, $d_1 = 0.25$ in., and $d_2 = 3.0$ in.

17. Using the information in Problem 16, compute F_2 when $F_1 = 40.0$ kg, $d_1 = 25.0$ mm, and $d_2 = 12.5$ cm.

18. The horsepower (H.P.) required to move an automobile through the air on a level road is proportional to the square of the speed s of the car. H.P. $= ks^2$, where s is the number of miles per hour and k is a constant determined by the aerodynamic characteristics of the particular car (i.e., good streamlining—small value of k; poor streamlining—large

value of k). Find the horsepower required for a certain car with $k = 0.0167$ and $s = 6.0 \times 10$ mph.

19. For a test-car traveling 85 mph, 88% of the horsepower developed by the engine is used to overcome air resistance. Compute the horsepower required to overcome the air resistance when this car is traveling 85 mph if $k = 0.0225$. (Use H.P. $= ks^2$ from Problem 18.)

20. The formula $V = ALN/144$ cu ft/min is used to calculate the theoretical volume of gas (in cubic feet) pumped per minute by a one-cylinder compressor where $A =$ the area of the piston (in square inches), $L =$ the length of the stroke (in feet), and $N =$ the number of pumping strokes per minute. Compute the volume of gas pumped in 1 hr by a one-cylinder compressor having a piston area of 5.50 sq in., a stroke of 6.00 in., and 1200 pumping strokes/min.

21. Using the information in Problem 20, compute the volume of gas pumped in 24 hr if $A = 10.2$ sq in., $L = 8.0$ in., and $N = 1400$ strokes/min.

22. The formula $S = 180wd^2/L$ is used to calculate the safe distributed load S (in pounds) supported by a wooden beam where $w =$ the number of inches of width, $d =$ the number of inches in depth, and $L =$ the number of feet of length of the beam. Compute the safe distributed load for a beam if $w = 8.0$ in., $d = 16$ in., and $L = 12$ ft.

23. Using the information in Problem 22, compare the results obtained if the width of the beam is doubled in contrast to doubling the depth of the beam.

24. The velocity V of sound (traveling through air) for a given air temperature is given by

$$V = 1090 + 1.14 \, (F - 32)$$

where $V =$ velocity in feet per second and $F =$ number of degrees Fahrenheit. Compute the velocity of sound when the temperature is:
(a) 0°F.
(b) 32°F.
(c) 132°F.
(Assume three-figure accuracy for the formula and the data.)

25. Using the information in Problem 24 and the formula $F = \frac{9}{5}C + 32$ ($C =$ Centigrade temperature and $F =$ Fahrenheit temperature), compute the velocity of sound when the temperature is:
(a) 80°C.
(b) -40°C.
(Assume three-figure accuracy for the formula and the data.)

4 Working with Equations and Formulas

4.1 BASIC OPERATIONS WITH EQUATIONS

In this section we shall undertake a systematic study of the mechanics of working with equations so that we are able to use the numerous formulas encountered in the technical fields.

We have already had contact with several algebraic equations in the form of formulas. Whenever you write a statement of equality by using the equal symbol ($=$) between two quantities, you are writing an *equation*. In other words, *an equation is a statement of equality existing between two quantities*. The statements $25 + 3 = 28$, $(4)(5) = 20$, $A = lw$, and 2 nickels = 1 dime are equations.

All of the terms on the left side of the equal sign form the *left-hand member* of the equation; all of the terms on the right of the equal sign make up the *right-hand member* of the equation. In practice, we refer to these members as simply the left side and right side of the equation.

Identities

Some equations are *identities*. These statements of equality are true regardless of what quantities the numbers or letters represent. For example,

$$\tfrac{6}{12} = \tfrac{1}{2}$$
$$(x + y)(x - y) = x^2 - y^2$$

Regardless of the values assigned to x and y, the statement of equality is true. (Try substituting several different values.)

Conditional Equations

A conditional equation is one in which the equality depends upon both members possessing some common property, such as weight, density, speed, monetary value, volume, numerical value, length, or area, which allows the equality to hold. Take for example the statement 2 nickels = 1 dime. This equality holds only for the comparison of the monetary value of both members. If the coefficient 2 is replaced with an unknown quantity such as x, the statement of equality "x nickels = 1 dime" rests upon the condition that there is some x (number) of nickels that equals the monetary value of 1 dime.

Because some equalities are conditional (i.e., their validity rests upon some condition), approach each equation with the idea that "solving an equation" is an attempt to find a value or values for the unknown terms or factors so that a condition of equality can be recognized. A statement of equality really asks the question: "Is there a value for x, or y, or z, etc., which will make the statement true?" To answer this question, you must "solve the equation" for x (or y, or z, etc.).

Fundamental Operations

The factors and terms that make up the members of an equation or a formula are mathematically related to each other by one or more of the operations of addition, subtraction, multiplication, and division. In the equation $(3x/4) - 2 = 0$, think of the 2 as being related (connected) to the factor x by subtraction, the 3 as being related to the x by multiplication, and the 4 as being related to the x by division. In solving this equation for x, you need to separate x from all other factors and terms. To do this, you must eliminate the 2, the 4, and the 3 from the left member, leaving x by itself. Each algebraic equation usually contains enough information to allow you to solve for the unknown quantity if you know how to read from the statement the steps that unlock the unknown factor.

The two sides of an equation or formula may be compared with the two pans of a simple beam balance (Figure 4-1). If the pans are in balance, you can do any

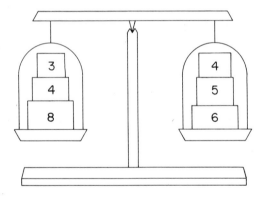

Figure 4-1

combination of the following and the pans will still contain equal weights and therefore be in balance:

1. Add equal weight to each pan.
2. Subtract equal weight from each pan.
3. Multiply the weight on each pan by the same factor.
4. Divide the weight on each pan by the same factor.

Thus, if a formula (or equation) is true, you can do any combination of the following and the resulting statements of equality will also be true:

1. Add the same number or amount to each side.
2. Subtract equal numbers or amounts from each side.
3. Multiply each side by the same factor.
4. Divide each side by the same (nonzero) factor.

When more than one of the above operations (1–4) are necessary, it is usually best to perform them in the order given.

example   Solve the statement

$$3x - 4 = 11 \tag{1}$$

Add 4 to each member:

$$3x = 15 \tag{2}$$

Divide both members by 3:

$$x = 5 \tag{3}$$

To see if $x = 5$ is a solution of the original equation, substitute 5 for x in equation (1):

$$\checkmark, \quad 3(5) - 4 = 11 \tag{4}$$

(The $\checkmark$ means checking.) Simplify each side independently of the other:

$$15 - 4 = 11$$
$$11 = 11 \tag{5}$$

example   Solve for R and check:

$$\frac{25R - 14}{6} = 3R \tag{1}$$

Multiply both members by 6:

$$25R - 14 = 18R \tag{2}$$

Subtract $18R$ from each member:

$$7R - 14 = 0 \tag{3}$$

Add 14 to each member:

$$7R = 14 \tag{4}$$

Divide each member by 7:

$$R = 2 \tag{5}$$

Check in the original:

$$\frac{25(2) - 14}{6} = 3(2) \qquad (6)$$

Simplify each member:

$$\frac{50 - 14}{6} = 6 \qquad (7)$$

$$\frac{36}{6} = 6 \qquad (8)$$

$$6 = 6 \qquad (9)$$

(Reduction to an identity is a necessary part of checking.)

Notice the form used in solving the next three examples. All equations are numbered and the "steps" are given.

example   Solve $(W/5) - 3 = 4$, for W and check.

In this column state each mathematical operation used to obtain the next equation. Call these statements "steps."		$\dfrac{W}{5} - 3 = 4$ (1)
Add the 3 to each side of equation (1) to obtain equation (2).	$+3,$	$\dfrac{W}{5} = 7$ (2)
Multiply both members of equation (2) by 5 to obtain equation (3).	$\cdot 5$	$W = 35$ (3)
Check	$\checkmark$	$\frac{35}{5} - 3 = 4$ (4)
	Simplify	$7 - 3 = 4$
		$4 = 4$ (5)

example   Solve and check:

$$\frac{2N + 5}{7} = 3 \qquad (1)$$

$$\cdot 7, \qquad 2N + 5 = 21 \qquad (2)$$

$$-5, \qquad 2N = 16 \qquad (3)$$

$$\div 2, \qquad N = 8 \qquad (4)$$

$$\checkmark, \qquad \frac{2(8) + 5}{7} = 3 \qquad (5)$$

Simplify, $\dfrac{16 + 5}{7} = 3$ (6)

$\dfrac{21}{7} = 3$ (7)

$3 = 3$ (8)

example   Solve and check :

$\dfrac{21}{Z} - 5 = 2$ (1)

$+5,$ $\dfrac{21}{Z} = 7$ (2)

$\cdot Z,$ $21 = 7Z$ (3)

$\div 7,$ $3 = Z$ (4)

$\checkmark,$ $\dfrac{21}{3} - 5 = 2$ (5)

Simplify, $7 - 5 = 2$ (6)

$2 = 2$ (7)

EXERCISE 4.1A

Solve and check each equation, following the form and procedure shown in the previous examples.

1. $x + 3 = 12$
2. $x - 4 = 8$
3. $y + 3 = 11$
4. $z - 4 = 21$
5. $n + 5 = 2$
6. $w - 7 = 9$
7. $R + 5 = 12$
8. $7 + k = 15$
9. $P - 4 = 8$
10. $w - \frac{1}{2} = 3$
11. $Z + 0.512 = 3.237$
12. $W - 9 = 7$
13. $K + \frac{3}{4} = \frac{5}{6}$
14. $L - \frac{4}{5} = \frac{2}{3}$
15. $12 = R - 4$

16. $N - 2.13 = 5.89$
17. $14.60 = P + 3.33$
18. $0 = T - 8.8$
19. $15.09 = D - 3.81$
20. $845.23 = 404.08 + W$
21. $3x = 45$
22. $\dfrac{y}{4} = 13$
23. $5y = 35$
24. $3v = 18$
25. $\dfrac{u}{6} = 40$
26. $3r - 5 = 13$

27. $\dfrac{x}{4} + 2 = 5$

28. $\dfrac{4x}{5} = 8$

29. $\dfrac{3y}{7} = 12$

30. $\dfrac{2W}{3} - 5 = 19$

31. $6K - 3 = 15$
32. $7N = 42$
33. $12 = 2x$
34. $5N - 4 = 21$
35. $3.5x = 70$
36. $\frac{1}{8}y = 2$

37. $18 = \dfrac{x}{5}$

38. $x + 3x = 40$
39. $2y - 4y = 16$

40. $-9 = \dfrac{360}{N}$

41. $5x = 2x + 12$
42. $8y = 10 + 3y$

43. $\dfrac{3x}{5} + 2 = 14$

44. $\dfrac{4x}{9} + 6 = -2$

45. $\dfrac{3R}{4} + 3 = 18$

46. $\dfrac{5W}{3} - 4 = 6$

47. $\dfrac{N}{3} + 5 = 10$

48. $3 + N = 12$

49. $12 = \dfrac{3K}{4}$

50. $\dfrac{N}{7} + 3 = 7$

51. $\dfrac{6y}{4} - 8 = 7$

52. $2x + 4 = x + 8$

53. $11 = \dfrac{28}{y} - 3$

54. $81 = 60 + 7W$

55. $\dfrac{49}{x} - 4 = 3$

56. $\dfrac{x - 4}{5} = 6$ (*Hint:* Eliminate the 5 first)

57. $\dfrac{3y + 6}{2} = 18$

58. $\dfrac{4r - 1}{8} = 0$

59. $\dfrac{3W - 5}{7} = \dfrac{4}{7}$

60. $kx = 12$ (*Hint:* What step would you perform if k was a 2? Do the same with k; k is a constant.)

Solving Equations with Literal Coefficients and Literal Terms

Up to this point* we have only used expressions with known coefficients. Frequently, we encounter algebraic expressions such as ax, ky, and bt^2. We shall now make a distinction between the roles played by different letters of the alphabet. It has proved helpful to use the letters from the beginning of the alphabet, a, b, c, d, etc., to represent unknown constants in an expression, and the letters near the

* Except for Problem 60, Exercise 4.1A.

end of the alphabet, p, q, r, s, t, z, etc., to represent "variable" factors in an expr sion.

Frequently the values for both the constant and variable factors may not l known and it is at first confusing to distinguish between them. Recall the equatio from Problem 18, Exercise 3.6, H.P. $= ks^2$, where k was a constant determined experimentally from aerodynamic data for each car. Before a particular car is tested, the value of the constant k is unknown. Once the value for k is determined, it does not change until the car is physically altered, such as changing its weight, design, etc. In contrast, s is the speed of the car and can vary from zero up to the maximum speed of the car. When we use expressions such as ks^2, we call k the literal coefficient and s the variable or "unknown" part.

Consider the following examples:

example   Solve for x:

$$ax = 12 \tag{1}$$

$$\div a, \qquad \frac{\cancel{a}x}{\cancel{a}} = \frac{12}{a} \tag{2}$$

$$x = \frac{12}{a} \tag{3}$$

$$\checkmark, \qquad \cancel{a}\left(\frac{12}{\cancel{a}}\right) = 12 \tag{4}$$

$$12 = 12 \tag{5}$$

example   Solve for x:

$$ax - b = c \tag{1}$$

$$+b, \qquad ax = c + b \tag{2}$$

$$\div a, \qquad \frac{\cancel{a}x}{\cancel{a}} = \frac{c+b}{a} \qquad \left(\text{or } \frac{c}{a} + \frac{b}{a}\right) \tag{3}$$

$$\checkmark, \qquad \cancel{a}\left(\frac{c+b}{\cancel{a}}\right) - b = c \tag{4}$$

$$\text{Remove ( ),} \qquad c + b - b = c \tag{5}$$

$$\text{C.T.,} \qquad c = c \tag{6}$$

example   Solve for y:

$$\frac{by}{c} + a = d \tag{1}$$

$$-a, \qquad \frac{by}{c} = d - a \qquad\qquad (2)$$

$$\cdot c, \qquad \cancel{c} \cdot \frac{by}{\cancel{c}} = c(d - a) \qquad (\text{or } cd - ca) \qquad (3)$$

$$\div b, \qquad \frac{\cancel{b}y}{\cancel{b}} = \frac{c(d - a)}{b} \qquad \left(\text{or } \frac{cd}{b} - \frac{ca}{b}\right) \qquad (4)$$

$$y = \frac{c}{b}(d - a) \qquad\qquad (5)$$

$$\checkmark, \qquad \frac{\cancel{b}}{\cancel{c}}\left(\frac{\cancel{c}}{\cancel{b}}\right)(d - a) + a = d \qquad (6)$$

Remove ( ), $\qquad d - a + a = d \qquad\qquad (7)$

C.T., $\qquad d = d \qquad\qquad (8)$

EXERCISE 4.1B

Solve Problems 1–8 for the variable involved. Number all equations, give steps, and check.

1. $ky = 15$
2. $ax - 3 = 12$
3. $br + 7 = 3$
4. $ct - 2 = 4$
5. $dx - 3 = 0$
6. $3ar = 6a$
7. $2cz = 8c$
8. $kw - 22 = 1$

Solve for the indicated factor in Problems 9–28. Number equations, give steps, and check.

9. $5S = 2T$; solve for S.
10. Solve the equation in Problem 9 for T.
11. $ar - 5 = W + 3$; solve for W.
12. Solve the equation in Problem 11 for r.
13. $8T - 4 = 2t + 2$; solve for T.
14. Solve the equation in Problem 13 for t.
15. $(2L + W)/6 = 1$; solve for W.
16. Solve the equation in Problem 15 for L.
17. $(kr - P)/9 = 2$; solve for r.
18. Solve the equation in Problem 17 for P.
19. $I = E/R$; solve for E (Ohms law).
20. Solve the equation in Problem 19 for R.
21. $A = lw$; solve for l.
22. Solve the equation in Problem 21 for w.
23. $A = \frac{1}{2}(b_1 + b_2)h$; solve for h.
24. Solve $S = Pd/2t$ for d (bursting strength of cast iron pipe).

25. Solve the equation in Problem 24 for P.
26. Solve the equation in Problem 24 for t.
27. Solve $R_x = R_1 R_3 / R_2$ for R_3 (formula from Wheatstone bridge).
28. Solve the equation in Problem 27 for R_2.

4.2 RATIO AND PROPORTION

You have already used numerous ratios. In Chapter 2, you solved several geometric problems by the use of proportions and the slide rule. At this point, we shall discuss the algebraic solution of a proportion.

Consider Figure 4-2 which shows the intersection of a railroad, an aqueduct, a highway, and a powerline, with the highway and powerline running parallel to each other. A draftsman needed to know distances S_4 and S_6. Remembering that

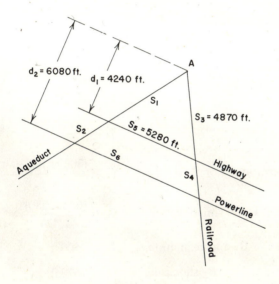

Figure 4-2

parallel lines cut all transversals proportionally and that corresponding parts of similar triangles are proportional, he set up the following proportions. (A proportion is a statement of equality between two ratios.)

$$\text{A.} \quad \frac{S_3}{S_4} = \frac{d_1}{(d_2 - d_1)} \qquad \text{B.} \quad \frac{S_6}{S_5} = \frac{d_2}{d_1}$$

To solve proportion A for S_4:

$$\frac{S_3}{S_4} = \frac{d_1}{d_2 - d_1} \qquad (1)$$

$\cdot (S_4),$
$$S_3 = \frac{d_1 S_4}{d_2 - d_1} \qquad (2)$$

$\cdot (d_2 - d_1),$
$$S_3(d_2 - d_1) = d_1 S_4 \qquad (3)$$

$\div d_1,$
$$\frac{S_3(d_2 - d_1)}{d_1} = S_4 \qquad (4)$$

Substituting for $S_3, d_1, d_2,$
$$\frac{(4870 \text{ ft})(6080 - 4240) \text{ ft}}{4240 \text{ ft}} = S_4 \qquad (5)$$

Upon simplification,
$$\frac{(487)(1840) \text{ ft}}{424} = S_4 \qquad (6)$$

By approximation,
$$S_4 \approx 2000 \text{ ft} \qquad (7)$$

By slide rule,
$$S_4 = 2110 \text{ ft} \quad \checkmark \quad (8)$$

To solve proportion B for S_6:

$$\frac{S_6}{S_5} = \frac{d_2}{d_1} \qquad (1)$$

$\cdot (S_5),$
$$S_6 = \frac{d_2 S_5}{d_1} \qquad (2)$$

Substituting for $S_5, d_1, d_2,$
$$S_6 = \frac{(6080 \text{ ft})(5280 \text{ ft})}{4240 \text{ ft}} \qquad (3)$$

Upon simplification,
$$S_6 = \frac{(608)(5280) \text{ ft}}{424} \qquad (4)$$

By approximation,
$$S_6 \approx 7500 \text{ ft} \qquad (5)$$

By slide rule,
$$S_6 = 7570 \text{ ft} \quad \checkmark \qquad (6)$$

Now observe that in proportion A, the draftsman set up the unknown factor S_4 in the denominator; in proportion B, he set up the unknown factor S_6 in the numerator. He could just as easily have set up A with S_4 in the numerator which would have saved several algebraic steps. Therefore, it is good practice to set up the proportion with the unknown factor in the numerator.

Ratio

To be very specific, a ratio is a number (without a dimension) that expresses the numerical relationship between two like quantities when the two quantities are compared as a quotient. For example, one design (No. 1) for a structural member of a bridge requires 480 rivets. An alternate design (No. 2) requires only 120 rivets. You can express a mathematical relationship between the two quantities by stating the difference in magnitude of the two quantities as: design No. 2 requires 360 fewer rivets than design No. 1, or design No. 1 requires 360 more rivets than design No. 2. Another way of expressing a relationship between the two quantities is by writing the ratios:

$$\frac{480 \text{ rivets}}{120 \text{ rivets}} = \frac{4}{1} \quad \text{and} \quad \frac{120 \text{ rivets}}{480 \text{ rivets}} = \frac{1}{4}$$

The first ratio tells us that there are 4 times as many rivets in design No. 1 as in design No. 2. The second ratio tells us that there are only $\frac{1}{4}$ as many rivets in design No. 2 as in design No. 1.

It is common practice to express fractional ratios such as $\frac{1}{4}$ as 0.25 to 1, $\frac{3}{5}$ as 0.6 to 1, $\frac{7}{8}$ as 0.875 to 1, and $\frac{9}{5}$ as 1.8 to 1.

For calculation purposes, it is not necessary that all ratios in fraction form be expressed as a decimal number compared to 1. However, it is essential for the technician to understand the various forms used to express ratios.

The Colon (:)

The colon is frequently used between two numbers instead of the word "to." For example, the ratio "3 to 1" is written $3:1$, "0.25 to 1" is written $0.25:1$, and "$\frac{5}{4}$" is written $5:4$ or $1.25:1$. Proportions are sometimes written which use the colon such as $3:6 = 1:2$ (read 3 is to 6 as 1 is to 2). Sometimes the equal sign is replaced with the double colon (: :)—$3:6 :: 1:2$ (reads the same, 3 is to 6 as 1 is to 2).

Consider a proportion written in three equivalent forms:

$$\text{A.} \quad \frac{5}{2} = \frac{x}{8} \qquad \text{B.} \quad 5:2 = x:8 \qquad \text{C.} \quad 5:2 :: x:8$$

Let us solve each of these in turn and see how the solutions of B and C bring us back to the procedure used in A.

$$\text{A.} \quad \frac{5}{2} = \frac{x}{8} \tag{1}$$

$$\cdot 2, \ \cdot 8, \qquad 5 \cdot 8 = 2 \cdot x \qquad \text{(This step is sometimes called} \tag{2}$$
clearing fractions because you
eliminate all denominators.)

$$\div 2, \qquad \frac{40}{2} = x \tag{3}$$

or,

$$20 = x \quad \checkmark$$

The key to the solution of form B and C comes from equation (2), i.e., the statement $5 \cdot 8 = 2 \cdot x$. To obtain this result, multiply the factors as shown by the arrows.

B. $5:2 = x:8$ which gives us $5 \cdot 8 = 2 \cdot x$

C. $5:2 :: x:8$ :: is replaced with $=$

$$5 \cdot 8 = 2 \cdot x$$

From here on the solutions are the same as for form A. The terms *means* (inner) and *extremes* (outer) are applied to the terms in proportions B and C.

$$\overset{\text{means}}{5:2} \; = \; x:8 \quad \text{and} \quad \overset{\text{means}}{5:2} \; :: \; x:8$$

$$\underset{\text{extremes}}{5 \cdot 8 = 2 \cdot x} \qquad\qquad \underset{\text{extremes}}{5 \cdot 8 = 2 \cdot x}$$

Note that the resulting statements equate the product of the means to the product of the extremes.

If the means (inside terms) of the proportion are equal as in the proportion $R:X = X:N$ or $R/X = X/N$, then the value of X is called the *mean proportional* between R and N. The solution of this proportion is $X = \sqrt{RN}$; i.e. the mean proportional between two numbers is the square root of their product.

EXERCISE 4.2

Copy and complete the following table (Problems 1–8) by writing ratios for the given quantities in common fraction form, colon form, decimal fraction to 1 form, and decimal colon form. Study the example first.

Quantities Compared	Fraction Form	Colon Form	Decimal to 1	Decimal Colon Form
Example. 6 in. with 4 ft	$\frac{6}{48} = \frac{1}{8}$	1:8	0.125 to 1	0.125:1
1. 3 pt with 6 gal				
2. 8 ft with 12 yd				
3. 1 sq in. with 1 sq ft				
4. 5 ohms with 2 ohms				
5. 6 adults with 3 children				
6. 15 oz with 1 lb				
7. Perimeter of a 5-in. square with perimeter of an 18-inch square				
8. 9 cu ft with 1 cu yd				

In Problems 9–14, write the given proportions in algebraic form, using first, the form $a/b = c/d$ and second the form $a:b = c:d$. Then solve each proportion.

9. Five is to twelve as x is to one hundred twenty.
10. Three compares to ten as N compares to fifteen.
11. Seven is to one as y is to three.
12. z is to four as six is to one.
13. Nine is to K as three is to one.
14. Five is to twenty-two as twenty-five hundredths is to W.

In Problems 15–18, write the given proportions in sentence form.

15. $3:5 = x:9$
16. $4:9 :: 12:R$

17. $11:J = 3:42$
18. $W:R :: z:T$

In Problems 19–24, write proportions involving X (the mean proportion) between the two given numbers. Then solve the proportion for X and, with the slide rule, determine its value to three significant figures (except Problems 23 and 24). Assume the given numbers are exact.

19. 4 and 16
20. 3 and 27

21. 1 and 4
22. 5 and 12

23. y and $4y$
24. a and b

Continued Proportions

A *continued proportion* between six or more quantities is expressed by the proportion $a:b:c = x:y:z$. This is a condensed way of simultaneously expressing the proportions:

$$\frac{a}{b} = \frac{x}{y}, \quad \frac{a}{c} = \frac{x}{z}, \quad \frac{b}{c} = \frac{y}{z}, \quad \text{or} \quad \frac{a}{x} = \frac{b}{y} = \frac{c}{z}.$$

Sometimes the relative proportions of ingredients are stated as numeric ratios. For example, a cement, sand, gravel ratio of $1:2:4$ means

$$\frac{\text{parts of cement}}{1} = \frac{\text{parts of sand}}{2} = \frac{\text{parts of gravel}}{4}$$

or

$$\frac{\text{parts of cement}}{\text{parts of sand}} = \frac{1}{2}$$

and

$$\frac{\text{parts of sand}}{\text{parts of gravel}} = \frac{2}{4}$$

or, as practically applied to a mixture, simply 1 part cement, 2 parts sand, and 4 parts gravel.

People who work in the same technical areas frequently leave off half of a proportion and simply state the numerical parts of a ratio. For example, $1:2:4$ (in the concrete industry) would mean the proportion discussed above.

EXERCISE 4.2B

In Problems 1–4, write four separate proportions (fraction form) contained in each of the continued proportions.

1. $r:s:t = 3:4:5$
2. $w:x:y = 1:5:12$

3. $5:2:1 = K:L:M$
4. $1:3:5 = R:W:J$

Referring to Figure 4-3, complete the following proportions.

5. $a:d:g:j = b:$

6. $a:j = c:$

7. $a:b:c = d:$

8. $\dfrac{a}{a+d+g+j} = \dfrac{b}{\rule{1.5cm}{0.4pt}} = \dfrac{c}{\rule{1.5cm}{0.4pt}}$

9. If $a = 68, b = 60, c = 62, f = 24, h = 24$, and $j = 32$, determine the lengths of the remaining lettered segments. (Assume that given values are exact and leave answers exact.)

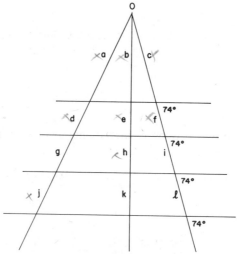

Figure 4-3

4.3 VARIATION

Before the modern technician can solve certain problems, he must first formulate them in mathematical language from observed and measured relationships. Consider the following example.

A technician was assigned to the operational design department of a large chemical company. One problem on which he worked required that he determine the optimum thickness of the plastic lining of a pipe. In addition, he was to determine the optimum pressure of the liquid being pumped through the pipe. During his study of this problem, he discovered the following important relationships:

1. Increasing the thickness of the plastic coating increased the service life of the pipe.

2. Increasing the pressure of the liquid in the pipe decreased the service life of the pipe.

The actual solution to this problem involves many other factors and mathematics beyond the scope of this text. However, let us carefully analyze each of the two statements because they contain the two basic mathematical relationships at the heart of problem solving.

Statement 1 says that the service life and plastic thickness increase or decrease together; they are in what is called *direct variation*. Statement 2 says that service life decreases when the operational pressure increases. This variation is called *inverse variation*. When the value of one quantity depends upon another quantity, the mathematical relationship between the two quantities is either a direct variation or an inverse variation.

One of the first steps in the development of algebraic statements of variation is to determine the type of variation between the involved quantities. The second step is to assign letters to represent the particular quantities. For Statement 1, we first determine that the relationship is a direct variation. Then we assign the letter S to stand for the service life of the pipe and the letter T to stand for thickness of the plastic. The direct variation statement can then be written as $S = k_1 T$. The k_1 is called the constant of variation or simply the proportionality constant. For Statement 2, we first conclude that the relationship is an inverse variation; then we assign the letter S to stand for service life and the letter P to stand for operational pressure. Then the inverse variation statement is written $S = k_2/P$.

examples

Word Statement	Algebraic Statement
R varies directly with W	$R = kW$
R varies inversely with W	$R = \dfrac{k}{W}$
N is directly proportional to P	$N = kP$
N is inversely proportional to P	$N = \dfrac{k}{P}$
Z varies as T	$Z = kT$
Z varies inversely as T	$Z = \dfrac{k}{T}$

EXERCISE 4.3A

In Problems 1–6, write each statement in algebraic form.

1. *R* varies directly with *W*.
2. *T* varies inversely as *Z*.
3. *P* is directly proportional to *Q*.
4. *N* is inversely proportional to *W*.
5. *Q* varies as *N*.
6. *S* varies as *P*.

In Problems 7–16, state whether the relationship is a direct or inverse variation and then write an equivalent algebraic statement.

7. *Z* increases as *W* increases.
8. *N* increases as *R* decreases.
9. *R* decreases as *S* decreases.
10. *Q* decreases as *Z* increases.
11. The quantities *P* and *Q* increase together.
12. The quantities *T* and *W* decrease together.
13. *P* diminishes in size as *T* grows larger.
14. The labor cost *C* per item is reduced as the volume *V* of finished items grows.
15. The pressure *P* of a gas flowing through a conduit drops as its velocity *V* is increased.
16. The strength *S* of fresh concrete increases with time *T*.

In Problems 17–24, state whether the relationship is a direct or inverse variation, assign appropriate letters to represent the specific quantities involved, and write an equivalent algebraic statement.

17. The force exerted by wind against a sail increases with the area of the sail.
18. The volume of a cylinder varies as its length.
19. Tire mileage is less at higher speeds.
20. The life of paint decreases with its exposure to direct sunlight.
21. The sag of a stretched wire can be diminished by increasing the tension on the wire.
22. The rate at which ice melts depends upon the amount of heat being absorbed by the ice.
23. The strength of a magnetic field around a wire becomes less as the current decreases.
24. The magnitude of an inscribed angle is measured by one-half its intercepted arc.

In Problems 25–30, assign appropriate letters to represent the specific quantities and write an equivalent algebraic statement.

25. The volume of a cylinder varies as the square of its diameter.
26. The volume of a sphere varies as the cube of its radius.
27. The volume of a cone varies as the square of the diameter of the base.
28. The cross-sectional area of a pipe varies as the square of the diameter.
29. The area of an equilateral triangle varies as the square of its altitude.
30. Wind resistance varies as the square of the velocity.

Determining the Type of Variation

In some problems, certain words give clues to the type of variation involved. The phrases "varies as," "varies directly with," " is directly proportional to," "is proportional to," and "increases with" show direct variation. The phrases "varies inversely," "is inversely proportional to," and "varies inversely with each other," show inverse variation.

Sometimes a problem does not contain any of these phrases. What might you look for then?

1. Consider the algebraic form of the direct variation $Y = kX$. If you solve for k, you have $Y/X = k$, which states that the ratio of Y/X is a constant. If you can tell that the ratio of two variable quantities is a constant, you can conclude that the relationship between the quantities is a direct variation.

2. Consider the algebraic form of the inverse variation $Y = k/X$. If you multiply the equation by X you have $XY = k$. This equation states that the product of the two quantities, X and Y, is a constant. When the product of two variable quantities is a constant, the two quantities vary inversely with each other.

example If the area of a rectangle is to remain constant, how are the length and width of the rectangle related? If you write $A = k = lw$ ($k =$ the constant area), then $l = k/w$ or $w = k/l$. These equations show that the length varies inversely with the width or that the width varies inversely with the length, respectively.

example If the ratio of the volume of a cylinder compared to its length is a constant, $V/l = k$, then the volume is directly proportional to the length.

EXERCISE 4.3B

Read each statement, assign appropriate letters for the involved quantities, state whether the variation is direct or inverse, and give reasons for your answer.

1. For a liquid being pumped through a pipe, the ratio of flow to pressure is a constant.
2. Distance equals velocity times the time, and the distance remains constant.

3. $\dfrac{\text{Circumference}}{\text{diameter}} = \pi$

4. $\dfrac{\text{Perimeter of a square}}{\text{edge}} = 4$

5. For any group of similar triangles, the ratio of altitude/base $= k$ (corresponding altitudes and bases).

6. $\dfrac{\text{Side of a regular hexagon}}{\text{diagonal}} = \dfrac{1}{2}$

7. Work = (force)(distance). (Consider work a constant.)
8. Pressure = (density)(depth). (Consider depth a constant.) $P = dD$
9. Do Problem 7 but consider distance a constant.
10. Do Problem 7 but consider force a constant.
11. Do Problem 8 but consider density a constant.
12. Do Problem 8 but consider pressure a constant.
13. Force = (pressure)(area). (Consider force a constant.)
14. Do Problem 13 but consider area a constant.
15. With no change in temperature, the pressure of a gas times the volume of the gas is a constant.
16. With no change in pressure, the ratio of the volume of a gas to its absolute temperature is a constant.

The Proportionality Constant

Consider the relationship $S = kA$, where S is the tensile strength of a bolt and A is the cross-sectional area of the bolt. If you solve this equation for k, you have $S/A = k$. This equation shows that the ratio S/A is a constant. However the particular constant obtained depends upon:

1. The particular units used to express tensile strength such as pounds/square inch, kilograms/square centimeter, and tons/square inch.

2. The type of material from which the bolt is made such as steel, cast iron, aluminum, and brass.

3. Other factors such as special treatment, hardening, crystallization, fatigue of the metal, and temperature.

Some relationships involve proportionality constants in which the constants cannot be altered. For example, for any circle, $C/D = \pi$. Regardless of the circle or the particular unit of length used for both C and D, π remains the same constant.

The constant of proportionality can be evaluated when values for the other factors are known.

example The perimeter P of a square is directly proportional to its edge e. If $P = 12$ cm when $e = 3$ cm, find the proportionality constant and then determine P when $e = 15$ yd.

First write an algebraic statement of the variation:

$$P = ke \tag{1}$$

Substitute values for P and e:

$$12 \text{ cm} = k\,(3 \text{ cm}) \tag{2}$$

$$\div 3 \text{ cm}, \qquad \frac{12 \text{ cm}}{3 \text{ cm}} = k \tag{3}$$

Upon simplifying, you have

$$4 = k \qquad (4)$$

Note that the proportionality constant $k = 4$ is independent of the unit of length used to express P and e. Then write the specific proportion:

$$P = 4e \qquad (5)$$

You can now solve the second part of the problem by using equation (5) and the given data $e = 15$ yd:

$$P = 4(15 \text{ yd}) \qquad (6)$$

$$P = 60 \text{ yd} \qquad (7)$$

example The pressure of a liquid on a submerged plate is directly proportional to the depth. Write an algebraic statement for this variation and evaluate k if the pressure is 43.3 lb/sq in. when the depth is 1.00×10^2 ft. Then determine the pressure at a depth of 2400 ft.

First assign letters for pressure and depth. Let $P =$ pressure in pounds/square inch and let $d =$ depth in feet. Then:

$$P = kd \qquad (1)$$

Substitute $P = 43.3$ lb/sq in. and $d = 100$ ft:

$$43.3 \text{ lb/sq in.} = (k)(100) \text{ ft} \qquad (2)$$

$$\div 100 \text{ ft}, \qquad \frac{43.3 \text{ lb/sq in.}}{100 \text{ ft}} = k \qquad (3)$$

Upon simplifying, you have:

$$\frac{0.433 \text{ lb/sq in.}}{1 \text{ ft}} = k \qquad (4)$$

Write the specific proportion:

$$P = \left(\frac{0.433 \text{ lb/sq in.}}{1 \text{ ft}} \right)(d) \qquad (5)$$

Note that the depth must be expressed in feet in order for pressure to be expressed in pounds/square inch. Now to find P when $d = 2400$ ft, substitute in equation (5):

$$P = \frac{(0.433 \text{ lb/sq in.})(2400 \text{ ft})}{\text{ft}} \qquad (6)$$

$$P = (0.433)(2400) \text{ lb/sq in.} \qquad (7)$$

$$P = 1040 \text{ lb/sq in.} \qquad \text{(at 2400-ft depth)} \qquad (8)$$

Return to equation (4). Once the proportionality constant is obtained, it is common practice to drop the units from it. (In this case, simply write $k = 0.433$.) However, when this is done, it is necessary to specify that the formula, $P = 0.433d$, yields pressure in pounds/square inch when d equals the number of feet of depth.

EXERCISE 4.3C

In the following problems: (a) assign letters for the involved quantities, (b) write algebraic statements for the variation, (c) evaluate the proportionality constant using the given data, and (d) write a specific algebraic statement of the proportionality using the value determined for k. When k depends upon particular units, specify in what units each quantity must be expressed. Also, if k depends upon a particular property of either quantity remaining constant, such as density, kind of material, or diameter, state the restrictions.

1. The perimeter of a regular hexagon varies as its diagonal. A hexagon with a diagonal of 140 mm has a perimeter of 420 mm.
2. The weight of a metal rod is directly proportional to its length. A 1.00-in. diameter rod with a length of 18.0 in. weighs 4.00 lb.
3. The average of three measurements is proportional to their sum. The average of three specific measurements is 123 mm when their sum is 369 mm.
4. A person's age in months is directly proportional to his age in years. A person 100 yr old is 1200 mo old.
5. The weight of an $8\frac{1}{2} \times 11$-in. sheet of typing paper varies as its thickness. A sheet with a thickness measuring 0.14 mm weighs 2.8 g.
6. The cost of $8\frac{1}{2} \times 11$-in. typing paper varies inversely with its thickness. Paper measuring 0.008 in. costs \$0.005/sheet.
7. Within certain limits, the viscosity rating of oil is inversely proportional to its absolute temperature. At 400° absolute, the viscosity rating of a special oil is 20.
8. When the temperature remains constant, the density of a gas varies as its pressure. A gas under 6000 g/sq cm pressure has a density of 24 g/ℓ.
9. The electrical resistance of a wire varies with its length. A 550-ft wire has a resistance of 99 ohms.
10. The electrical resistance of a wire varies inversely with its cross-sectional area and directly with its length; 1000 ft of copper wire, whose cross-sectional area is 300,000 circular mils, has a resistance of 36 ohms.
11. The number of units of heat generated by an electric-heating coil is directly proportional to the time of current flow through the coil. For a certain current, 400 B.t.u. of heat was generated in 250 sec.
12. The strength of a horizontal wooden beam (rectangular cross section) varies inversely with the distance between supports. A beam 12 ft long and supported at each end has a safe load capacity of 840 lb.

Combined Variation

Going back to our discussion about the service life of plastic-lined pipe, $S = k_1 t$ and $S = k_2/P$, we could combine these two statements as $S = kt/P$. This equation now reads: "The service life S of the pipe is directly proportional to the thickness t of the plastic lining and inversely proportional to the operational pressure P. The proportionality constant k is now a different constant than either k_1 or k_2. Note that S is now mathematically related to both t and P in the same statement.

examples

$$y = \frac{kr}{w}$$

"y varies directly with r and inversely with w."

$$R = \frac{kV}{t}$$

"R varies directly as V and inversely as t."

Joint Variation

Frequently a quantity varies as a product of two other quantities. For example, $Y = kXZ$. When this occurs, we say that the quantity Y varies *jointly* as X and Z.

examples

$$T = kRW$$

"T varies jointly as R and W."

$$A = klw$$

"A varies jointly with l and w."

Combined Joint and Inverse Variation

For the combined joint and inverse variation

$$R = \frac{kWZ^2}{t}$$

you would read: "R varies jointly as W and Z^2 and inversely with t." For

$$W = \frac{kL^2 r}{p^2}$$

you would read: "W varies jointly as L^2 and r and inversely as p^2." For

$$C = \frac{kn^2t}{q}$$

you would read: "C varies jointly as n squared and t and inversely as q."

example   The safe load (uniformly distributed) on a beam supported at each end varies jointly as its width and the square of its depth and inversely as the distance between the supports. If a 6 by 12-in. wooden beam, 18 ft long, will carry a safe load of 2400 lb, what will be the safe load for a 2 by 8-in. beam, 10 ft long (same kind of wood)?

1. Assign letters for the quantities involved. Let S = safe load in pounds, W = width of beam in inches, d = depth of beam in inches, and L = length of beam in feet.

2. Write the variation in algebraic form:

$$S = \frac{kWd^2}{L}$$

3. Determine the proportionality constant by substituting given data for S, W, d and L:

$$2400 \text{ lb} = \frac{k(6 \text{ in.})(12 \text{ in.})^2}{18 \text{ ft}}$$

$$k = \frac{(50 \text{ lb})(\text{ft})}{\text{in.}^3}$$

4. Write the specific variation formula, using the computed value for k:

$$S = \left(\frac{(50 \text{ lb})(\text{ft})}{\text{in.}^3}\right)\left(\frac{Wd^2}{L}\right)$$

5. Solve for S when $W = 2$ in., $d = 8$ in., and $L = 10$ ft:

$$S = \left(\frac{50 \text{ lb } \cancel{\text{ft}}}{\text{in.}^3}\right)\left(\frac{(2 \text{ in.})(8 \text{ in.})^2}{10 \cancel{\text{ft}}}\right)$$

$$= \frac{(50 \text{ lb})}{\cancel{\text{in.}^3}} \frac{(2)(64) \cancel{\text{in.}^3}}{\cancel{10}}^{5}$$

$$S = 640 \text{ lb}$$

The units

$$\frac{(\text{lb})(\text{ft})}{\text{in.}^3}$$

may be dropped from the proportionality constant if you specify that

$$S = \frac{50Wd^2}{L}$$

gives the safe load in pounds where W is the number of inches of width, d is the number of inches of depth, and L is the number of feet of length.

Variation formulas are general until the proportionality constant is determined. In the last example, the formula in Step 2 is valid for any material and for any dimensions expressed in suitable units. The particular proportionality constant obtained will depend upon the material and dimensions selected. Once these are determined, the variation formula (as in Step 4) becomes specific and is only good for beams of like material using the specified dimensional units.

Dimensional Soundness

One of the most important habits a technician can form is that of checking *each* proportionality constant to see that it is *dimensionally correct*. Dimensionally correct does not mean that the proper units are annexed after a numeric value for the constant is obtained. (This is a poor and costly practice which should be avoided.) It means that proper dimensions accompany all factors involved in the algebraic solution for k and that these units properly express the type or kind of proportionality constant represented by k. For example, a velocity constant should be expressed in units that represent linear distance/unit of time.

Algebraic operations with dimensional units are performed in the same way as with other literal factors or terms. Dimensional units can only be eliminated or changed by performing suitable algebraic operations.

example   The electrical resistance of a wire is directly proportional to its length and inversely proportional to the square of its diameter. If 400 ft of a particular wire, with a diameter of 0.12 in., has a resistance of 0.04 ohm, determine a proportionality constant for the relationship between resistance, length, and diameter; analyze the equation for dimensional soundness.

1. Let R = the resistance of the wire, L = the length of the wire, and k = the proportionality constant.

2.
$$R = \frac{kL}{d^2}$$

3. Substitute the given data for R, L, and d;

$$0.04 \text{ ohm} = \frac{(k)(400 \text{ ft})}{(0.12 \text{ in.})^2}$$

4. Solve this equation for k:

$$k = \frac{(0.04 \text{ ohm})(0.12 \text{ in.})^2}{400 \text{ ft}}$$

5. Simplify:

$$k = \frac{1.44 \times 10^{-6} \, (\text{ohm})(\text{in.}^2)}{\text{ft}}$$

6. It is sometimes easier to solve the general formula in step 2 for k first, obtaining $k - Rd^2/L$, and then to substitute for R, d, and L, obtaining the result in step 4. To obtain a clearer picture of the dimensional soundness of this constant, substitute this value for k in step 2;

$$R = \frac{1.44 \times 10^{-6} \, (\text{ohm})(\text{in.}^2)(L)}{\text{ft} \, (d^2)}$$

7. Then note that for L expressed in feet and d in inches, the feet units cancel each other as do the square inch units, leaving just

$$R = \frac{(1.44 \times 10^{-6})(\text{the number } L)}{(\text{the number } d)^2} \text{ohm}$$

Resistance is properly expressed in ohms so you can conclude that you have a dimensionally sound formula. After evaluating the constant and determining its dimensional soundness, it is permissible to drop all dimensions from the formula if you specify the units in which the various quantities are expressed. Therefore, to complete the example, write:

$$R = \frac{1.44 \times 10^{-6}L}{d^2}$$

where

$$R = \text{resistance of wire in ohms}$$
$$L = \text{the number of feet in length}$$
$$d = \text{the number of inches in diameter}$$

The following table lists some important dimensional units.

Quantity	Computed by	Expressed by
Area	length × length	$(ft)(ft)^* = ft^2$
Volume	length × length × length	$(ft)(ft)(ft) = ft^3$
Volume	area × length	$(ft^2)(ft) = ft^3$
Average velocity	$\dfrac{\text{distance}}{\text{time}}$	$ft \div sec = ft/sec$
Average acceleration	$\dfrac{\text{change in velocity}}{\text{time}}$	$\dfrac{ft}{sec} \div sec = \dfrac{ft}{sec^2}$
Pressure	$\dfrac{\text{force}}{\text{area}}$	$lb \div ft^2 = \dfrac{lb}{ft^2}$.

* Or other appropriate units.

EXERCISE 4.3D

In Problems 1–8, write the given variation statement in words.

1. $R = \dfrac{ks}{t}$

2. $W = kTZ$

3. $A = kLW$

4. $V = \dfrac{kT}{P}$

5. $S = \dfrac{kPT}{W^2}$

6. $L = \dfrac{kVT}{R^2}$

7. $H = \dfrac{kR^2S}{T}$

8. $C = \dfrac{klwh}{n}$

In Problems 9–14, write an equivalent algebraic statement for the given statement of variation.

9. *R* varies jointly with *W* and *T*.
10. *Z* varies jointly as *S* and *P* and inversely as *R*.
11. *C* varies directly with *t* and inversely as P^2.
12. *R* varies directly as the square of *t* and inversely with *W*.
13. *W* varies jointly as *S* and T^2 and inversely with r^3.
14. *d* varies jointly as *S* and the square root of *W* and inversely as t^2.

In Problems 15–22: (a) write the given statement in algebraic form; (b) evaluate the proportionality constant using the given data; (c) then using this value for k, write the specific proportionality between the given quantities, and (d) use this Proportion to solve for the indicated quantity.

15. *y* is directly proportional to *x*. Find *y* when $x = 40$, given that $y = 270$ when $x = 9$.
16. *T* varies inversely with *W*. If $T = 12$ when $W = 8.0$, determine *T* when $W = 420$.

17. Z varies directly with R and inversely with Y. If $Z = 84$ when $R = 4$ and $Y = 9$, determine Z when $R = 144$ and $Y = 36$.

18. H varies directly with d and inversely as r. If $H = 24.4$ when $d = 105$ and $r = 52.5$, determine H when $d = 28.0$ and $r = 85.4$.

19. If Z varies directly as the square of d and $Z = 22.5$ when $d = 1.5$, determine Z when $d = 5.5$.

20. If r varies directly as w and inversely as the square of t, and $r = 42.3$ when $w = 56.4$ and $t = 20.0$, determine r when $w = 67.5$ and $t = 15.0$.

21. The quantity K varies jointly as s and t and inversely as the square root of d. If $K = 71.60$ when $s = 4.000$, $t = 0.8950$, and $d = 64.00$, determine K when $s = 5.10$, $t = 1.02$, and $d = 2.25$.

22. If N varies jointly as x, y, and z and inversely as r^2, and if $N = 6.30$ when $x = 14.4$, $y = 10.5$, $z = 6.00$, and $r = 12.0$, determine N when $x = 0.450$, $y = 1.11$, $z = 10.4$, and $r = 0.850$.

In Problems 23–28: (a) assign letters for the involved quantities; (b) write an algebraic statement for the variation; (c) evaluate the proportionality constant, retaining all dimensional units; (d) write a specific algebraic statement of the proportionality using the value for k, retaining dimensional units, and analyze the dimensional soundness of the formula; (e) rewrite the formula, dropping all dimensional units, and specify the necessary units for each quantity involved; and (f) use the formula in part (d) to solve for the required quantity.

23. The force needed to stretch a spring is proportional to the amount the spring is stretched. If 80 lb will stretch a spring 16 in. what force is required to stretch the spring 24 in.?

24. The loss in pressure of a liquid flowing through a pipe varies as the length and inversely as the diameter of the pipe. If a water pipe 240 ft long and 12 in. in diameter results in a drop in water pressure of 0.80 lb/sq in., find the drop in pressure for a water pipe 880 ft long whose diameter is 22 in.

25. The capacity of a rectangular coal bin varies jointly as the length, width, and height of the bin. If a bin 40.0 ft long, 20.0 ft wide, and filled to a depth of 12.0 ft holds 320 tons, how many tons of the same coal can be stored in a rectangular bin 82 ft long, 46 ft wide, and 18 ft deep?

26. The kinetic energy contained by a moving object varies as the square of its speed. If the kinetic energy of an automobile traveling 22 ft/sec (15 mph) is 242,000 ft-lb, what is the kinetic energy when the car is traveling 88 ft/sec?

27. The number of trees that can be planted per acre varies inversely as the square of the distance between them. If 43,560 seedlings/acre can be planted 1.0 ft apart, how many seedlings/acre will remain when thinned to 5.0 ft apart? (Assume distances are exact.)

28. The time required to dig a canal varies jointly as its length, width, and square of its depth and inversely with the number of machines used. If it takes 1600 hr to dig 1 mile of a canal that is 80.0 ft wide and 20.0 ft deep using 50 earth-moving units, how long will it take to dig 7.2 miles of a canal which is 96.0 ft wide and 30.0 ft deep using 180 earth-moving units?

4.4 LINEAR EQUATIONS

The word linear means in line or pertaining to a line. We have already worked with linear relationships expressed by such direct variation statements as $y = \frac{1}{2}x$, $y = x$, and $y = 4x$. Figure 4-4 shows the graphic representation of these three linear equations. Suppose that we change each of these equations by adding the constant 5 to the right-hand members: $y = \frac{1}{2}x + 5$, $y = x + 5$, and $y = 4x + 5$.

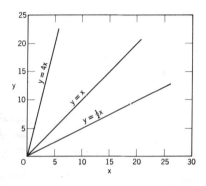

Figure 4-4

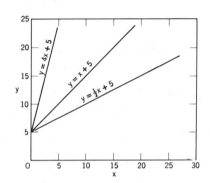

Figure 4-5

What does this do to the linear relationship between y and x? Figure 4-5 shows the graph of these equations. Observe that we still obtain straight lines. Adding the constant 5 does not destroy the linear relationship, but it does shift the lines 5 units in the positive y direction.

How can you tell when a linear relationship exists between two quantities without graphing the equation? If you solve the equation for one of the variable quantities in terms of the other, you should have a statement of direct variation in which the exponents of the two quantities are always ones (usually not written).

For example, if you solve $2T - 4 = 8R$ for T, you have $T = 4R + 2$. This shows that the relationship between T and R is linear, because T varies directly as R and because both variable quantities T and R have exponents of 1. The equations $2T - 4 = 8R$ and $T = 4R + 2$ are both linear equations. As stated in Section 4.1, any solution of $T = 4R + 2$ will also be a solution of $2T - 4 = 8R$; the addition and division operations performed on $2T - 4 = 8R$ to obtain $T = 4R + 2$ do not change the relationship between T and R. Therefore, the two equations are *equivalent*. Equivalent linear equations have the same solution or solutions and their graphs are the same straight line. Figure 4-6 shows the graph of both $2T - 4 = 8R$ and $T = 4R + 2$.

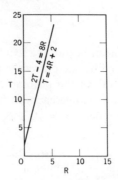

Figure 4-6

example   Starting with $3Z - 9W = 18$, obtain an equivalent equation showing that Z is directly proportional to W.

$$3Z - 9W = 18 \tag{1}$$

$$+\, 9W, \qquad 3Z = 9W + 18 \tag{2}$$

$$\div\, 3, \qquad Z = 3W + 6 \tag{3}$$

To check, substitute $3W + 6$ for Z in equation (1):

$$3(3W + 6) - 9W = 18 \tag{4}$$

Remove ( ) and C.T.:

$$9W + 18 - 9W = 18$$

$$18 = 18 \tag{5}$$

Therefore, equations (1) and (3) are equivalent.*

example   Starting with $4X - 12 = 20Y$, obtain an equivalent equation which shows that X is directly proportional to Y.

$$4X - 12 = 20Y \tag{1}$$

$$+\, 12, \qquad 4X = 20Y + 12 \tag{2}$$

$$\div\, 4, \qquad X = 5Y + 3 \tag{3}$$

To check, substitute $5Y + 3$ for X in equation (1):

$$4(5Y + 3) - 12 = 20Y \tag{4}$$

---

* Actually, equations (1), (2), and (3) are all equivalent, which can be verified by substituting $(3W + 6)$ for Z in equations (1) and (2).

Remove ( ),

$$(), \qquad 20Y + 12 - 12 = 20Y \qquad (5)$$

C.T.,

$$20Y = 20Y \qquad (6)$$

Therefore, equation (3) is the required equation.

Dependent and Independent Variables

When the value of one quantity depends upon the value of another, the first quantity is called the *dependent variable* and the second quantity is called the *independent variable*. How can you tell which variable quantity is the dependent variable and which is the independent variable?

1. The subject of the statement of variation tells which is the dependent variable and which is the independent variable. For example, if Y varies directly as X, the Y value depends upon the X value. Therefore, Y is the dependent variable, whereas X is the independent variable.

2. The dependent variable can sometimes be determined from the form of the equation or formula itself. Consider the equation used in the last example, $4X - 12 = 20Y$. From this equation, you cannot tell which variable is dependent and which is independent unless you know something else about the quantities represented by X and Y. In this example, you ended up with the equation $X = 5Y + 3$. This form of the equation shows that X varies directly as Y; therefore, X is the dependent variable and Y is the independent variable. The equation $4X - 12 = 20Y$ can also be solved for Y, obtaining $Y = 1/5X - 3/5$. In this form, X is the independent variable and Y is the dependent variable.

3. You can sometimes determine the role of the two variables from known information about the two quantities. For example, after the first year the value of a car varies inversely as its age, i.e., $V = k/a$ (V = value in dollars and a = age in years). Because the value decreases as the age increases, and the age does not depend upon the value, you cannot change its age by altering its value. So you know that V is the dependent variable and a is the independent variable.

EXERCISE 4.4A

1. What kind of graph is obtained by graphing a linear equation?
2. What is the relationship called between two quantities whose graph is a straight line?
3. What kind of variation is expressed by a linear equation?
4. What exponents do the variable quantities in a linear equation have?
5. True or false: The exponents referred to in Problem 4 must always be written on the quantities.

6. What are "understood" exponents?

7. True or false: Adding a constant to the right or left member of a linear equation will not alter the direct relationship between the two variables.

8. True or false: Equivalent equations will have the same graph but different solutions. Explain.

9. For the following algebraic statements of variation, write the word linear after those which express a linear relationship. (You may wish to change some of the equations to equivalent equations before deciding on the answer.)

   (a) $Z = 5t - 3$
   (b) $5Z = kt^2$

   (c) $N = \dfrac{k}{W}$

   (d) $R^2 = kt$
   (e) $A = lw$
   (f) $V = lwh$
   (g) $2Y - 6 = 8X + 4$ (Consider Y the dependent variable.)

   (h) $\dfrac{R}{S} = k$

   (i) $Wt = k$

10. For each problem referred to below, state whether or not the relationship between the variable quantities is linear and state which is the dependent variable and which is the independent.
   (a)   Problem 15, Exercise 4.3D
   (b)   Problem 16, Exercise 4.3D
   (c)   Problem 23, Exercise 4.3D

11. Starting with the equation $2r - 4s - 12 = 0$, obtain equivalent linear equations, showing: (a) that r is directly proportional to s, and (b) that s is directly proportional to r.

12. Starting with the equation $6W - 24 = 30T + 12$, obtain equivalent linear equations showing: (a) W as the independent variable, and (*b*) T as the independent variable.

13. Show that the equations $2x - 4y = 8$ and $y = \frac{1}{2}x - 2$ are equivalent.

14. If the two linear equations in Problem 13 were graphed, what would the graph look like?

15. In relating the heat content H of a certain fuel with the carbon content C of the fuel, which do you think would most likely be the dependent* variable? Give reasons for your answer.

16. In an equation relating the frequency f and the tension T of a vibrating musical wire, which do you think would be the dependent* variable? Give reasons for your answer.

17. When two people ride a teeter-totter (see-saw), one goes up while the other goes down. Could their relative positions be expressed by a linear equation? Justify your answer.

---

  * Either variable may be made the dependent variable mathematically. However, it is frequently helpful to establish any physical dependence that exists.

18. For a constant wind, the relationship between the area A of a sail and the force F of the air against the sail can be expressed by a linear equation. Write a linear equation showing this relationship; underline the independent variable, giving reasons for your choice.

Solution(s) of an Equation

A linear equation may have only one solution or it may have an infinite number of solutions. A solution is any value(s) for the variable(s) which satisfies the relationship expressed by the equation. Sometimes there are physical limitations which restrict the number of solutions of a linear equation.

The equation $Y = 5$ has only one solution—the number 5. However, the equation $Y = 3x$ has an infinite number of solutions. A few of the solutions are:

$$(x = 0, Y = 0), \qquad (x = 1, Y = 3)$$
$$(x = -1, Y = -3), \qquad (x = 100, Y = 300)$$

example   The equation $h = -16t^2 + 96t$, when used to express the height of an object (thrown vertically upward) from the ground as a function of the time t, has an infinite number of solutions. However, we must rule out negative values for time and all values for time greater than $t = 6$ sec, because at $t = 6$ sec the object has returned to the ground and the equation no longer expresses its position in relation to time.

Solution of Simultaneous Linear Equations

Technicians frequently encounter a situation in which they know two different relationships between two quantities but need a solution that satisfies both relationships. For example, specifications for the design of a balanced connecting-rod bearing with two inserts require that the combined weights of the two inserts be 80.0 g and that the difference in the weights of the two inserts be 22.0 g. What is the required weight of each insert? If you let $x =$ the weight (in grams) of the heavier insert and $y =$ the weight (in grams) of the lighter insert, then you know two things about x and y. First, $x + y = 80.0$ g and second, $x - y = 22.0$ g.

Solving a pair of equations for a solution that will satisfy both relationships is called solving *simultaneous equations*. To obtain a solution to the problem, add the two equations together.

$$
\begin{array}{ll}
x + y = 80.0\text{ g} & (1) \\
\underline{x - y = 22.0\text{ g}} & (2) \\
2x = 102.0\text{ g} & (3)
\end{array}
$$

Then solving equation (3) for x, you have $x = 51.0$ g.

Now that you know that the weight of the heavier insert is 51.0 g and that the other bearing must be 22 g lighter, you can subtract 22.0 g from 51.0 g, obtaining

29.0 g for the weight of the lighter insert. Do these weights satisfy the conditions of both equations?

$$
\left.
\begin{aligned}
51.0\,\text{g} + 29.0\,\text{g} &= 80.0\,\text{g} \\
51.0\,\text{g} - 29.0\,\text{g} &= 22.0\,\text{g}
\end{aligned}
\right\} \text{Both are satisfied.}
$$

Solution by Addition and Substitution

A combination of the addition and the substitution methods was used to solve the last example. The combination of these two methods is frequently used in solving simultaneous equations. When you attempt to solve a pair of simultaneous equations, you are looking for a solution that satisfies both equations. This search has a counterpart in the graphic interpretation of the system of equations.

Note that lines L_1 and L_3 in Figure 4-7 are parallel, whereas L_1 and L_2 intersect and L_2 and L_3 intersect. If the lines (obtained by graphing the two linear

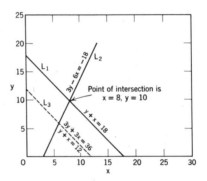

Figure 4-7

equations) intersect, they intersect at only one point. The values of the two variables represented by the coordinates of the point of intersection constitute the solution. In short, if the lines intersect, they have both an algebraic and graphic solution. If the lines are parallel, such as L_1 and L_3, then they do not intersect; the equations of these lines do not have a simultaneous solution. Line L_3 is represented by two equations: $3y + 3x = 36$ and $y + x = 12$. Because these two equations are equivalent, their respective graphs turn out to be the same identical line.

With this as a background, let us consider several examples, starting with the equations for lines L_1 and L_2.

example Solve by the combined addition and substitution method:

$$y + x = + 18 \qquad L_1 \qquad (1)$$

$$3y - 6x = - 18 \qquad L_2 \qquad (2)$$

To use an addition (or subtraction) step, you need the same coefficients on both x terms or both y terms. Suppose that you multiply equation (1) by 6, obtaining equation (3) below, and then take a look at the pair of equations:

$$6y + 6x = 108 \qquad (3)$$
$$3y - 6x = - 18 \qquad (2)$$

Now add them, obtaining

$$9y = 90 \qquad (4)$$

Solving equation (4) for y, you have $y = 10$. Now substitute this value for y in either equation (1) or (2). If you pick equation (1), it becomes

$$10 + x = 18$$

Solving this for x, you have

$$x = 8 \qquad (5)$$

To check, substitute $x = 8$ and $y = 10$ in both equations (1) and (2):

$$10 + 8 = 18 \qquad (1)$$

$$18 = 18 \checkmark$$

$$3(10) - 6(8) = - 18 \qquad (2)$$

$$30 - 48 = - 18$$

$$- 18 = - 18 \checkmark$$

Both are satisfied by $x = 8$ and $y = 10$. Note that the solution checks with the graph in Figure 4-7.

example Solve the following system of equations:

$$3y - 3x = 3 \qquad (1)$$

$$2y + 2x = 26 \qquad (2)$$

To obtain the same coefficients on the x terms, multiply equation (1) through by the factor 2 and equation (2) through by the factor 3. This gives:

$$6y - 6x = 6 \qquad (3)$$

$$6y + 6x = 78 \qquad (4)$$

You can add the two equations and eliminate the x terms, or you can subtract either of the equations from the other eliminating the y terms. To subtract equation (3) from equation (4), change the signs of every term in the subtrahend. Then equation (3) becomes equation (5), and equation (4) stays the same.

$$-6y + 6x = -6 \tag{5}$$

$$6y + 6x = 78 \tag{4}$$

Now add them, obtaining:

$$12x = 72 \tag{6}$$

Solve for x:

$$x = 6 \tag{7}$$

Instead of substituting $x = 6$ for x in equation (1) or (2), just add equations (3) and (4) as they stand, obtaining $12y = 84$ which yields $y = 7$. Now check to see if $x = 6$, $y = 7$ satisfy both equations (1) and (2):

$$3(7) - 3(6) = 3 \tag{1}$$

$$21 - 18 = 3$$

$$3 = 3\checkmark$$

$$2(7) + 2(6) = 26 \tag{2}$$

$$14 + 12 = 26$$

$$26 = 26 \checkmark$$

Therefore, the simultaneous solution is $x = 6$, $y = 7$.

example   Solve the following system of equations by the substitution method only:

$$x + y = 6 \tag{1}$$

$$2y - 2x = -4 \tag{2}$$

To solve a system of equations by the substitution method, you solve either of the equations for either of the variables. If you solve equation (1) for y, you obtain

$$y = -x + 6 \tag{3}$$

Then substitute this value for y in equation (2), obtaining:

$$2(-x + 6) - 2x = -4 \qquad (4)$$

Remove ( ), $\qquad -2x + 12 - 2x = -4 \qquad (5)$

C.T., $\qquad -4x + 12 = -4 \qquad (6)$

-12, $\qquad -4x = -16 \qquad (7)$

$\div(-4)$, $\qquad x = 4 \qquad (8)$

Now you can substitute $x = 4$ in either equation (1) or (2). If you pick (1):

$$4 + y = 6 \qquad (9)$$

-4, $\qquad y = 2 \qquad (10)$

Check with $x = 4$, $y = 2$ in equations (1) and (2):

$$4 + 2 = 6 \qquad (1)$$

$$6 = 6 \checkmark$$

$$2(2) - 2(4) = -4 \qquad (2)$$

$$4 - 8 = -4$$

$$-4 = -4 \checkmark$$

The solution is $x = 4$, $y = 2$.

example   Return now to the equations for L_1 and L_3 in Figure 4-7 to see what happens if you attempt to solve this system of equations:

$$y + x = 18 \qquad L_1 \qquad (1)$$

$$y + x = 12 \qquad L_3 \qquad (2)$$

If you subtract equation (2) from (1), you obtain $0 = 6$ which indicates that something is wrong. If you had found an algebraic solution for some x value and some y value for this system of equatious, the lines would meet at some point, therefore L_1 would not be parallel to L_3. If the lines are parallel, the attempt to find an algebraic solution always ends up with an absurdity such as $0 = 6$.

Equations whose graphs are parallel lines are called *inconsistent* and there is no solution for them. Because the graphs of equivalent equations turn out to be the same line, such equations are called *dependent* equations (or usually just *equivalent* equations).

The following tips can help you gain insight into a system of equations.

1. If corresponding coefficients and constants of two equations are proportional, the equations are equivalent (one line—infinite number of solutions).

2. If corresponding coefficients of two equations are proportional but not the constants, the equations are inconsistent (parallel lines—no solution).

3. If corresponding coefficients of two equations are not proportional, the equations are independent (intersecting lines—one solution).

EXERCISE 4.4B   *Prime* II *problems.*

1. If you graph the system of equations in the example on page 148, what would the graph look like?

2. If you graph the system of equations in the second example on page 147, what would the graph look like?

   In Problems 3–16, determine by inspection whether the system is dependent, inconsistent or independent. Solve those that possess a common solution.

3. $x + y = 8$
   $x - y = 2$

4. $2x - y = 6$
   $3x + y = 4$

5. $x + 2y = 10$
   $4x + 8y = 2$

6. $3x - 3y = 6$
   $2x - 2y = -6$

7. $3y - 4x = 5$
   $4y + 4x = 9$

8. $7x - 3y = 10$
   $5x - 2y = 8$

9. $4x + 7y = 3$
   $6x - 5y = 20$

10. $8y - 5x = 18$
    $x - y = 0$

11. $9x + 8y = 77$
    $x - y = 1$

12. $8x - 5y = 58$
    $x + y = 4$

13. $9R - 13W = -3$
    $6R - 7W = 3$

14. $2x + 6y = 10$
    $3x + 9y = 15$

15. $7Z - 3R = 41$
    $4Z + 5R = 10$

16. $1.1W - 0.3R = -3.4$ (*Hint:* Multiply each equation through by 10.)
    $0.8W + R = 2.4$

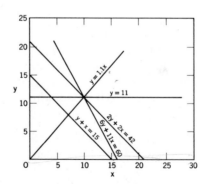

Figure 4-8

17. If you were to draw five more straight lines through the point where $x = 8$ and $y = 10$ in Figure 4-7, and were then able to determine the linear equation for each line, would all the lines through this point represent a simultaneous system of equations? Explain.

18. Figure 4-8 shows the graph of several linear equations. List the sets of equations that have a common solution.

Challenge Problems

19. Two shims are to have a combined thickness of 0.070 in. One shim must be 2.5 times as thick as the other. What are their respective thicknesses?

20. The sum of two resistors in series must equal 414 ohms. The smaller resistor must have a resistance equal to 80.0% of the resistance of the larger resistor. What must be their respective ratings?

19. $X + \frac{y}{2} = .070$ $y = 2.5 X$

X

Prime number - divisible only by itself and one - evenly

5 Additional Topics in Algebra

5.1 FACTORING AND SPECIAL PRODUCTS

We have seen how the work involved in performing arithmetic operations can be simplified if the numbers are written in factored form, i.e., as a product of two or more factors. In a similar manner, the work on algebraic expressions can also be simplified if the expressions are in factored form. For example:

$$\textit{Fraction} \qquad \textit{Factored Form} \qquad \textit{Simplified Fraction}$$

$$\frac{68}{170} = \frac{\cancel{17} \cdot \cancel{2} \cdot 2}{\cancel{17} \cdot \cancel{2} \cdot 5} = \frac{2}{5}$$

$$\frac{5a + 15b}{7a + 21b} = \frac{5\cancel{(a + 3b)}}{7\cancel{(a + 3b)}} = \frac{5}{7}$$

When we factor a number such as 30 as equal to $5 \cdot 6$ or $5 \cdot 2 \cdot 3$, we are relying on previously memorized products. In a similar manner, we must memorize a few algebraic products. When learning to factor, it is also a great help to learn the names of the most frequently encountered products. Study the multiplication involved in each of the following products. Also learn the name and pattern or form of each product. Each name is derived from the type of product obtained.

1. Common monomial (one-term) factor:

$$\underline{a}(b + c) = \underline{a}b + \underline{a}c$$

$$\underline{5H}(R - T + Z) = \underline{5H}R - \underline{5H}T + \underline{5H}Z$$

2. Common binomial (two-term) factor:

$$\underline{(x - y)}(3R + 5K) = \underline{(x - y)}3R + \underline{(x - y)}5K$$

$$\underline{(3x - 2)}(8L - N + W) = \underline{(3x - 2)}(8L) + \underline{(3x - 2)}(- N) + \underline{(3x - 2)}(W)$$

3. The difference of two squares:

$$(x - y)(x + y) = x^2 + xy - xy - y^2 = \underline{x^2 - y^2}$$

$$(5W + 6L)(5W - 6L) = 25W^2 + 30WL - 30WL - 36L^2 = \underline{25W^2 - 36L^2}$$

4. Trinomial square:

$$(a + b)^2 = (a + b)(a + b) = \underline{a^2 + 2ab + b^2}$$

$$(2x - 3y)^2 = (2x - 3y)(2x - 3y) = \underline{4x^2 - 12xy + 9y^2}$$

5. Trinomial of the form (squared variable has a coefficient of 1):

$$(x + a)(x + b) = \underline{x^2 + (a + b)x + ab}$$

$$(y + 5)(y + 3) = y^2 + (5 + 3)y + 15 = \underline{y^2 + 8y + 15}$$

6. General trinomial:

$$(ax + b)(cx + d) = \underline{acx^2 + (ad + bc)x + bd}$$

$$(3x - 2)(2x + 3) = 3x \cdot 2x + (3 \cdot 3 - 2 \cdot 2)x - 2 \cdot 3 = \underline{6x^2 + 5x - 6}$$

7. Factoring by grouping:

$$(a + b)(x + y) = a(x + y) + b(x + y) = \underline{(ax + ay)} + \underline{(bx + by)}$$

$$= ax + ay + bx + by$$

$$(5L + R)(K - P) = 5L(K - P) + R(K - P) = \underline{(5KL - 5PL)} + \underline{(RK - RP)}$$

$$= 5KL - 5PL + RK - RP$$

EXERCISE 5.1A

Multiply Problems 1–10 by inspection

1. $3a(2 + b)$ $6a + 3ab$
2. $6x(3x - y)$ $18x^2 - 6xy$
3. $5(a - 1)$ $5a - 5$
4. $2a(3b - 5)$ $6ab - 10a$
5. $6x(1 - 4x)$

6. $3w(2R - w + 1)$
7. $6xz(2x + 3z - 1)$
8. $9R^2(R + S)$
9. $2N(R - K + 1)$ $2NR - 2NK + 2N$
10. $15z(3z - 4)$ $45z^2 - 60z$

For Problems 11–20, factor the answers to Problems 1–10. Factor each product in Problems 21–30 by removing the largest common factor.

21. $12x^2 - 6xy$ (6)
22. $16R^2N + 4R - 2N$ (2) $8r^2N + 2r - N$
23. $5ab - 5ac$ $(5a)(b - c)$
24. $12x^2z - 3xz^3$ $(3xz)$
25. $8xy - 16yz + 4y$ $(4y)$

26. $T^2 - T + TR$ (T)
27. $18R^4 + 9NR^2 + 3R^3$ $(3R^2)$
28. $15W - 3W^2 + W^3$ (w)
29. $24Z^5Y^4 - 18Z^2Y^2 + 6Z^3Y$ $(6Z^2y)$
30. $8Z^2W - 4ZW^2 - 40ZY$ $(4z)$

Solving Linear Equations by Factoring

Let us now consider a method of factoring in solving a few equations for a specific letter. Consider the formula $RT - 2T = 6$. To obtain an equivalent equation showing T as the dependent variable, proceed as shown in the following example.

example $RT - 2T = 6$

1. Factor T from both terms in the left member:

$$T(R - 2) = 6$$

2. Divide both members by the factor $(R - 2)$:

$$\frac{T(R - 2)}{(R - 2)} = \frac{6}{(R - 2)}$$

or simply

$$T = \frac{6}{R - 2}$$

example   Solve the formula $3RD - 6 = 8 - 2D$ for D, and evaluate when $R = 4$. This formula has a term containing the factor D in each member of the equation. You need to get all the terms containing D together on one side of the equation. To do this, either subtract $3RD$ from each side or add $2D$ to both sides. Let us add $2D$ to both members, obtaining

$$3RD - 6 + 2D = 8 \tag{1}$$

$$+6, \quad 3RD + 2D = 14 \tag{2}$$

$$\text{Factor,} \quad D(3R + 2) = 14 \tag{3}$$

$$\div (3R + 2), \quad D = \frac{14}{3R + 2} \tag{4}$$

To evaluate D when $R = 4$, substitute 4 for R:

$$D = \frac{14}{3(4) + 2} = \frac{14}{12 + 2} = \frac{14}{14} = 1 \tag{5}$$

As a check, go back to the original equation and substitute $D = 1$ and $R = 4$:

$$3(4)(1) - 6 = 8 - 2(1) \tag{6}$$

$$12 - 6 = 8 - 2$$

$$6 = 6\checkmark$$

EXERCISE 5.1B

Solve the following formulas for the specified factor, evaluate using the given data and check.

1. $2WN - 3W = 12$ Solve for W; evaluate with $N = 5$.
2. $5xy - 2y = 6x^2$ Solve for y; evaluate with $x = 1$.
3. $2RW - 4W = 6 - R$ Solve for R; evaluate with $W = 2$.
4. $T - RT = 12$ Solve for T; evaluate with $R = -5$.
5. $KL = 2L - 5$ Solve for L; evaluate with $K = 3$.
6. $WN^2 = 3N - 4W + 6$ Solve for W; evaluate with $N = -2$.
7. $TS^2 - TS = T + 3 + S$ Solve for T; evaluate with $S = -1$.
8. $5L + 6H = 12 - L + 2H$ Solve for L; evaluate with $H = 6$.

9. $\dfrac{R + 3}{R} = J$ Solve for R; evaluate with $J = -2$. (*Hint:* Suppose that the R in the denominator was a 5.)

10. $\dfrac{6W - 3R}{W} = 2R$ Solve for R; evaluate with $W = -4$.

In Problems 11–50, multiply by inspection only. (You may refer to the products on page 152 or 153.)

11. $(r + s)(r - s)$
12. $(5 - a)(5 + a)$
13. $(x - y)(x + y)$
14. $(2 + Z)(2 - Z)$
15. $(R - K)(R + K)$
16. $(T - W)(T + W)$
17. $(5 - L)(5 + L)$
18. $(3 - R)(3 + R)$
19. $(H + 1)(H - 1)$
20. $(W + 3)(W - 3)$
21. $(5L - 2P)(5L + 2P)$
22. $(6a - 5)(6a + 5)$
23. $(a + b)^2$
24. $(x - y)^2$

25. $(x + 3)^2$
26. $(3 - y)^2$
27. $(2x + 5)^2$
28. $(r + s)^2$
29. $(T - W)^2$
30. $(a + 3b)^2$
31. $(S - N)^2$
32. $(x + 12)(x - 3)$
33. $(y + 1)(y - 2)$
34. $(x - 5)(x + 6)$
35. $(R - 4)(R + 1)$
36. $(R + 5)(R + 7)$
37. $(Z - 3)(Z - 8)$
38. $(H + 3)(H - 9)$

39. $(W + 1)(W + 9)$
40. $(L - 5)(L + 7)$
41. $(3X + 1)(X + 4)$
42. $(3n + 2)(n - 5)$
43. $(5x - 3)(2x + 4)$
44. $(3n - 5)(2n - 2)$
45. $(8x - 5)(3x + 4)$
46. $(2x - 3)(5x + 7)$
47. $(3R + 8)(2R - 5)$
48. $(7K - 3)(K + 9)$
49. $(11L + 4)(3L + 5)$
50. $(9R - 6)(3R + 7)$

In Problems 52–60, follow the form shown in Problem 51.

51. $(a + b)(c + d) = (a + b)c + (a + b)d = ac + bc + ad + bd$
52. $(x + y)(a + b)$
53. $(m + w)(r - s)$
54. $(x - 4)(v + 3)$
55. $(V + K)(R - T)$
56. $(2R - 3)(4 - 5K)$

57. $(3R - 4P)(5N + 2W)$
58. $(2L - 3Z)(5x + 2y)$
59. $(12N + 3K)(Z - P)$
60. $(6L - 3W)(8R + 2K)$

In Problems 61–118, factor each product to prime factors.

61. $x^2 - y^2$
62. $16N^2 - 25$
63. $R^4 - 1$ (3 factors)
64. $81W^2 - 9T^2$
65. $121 - 81W^2$
66. $625Z^4 - T^4$ (3 factors)
67. $Z^8 - 1$ (4 factors)
68. $1 - w^8$ (4 factors)
69. $25x^2 - 49y^2$
70. $49W^2 - 64R^2$

71. $x^2 + 2xy + y^2$
72. $a^2 - 2ab + b^2$
73. $r^2 + 2rs + s^2$
74. $W^2 - 2WK + K^2$
75. $9x^2 + 12x + 4$
76. $4y^2 - 12y + 9$
77. $b^2 + 8b + 16$
78. $a^2 - 12a + 36$
79. $49x^2 - 56x + 16$
80. $16y^2 + 56y + 49$

81. $18x^2 - 48xy + 32y^2$ (*Hint:* Remove a common factor first.)
82. $75y^2 - 90wy + 27w^2$
83. $x^2 + 7x + 12$ (*Hint:* Two factors of 12 must add up to 7.)
84. $a^2 + 17a + 72$

85. $N^2 - N - 6$
86. $x^2 + 4x + 3$
87. $y^2 + 12y + 35$
88. $x^2 - 7x + 10$
89. $m^2 - 5m + 6$
90. $Z^2 + 9Z - 10$

91. $2x^2 - 2x - 60$
92. $3m^2 - 15m + 18$
93. $bx + by + cx + cy$
94. $5rx - 30x + 11ry - 66y$
95. $WZ - 6NR - 3RZ + 2NW$
96. $6KR - LT - 3RL + 2KT$
97. $12HZ - 10RW + 15RZ - 8HW$

98. $12NZ - 6NR + 10N - 6KZ + 3KR - 5K$ (Try 3 groups of 2 terms each and then 2 groups with 3 terms.)
99. $10WL - 15LZ + 10L - 12WT + 18TZ - 12T$
100. $6xz - 4yz + 6xw - 4yw$
101. $10RW + JK - 2RJ - 5KW$
102. $35xr - 12kw - 20rw + 21xk$

MISCELLANEOUS PROBLEMS

103. $7R - 14W$
104. $R^2 - 9Z^2$
105. $3mbx + 2mb - 3pqx - 2pq$
106. $W^2 + 2WD + D^2$
107. $8RS - 8WS - 7RT + 7WT$
108. $m^4 - 1$
109. $ab + a + b + 1$
110. $5a^2 - 15ab - 6ad + 18bd$

111. $(x - 1)^3 - 9(x - 1)^2 + 8(x - 1)$
112. $9(x + 1) + b(x + 1) + 9(x - 1) + b(x - 1)$
113. $x^2 - \frac{1}{4}$ (*Hint:* $(\frac{1}{2})^2 = \frac{1}{4}$)
114. $(Z^2 - 1)^2 + (Z^2 - 1)$
115. $625Z^4 - T^2$
116. $TR^2 - 4R + R^2T$
117. $15ax^3 - ax^2 - 6ax$
118. $1 + x + y + xy$

5.2   QUADRATIC EQUATIONS

The algebraic statement of variation showing that one quantity varies directly as the square of another quantity is called a quadratic equation. The word quadratic means second degree; i.e., the exponent on the independent variable is a 2.

Quadratic (second-degree) variables appear in numerous formulas and relationships. For example, the algebraic statement showing that wind resistance varies as the square of the velocity V is a quadratic equation: $R = kV^2$.

We shall limit our discussion to the general quadratic equation of the form $y = ax^2 + bx + c$, where a is the coefficient of x^2, b is the coefficient of x, and c is the constant. The $ax^2 + bx + c$ member is sometimes called the quadratic member. Frequently the name "quadratic function" is given to the quadratic member. For example, considering the equation $R = kV^2$, we could say that R is a quadratic function of V.

The quadratic member may have one, two, or three term(s) as follows:

$$y = ax^2 \qquad (b \text{ and } c \text{ both } 0)$$

$$y = ax^2 + bx \qquad (c = 0)$$

$$y = ax^2 + c \qquad (b = 0)$$

$$y = ax^2 + bx + c$$

Roots of a Quadratic Equation

By *root*, we mean the value or values of *the independent variable* which makes the function (quadratic member) equal to zero. For example, in $y = (x - 2)(x + 5) = x^2 + 3x - 10$, 2 and -5 are roots because for these two values of x, $y = x^2 + 3x - 10 = 0$. Roots are special solutions of an equation; however, all solutions are not roots. A solution of a quadratic equation consists of a value for the independent variable *and* a value for the dependent variable. For example, $x = 1$ and $y = -6$ constitute a solution of the equation: $y = x^2 + 3x - 10$; because with $y = -6$ and $x = 1$, the equation reduces to the identity $-6 = -6$. However, $x = 1$ and $y = -6$ are not roots.

Roots by Factoring

We have already covered the factoring necessary to determine the roots of those quadratic functions that can be factored.

example   Solve for the roots of the quadratic equation $y = x^2 - x - 6$. Because a root is defined as a value of the independent variable that will make the quadratic member equal to zero, start by setting this member equal to zero. (*Note:* $y = 0$.)

$$x^2 - x - 6 = 0 \qquad (1)$$

$$\text{Factor,} \quad (x - 3)(x + 2) = 0 \qquad (2)$$

If the product of two factors is zero, then at least one of the factors must be zero. Set each factor equal to zero:

$$x - 3 = 0 \quad \text{and} \quad x + 2 = 0$$

Solving for x, you have the roots

$$x = 3 \quad \text{and} \quad x = -2$$

As a check, substitute in turn $x = 3$ and $x = -2$ into equation (1) to see that these values do indeed give the function a zero value:

$$3^2 - 3 - 6 = 0 \quad \text{and} \quad (-2)^2 - (-2) - 6 = 0$$
$$9 - 3 - 6 = 0 \qquad\qquad\qquad 4 + 2 - 6 = 0$$
$$0 = 0 \checkmark \qquad\qquad\qquad\qquad 0 = 0 \checkmark$$

example   Solve for the roots of the following equation:

$$y = 6x^2 - x - 12 \tag{1}$$

Set $y = 0$:

$$6x^2 - x - 12 = 0 \tag{2}$$

Factor:

$$(2x - 3)(3x + 4) = 0 \tag{3}$$

Set each factor equal to zero and solve for x:

$$2x - 3 = 0 \quad\quad 3x + 4 = 0 \tag{4}$$
$$2x = 3 \quad\quad\quad 3x = -4 \tag{5}$$
$$x = \frac{3}{2} \quad\quad\quad x = \frac{-4}{3} \tag{6}$$

Roots are

$$x = \frac{3}{2} \quad \text{and} \quad x = \frac{-4}{3} \tag{7}$$

Check:

$$6\left(\frac{3}{2}\right)^2 - \left(\frac{3}{2}\right) - 12 = 0 \quad \text{and} \quad 6\left(\frac{-4}{3}\right)^2 - \left(\frac{-4}{3}\right) - 12 = 0$$
$$6(\tfrac{9}{4}) - \tfrac{3}{2} - 12 = 0 \qquad\qquad 6(\tfrac{16}{9}) + \tfrac{4}{3} - 12 = 0$$
$$\tfrac{27}{2} - \tfrac{3}{2} - 12 = 0 \qquad\qquad\qquad \tfrac{32}{3} + \tfrac{4}{3} - 12 = 0$$
$$\tfrac{24}{2} - 12 = 0 \qquad\qquad\qquad\qquad \tfrac{36}{3} - 12 = 0 \tag{8}$$
$$12 - 12 = 0 \qquad\qquad\qquad\qquad 12 - 12 = 0$$
$$0 = 0 \checkmark \qquad\qquad\qquad\qquad\qquad 0 = 0 \checkmark$$

EXERCISE 5.2A

assume all letters = 0

In Problems 1–8, determine the roots of the equations by inspection.

1. $y = (x - 1)(x + 1)$
2. $z = (t + 3)(t - 4)$ $(-3, 4)$
3. $H = (s - 5)(s + 2)$
4. $V = (T - 4)(T + 8)$

5. $Z = (P - 7)(P + 4)$
6. $Z = (2x - 1)(x + 3)$
7. $W = (3x + 6)(x - 1)$
8. $Y = (2T - 3)(T + 1)$

Solve for the roots of the quadratic equations in Problems 9–16 by factoring; check.

9. $y = x^2 - 2x + 1$
10. $S = t^2 - 2t - 15$
11. $H = 2r^2 - 5r - 12$
12. $N = 2s^2 + 7s + 5$

13. $R = q^2 - 8q + 15$
14. $S = 2V^2 + 10V + 12$
15. $L = 3T^2 - 3T - 6$
16. $W = 6P^2 - 24$

The Quadratic Formula

The quadratic formula* will give the roots of any quadratic function written in the form $ax^2 + bx + c = 0$. This function has roots x_1 and x_2 as follows:

$$x_1 = \frac{-b + \sqrt{b^2 - 4ac}}{2a} \qquad x_2 = \frac{-b - \sqrt{b^2 - 4ac}}{2a}$$

Usually these two parts of the formula are combined as

$$x = \frac{-b \pm \sqrt{b^2 - 4ac}}{2a}$$

example Determine the roots of $5x^2 + 3x - 7 = 0$, using the quadratic formula. In this function $a = 5, b = 3$, and $c = -7$. Substituting these values in the formula, you obtain

$$x = \frac{-3 \pm \sqrt{3^2 - 4(5)(-7)}}{(2)(5)}$$

$$x = \frac{-3 \pm \sqrt{9 + 140}}{10}$$

$$x = \frac{-3 \pm \sqrt{149}}{10}$$

---

* The development of this formula can be found in any standard algebra text. See "quadratic formula" in the index.

The $\pm$ sign in front of the $\sqrt{}$ symbol means that there are two values: one a plus value and the other a minus value. Therefore, you can write the two roots as

$$x_1 = \frac{-3 + \sqrt{149}}{10} \qquad x_2 = \frac{-3 - \sqrt{149}}{10}$$

These roots are called *irrational* roots because 149 is not a perfect square and its square root is an irrational number. The $\sqrt{149} \approx 12.2$. Therefore, the two roots are approximately

$$\frac{-3 + 12.2}{10} = \frac{9.2}{10} = 0.92$$

and

$$\frac{-3 - 12.2}{10} = \frac{-15.2}{10} = -1.52$$

example Use the quadratic formula to determine the roots of $x^2 - 3x - 10 = 0$ with $a = 1, b = -3$, and $c = -10$. Then:

$$x = \frac{-(-3) \pm \sqrt{(-3)^2 - 4(1)(-10)}}{2(1)}$$

$$x = \frac{3 \pm \sqrt{9 + 40}}{2} = \frac{3 \pm \sqrt{49}}{2} = \frac{3 \pm 7}{2}$$

The two roots are

$$x_1 = \frac{3 + 7}{2} = \frac{10}{2} = 5$$

and

$$x_2 = \frac{3 - 7}{2} = \frac{-4}{2} = -2$$

When you obtain rational roots such as 5 and -2, you know that the quadratic function could have been factored:

$$x^2 - 3x - 10 = (x - 5)(x + 2) = 0$$

example   Use the quadratic formula to determine the roots of $2x^2 + 3x + 5 = 0$ with $a = 2$, $b = 3$, and $c = 5$. Then:

$$x = \frac{-3 \pm \sqrt{3^2 - 4(2)(5)}}{4} = \frac{-3 \pm \sqrt{9 - 40}}{4}$$

$$x = \frac{-3 \pm \sqrt{-31}}{4}$$

$$x_1 = \frac{-3 + \sqrt{-31}}{4}$$

$$x_2 = \frac{-3 - \sqrt{-31}}{4}$$

The square root of a negative number is called an imaginary number and the roots are called imaginary roots. In other words, there are no real numbers which will make this function zero.

example   Solve the following function for its roots, using the quadratic formula: $4x^2 - 12x + 9 = 0$ with $a = 4$, $b = -12$, and $c = 9$. Then:

$$x = \frac{-(-12) \pm \sqrt{(-12)^2 - 4(4)(9)}}{2(4)}$$

$$x = \frac{12 \pm \sqrt{144 - 144}}{8}$$

$$x = \frac{12 \pm \sqrt{0}}{8}$$

$$x_1 = \frac{12 + 0}{8} = \frac{12}{8} = 1\tfrac{1}{2}$$

$$x_2 = \frac{12 - 0}{8} = \frac{12}{8} = 1\tfrac{1}{2}$$

When the quantity under the radical is zero, you end up with two equal roots (called a double root).

example   Develop a quadratic function of x having roots of 3 and 5. If 3 and 5 are roots, then $(x - 3)$ and $(x - 5)$ are factors; their product $x^2 - 8x + 15$ is the required function.

example   Develop a quadratic function of t having roots of 4 and -7. If 4 and -7 are roots, then $(t - 4)$ and $(t + 7)$ are factors; their product $t^2 + 3t - 28$ is the required function.

EXERCISE 5.2B

In Problems 1–6, determine a quadratic function of the indicated variable having the given roots.

1. Roots of 1 and -1 (x)
2. Roots of 2 and -3 (r)
3. Roots of 2 and 2 (t)
4. Double root of 5 (w)
5. Roots of -5 and 5 (s)
6. Roots of $\frac{1}{2}$ and 2 (z)

In Problems 7–16, use the quadratic formula to determine the roots for the following functions. $x = \frac{-b \pm \sqrt{b^2 - 4ac}}{2a}$

7. $x^2 + 4x + 4 = 0$
8. $x^2 + 7x + 12 = 0$
9. $3x^2 - 3x - 5 = 0$
10. $5x^2 + 2x - 6 = 0$
11. $3x^2 + 4x + 5 = 0$
12. $4x^2 + 5x + 2 = 0$
13. $4x^2 - 4x - 3 = 0$
14. $10x^2 + 13x - 3 = 0$
15. $2t^2 - 16t + 32 = 0$
16. $9v^2 + 12v + 4 = 0$

5.3   ALGEBRAIC FRACTIONS

Three areas seem to cause most of the difficulties students have with algebraic fractions: (1) proper handling of signs, (2) obtaining the lowest common denominator, and (3) simplifying complex fractions. We shall start our work with a close look at these three topics.

Signs

The negative of a number or of an algebraic expression can be obtained by multiplying the number or expression by a minus one (-1). The negative of 5 is $(-1)(5) = -5$; the negative of -638 is $(-1)(-638) = +638$; the negative of $(x - 3)$ is $(-1)(x - 3) = (-x + 3)$ or $(3 - x)$. The negative of $(x - y)$ is $(y - x)$ and not $(x + y)$. The terms $(x - y)$ and $(x + y)$ are called *conjugates* of each other and are not negatives of each other. It is easy to make the mistake of thinking that expressions such as $(2 - x)$ and $(2 + x)$ are negatives. The negative of $(2 - x)$ is $(-2 + x)$ or $(x - 2)$.

The negative of a fraction can be obtained by multiplying the fraction by a minus one (-1). For example, the negative of

$$\frac{3}{x} = (-1)\left(\frac{3}{x}\right) = \frac{-3}{x}.$$

The negative of

$$\frac{5}{x-1} = (-1)\left(\frac{5}{x-1}\right) = -\frac{5}{x-1}.$$

Consider the indicated quotient $-\frac{12}{3}$ which equals -4. In fact, the fractions $-\frac{12}{3}, \frac{-12}{3}$, and $\frac{12}{-3}$ all equal -4. This tells you that you can change the sign of a fraction by changing the sign of either the numerator, the denominator, or the fraction itself.

Now instead of changing the sign of the fraction only once, change its sign twice. Again start with $-\frac{12}{3}$ and change the sign of both the numerator and denominator. You obtain $-\frac{-12}{-3}$ which still equals -4. Or you could change the sign in front of the fraction and either the sign of the numerator or the denominator, giving, respectively,

$$+\frac{-12}{3} = +\frac{12}{-3} = -4$$

Let us now consider sign changes of algebraic fractions.

example Determine the necessary plus or minus sign for the (?) to make the fractions equivalent.

$$-\frac{x}{(5-x)} = (?)\frac{-x}{(5-x)} = (?)\frac{-x}{(x-5)}$$

$$= (?)\frac{-x}{-(x-5)} = (?)\frac{-x}{-(5-x)}$$

In this kind of problem, it is best to work from the first fraction each time to avoid carrying a possible mistake all the way to the end fraction. Try the problem before looking at the answers.* (*Hint:* An even number of sign changes is equivalent to no sign change.)

Note the role that sign changes play in reducing the following fractions:

$$\frac{(x-y)}{(y-x)} = -\frac{(x-y)}{(x-y)} = -1$$

You can change the sign of the numerator or the denominator if you also

---

* Reading from right to left, the signs are: $+ - + -$.

change the sign of the fraction:

$$\frac{(a + b)(a - b)}{(b - a)} = -\frac{(a + b)(a - b)}{(a - b)} = -(a + b)$$

It is not necessary to make both sign changes physically because any factor divided by its negative gives a -1 quotient; i.e., $(y - x)/(x - y) = -1$ just as $\frac{-7}{7}$ or $\frac{7}{-7} = -1$. However, fewer mistakes are made when sign changes are actually made. By a sign change, we mean multiplying a factor by minus one. If a factor such as $(x - y + 5)$ has its sign changed, e.g., $-(x - y + 5)$ or $(-x + y - 5)$, it is only counted as one sign change, not three.

EXERCISE 5.3A

In Problems 1–7, write the negatives for each given quantity.

1. $8a$
2. -3
3. $-5xy^2$

4. $a - x$
5. $-(a - x)$

6. $-(-3)$
7. $-(-3x + 4y)$

In Problems 8–12 give three equivalent answers.

8. $\frac{1}{2}$

9. $-\frac{3}{x}$

10. $\frac{1}{x - 3}$

11. $\frac{x + 4}{-5}$

12. $\frac{x - y}{a - b}$

In Problems 13–14, determine the plus or minus sign for each (?) so that the fractions are equivalent.

13. $\dfrac{1}{x - y} = \dfrac{(?)1}{y - x} = \dfrac{-1}{(?)(y - x)} = (?)\dfrac{-1}{-y + x}$

14. $\dfrac{a - b}{c - d} = (?)\dfrac{b - a}{c - d} = (?)\dfrac{b - a}{d - c} = (?)\dfrac{-a + b}{-c + d}$

In Problems 15–24, reduce each fraction to lowest terms by eliminating all common factors from the numerator and denominator.

15. $\dfrac{-x}{2x}$

16. $\dfrac{3ab}{-ab}$ -3

17. $\dfrac{x - y}{-1}$

18. $-\dfrac{x - y}{1}$

19. $-\dfrac{x - y}{-1}$

20. $\dfrac{x^2 - y^2}{x - y}$

21. $\dfrac{r-s}{s^2-r^2}$ 22. $-\dfrac{w-s}{s-w}$ 23. $\dfrac{6x^2-13x+6}{(2-3x)}$ 24. $\dfrac{(b-3a)}{3a^2+2ab-b^2}$

In Problems 25–28, make the necessary sign changes in the second fraction so that the given fractions have a common denominator; then add the fractions.

25. $\dfrac{5}{x-y}$, $\dfrac{3}{y-x}$ $\dfrac{-3}{-y+x}$

27. $\dfrac{14y}{S+t}$, $\dfrac{8k}{-S-t}$ $\dfrac{-8x}{-S+t}$ $=\dfrac{14y-8x}{S+t}$

26. $\dfrac{7x}{a-b}$, $\dfrac{2x}{b-a}$

28. $\dfrac{3x}{5-y}$, $\dfrac{-8x}{y-5}$

Addition and Subtraction

To add fractions having a common denominator, add their numerators together and write this sum over the common denominator. It is standard practice to reduce all fractions to lowest terms.

example   Add:

$$\frac{3x}{5-y}+\frac{7x}{5-y}+\frac{4x}{5-y}-\frac{2x}{5-y}$$

$$=\frac{3x+7x+4x-2x}{5-y}$$

$$=\frac{12x}{5-y}$$

example   This example contains a common mistake which causes much difficulty.
Add:

Contains a mistake
$$\begin{cases} \dfrac{5-x}{a+b}+\dfrac{4x-7}{a+b}-\dfrac{x-2}{a+b} \\[2mm] \dfrac{5-x+4x-7-x-2}{a+b}=\dfrac{2x-4}{a+b} \end{cases}$$

Correct
$$\begin{cases} \dfrac{5-x}{a+b}+\dfrac{4x-7}{a+b}-\dfrac{x-2}{a+b} \\[2mm] \dfrac{5-x+4x-7-x+2}{a+b}=\dfrac{2x}{a+b} \end{cases}$$

The mistake occurs in moving or changing the negative sign in front of the last fraction to the numerator but failing to change the sign of the entire numerator (specifically, the sign of the -2).

EXERCISE 5.3B

In Problems 1–8, make the necessary sign changes to obtain equivalent fractions having a common denominator. Then add the fractions and simplify.

1. $\dfrac{5}{x-y} + \dfrac{3}{x-y} + \dfrac{2}{y-x}$

5. $\dfrac{3W-P}{W-2P} - \dfrac{2W-4P}{2P-W} - \dfrac{2W+P}{W-2P}$

2. $\dfrac{3x-y}{12} - \left(\dfrac{2x+y}{12}\right) + \dfrac{x-y}{-12}$

6. $\dfrac{2r+s}{r+s} - \dfrac{2r+3s}{s+r} - \dfrac{r+s}{s+r}$

3. $\dfrac{a-b}{a+b} - \dfrac{2a}{a+b} - \dfrac{3a-b}{a+b}$

7. $\dfrac{5x-2y}{x^2-y^2} + \dfrac{3x+2y}{y^2-x^2} - \dfrac{x-3y}{x^2-y^2}$

4. $\dfrac{5r-s}{-6} + \dfrac{r+2s}{-6} - \dfrac{2r-3s}{6}$

8. $\dfrac{3r}{a-r} - \dfrac{2a}{r-a} - \dfrac{7r}{a-r} + \dfrac{r-3a}{r-a}$

In Problems 9–14 subtract the second fraction from the first and simplify.

9. $\dfrac{5}{x-y}$, $\dfrac{3}{y-x}$

12. $\dfrac{3a-2b}{a-b}$, $\dfrac{5a-6b}{b-a}$

10. $\dfrac{2s+t}{t}$, $\dfrac{2(t+s)}{t}$

13. $\dfrac{4t+5s}{s-2t}$, $\dfrac{7s}{s-2t}$

11. $\dfrac{a-b}{b}$, $\dfrac{2a-b}{-b}$

14. $-\dfrac{2y+3}{y-5}$, $-\dfrac{3y-2}{y-5}$

Lowest Common Denominator

The lowest common denominator (LCD) becomes almost a necessity when adding algebraic fractions. Use the procedure of factoring all denominators to prime factors and then selecting one of each different factor as factors of the LCD. When a choice must be made between two identical factors having different exponents, use the one with the larger exponent.

example   Determine the LCD for the following fractions and then find their sum:

$$\frac{5}{x-y} + \frac{6}{x+y} + \frac{3}{x^2-y^2}$$

Remember that $(x-y)$ and $(x+y)$ are factors of $x^2 - y^2$ and, therefore, $x^2 - y^2$ is the LCD for these fractions. Write the LCD in factored form under a vinculum as

$$\overline{(x-y)(x+y)}$$

Looking at the fraction $5/(x-y)$, you must multiply the denominator

$(x - y)$ by $(x + y)$ to obtain the LCD. Then you must multiply the numerator by this same factor. This can be indicated as

$$\frac{5(x + y)}{(x - y)(x + y)}$$

The fraction so far is exactly equal to $5/(x - y)$. Now look at the second fraction $6/(x + y)$. You must multiply its denominator by $(x - y)$ to obtain the LCD, and therefore multiply the numerator 6 by $(x - y)$ also. This can be shown as

$$\frac{5(x + y) + 6(x - y)}{(x - y)(x + y)}$$

Finally, you only need to add the numerator 3 to the sum of the numerators because the 3 already has the common denominator:

$$\frac{5(x + y) + 6(x - y) + 3}{(x - y)(x + y)}$$

Remove parentheses and collect terms:

$$\frac{11x - y + 3}{(x - y)(x + y)} \quad \text{or} \quad \frac{11x - y + 3}{x^2 - y^2}$$

example   Add the following:

$$\frac{3}{x^2 - 6x + 9} + \frac{4}{18 - 6x} + 5$$

First factor the denominators:

$$\frac{3}{(x - 3)^2} + \frac{4}{6(3 - x)} + \frac{5}{1}$$

The $(x - 3)$ is the negative of $(3 - x)$, so make two sign changes in the second fraction:

$$\frac{3}{(x - 3)^2} - \frac{4}{6(x - 3)} + \frac{5}{1}$$

With the LCD $= 6(x - 3)^2$, the sum becomes

$$\frac{3(6) - 4(x - 3) + 5(6)(x - 3)^2}{6(x - 3)^2}$$

Expand and collect terms in the numerator:

$$\frac{30x^2 - 184x + 300}{6(x - 3)^2} = \frac{15x^2 - 92x + 150}{3(x - 3)^2}$$

example This relatively simple problem still causes some difficulty. Combine the terms:

$$\frac{1}{a - 1} + 1$$

It may help you to think of the 1 as the fraction $\frac{1}{1}$. The LCD for these two fractions has the factors $(a - 1)$ and 1 so the LCD $= a - 1$:

$$\frac{1}{1} = \frac{a - 1}{a - 1}$$

The addition becomes

$$\frac{1}{a - 1} + \frac{a - 1}{a - 1} = \frac{1 + a - 1}{a - 1} = \frac{a}{a - 1}$$

EXERCISE 5.3C

In Problems 1–10, reduce to lowest terms. If the fraction is already prime, write prime for the answer.

1. $\dfrac{ab + ac}{a}$

2. $\dfrac{b - a}{a - b}$

3. $\dfrac{m - n + 1}{n - m - 1}$

4. $\dfrac{x}{x - x^2}$ $\dfrac{x}{x(1-x)} = \dfrac{1}{1-x}$

5. $\dfrac{y^2}{y^2 - x^2}$ *prime*

6. $\dfrac{(a + b)}{a^2 + 2ab + b^2}$

7. $\dfrac{x^2 - 9}{x^2 - x - 6}$

8. $\dfrac{x^3 - 2x^2 - 8x}{x^3 - x^2 - 6x}$

9. $\dfrac{a^2 + ab - 2b^2}{a^2 + 2ab - 3b^2}$

10. $\dfrac{r^3 - 7r^2 + 10r}{r^2 - 5r}$

In Problems 11–28, combine the fractions, obtaining the LCD using prime factors.

11. $\dfrac{2a + b}{12} + \dfrac{a - 4b}{18}$

12. $\dfrac{x - 3y}{20} + \dfrac{2x + y}{15}$

13. $\dfrac{1}{x} + 1$

14. $\dfrac{1}{a} - 1$

15. $\dfrac{1}{a} + \dfrac{1}{b}$

16. $\dfrac{1}{x} - \dfrac{1}{y} + 1$

17. $\dfrac{1}{x - 1} + \dfrac{1}{x + 1}$

18. $\dfrac{a}{x} + \dfrac{b}{y}$

19. $\dfrac{3}{x^2 - y^2} + \dfrac{1}{x - y}$

20. $\dfrac{7}{x^2 + 2xy + y^2} + \dfrac{5}{x + y}$

21. $\dfrac{1}{r_1} + \dfrac{1}{r_2}$

22. $5 - \dfrac{3x}{x - y}$

23. $\dfrac{1}{2x - 3} - \dfrac{1}{2x + 3}$

24. $\dfrac{3a + 2}{a + 1} + \dfrac{a^2 + 1}{1 - a^2}$

25. $\dfrac{4}{x + 1} - \dfrac{x - 2}{x^2 + x} + \dfrac{3x}{1 - x^2}$

26. $\dfrac{1}{x + 2} + \dfrac{2}{x - 3} + \dfrac{3}{x^2 - x - 6}$

27. $\dfrac{1}{x} + \dfrac{1}{y} + \dfrac{1}{z}$

28. $\dfrac{3}{x + 2} + \dfrac{4}{x^2 + 4x + 4} - \dfrac{5}{x^2 - 4}$

Multiplication and Division

The multiplication of two algebraic fractions is performed in the same manner as the multiplication of common fractions. It is best, however, to factor and reduce whenever possible before multiplying. For example:

$$\frac{x^2 + x - 6}{x^2 + 4x + 3} \cdot \frac{x^2 + 3x + 2}{x^2 - 2x}$$

$$= \frac{(x + 3)(x - 2)}{(x + 3)(x + 1)} \cdot \frac{(x + 2)(x + 1)}{(x)(x - 2)} = \frac{x + 2}{x} \quad \text{or} \quad 1 + \frac{2}{x}$$

When dividing by an algebraic fraction, first invert the divisor and change the operation sign to multiplication before factoring the terms of the fraction. Then cancel common factors and multiply.

example Divide:

$$\frac{x^2 + 2xy}{x^2 + 4y^2} \div \frac{x^2 - 4y^2}{xy - 2y^2}$$

First, invert the divisor and change the sign:

$$\frac{x^2 + 2xy}{x^2 + 4y^2} \cdot \frac{xy - 2y^2}{x^2 - 4y^2}$$

Second, factor each term to prime factors and cancel common factors:

$$\frac{(x)(x + 2y)}{x^2 + 4y^2} \cdot \frac{y(x - 2y)}{(x - 2y)(x + 2y)} = \frac{xy}{x^2 + 4y^2}$$

example   Divide $(a - 1/b)$ by $(b - 1/a)$. Before inverting the divisor, the terms of each factor must be combined:

$$a - \frac{1}{b} = \frac{ab - 1}{b} \qquad b - \frac{1}{a} = \frac{ab - 1}{a}$$

The problem becomes

$$\frac{ab - 1}{b} \div \frac{ab - 1}{a}$$

which gives

$$\frac{\cancel{ab - 1}}{b} \cdot \frac{a}{\cancel{ab - 1}} = \frac{a}{b}$$

EXERCISE 5.3D

Perform the following multiplications and divisions and simplify.

1. $\dfrac{7}{x + y} \cdot \dfrac{5}{x^2 - xy + y^2}$

2. $\dfrac{6}{a + b} \cdot \dfrac{-3}{a + b}$

3. $\dfrac{7}{a - b} \div \dfrac{-1}{b - a}$

4. $\dfrac{9a^3b}{6ab^2} \div \dfrac{ab^4}{15a^3}$

5. $\dfrac{x^2 - 2xy + y^2}{x^2 - y^2} \div \dfrac{x - y}{x + y}$

6. $\dfrac{2x^2 + 3x - 2}{2x^2 - 7x - 4} \div \dfrac{x^2 + 3x + 2}{x^2 - 3x - 4}$

7. $\dfrac{16a - 4}{5a - 5} \div \dfrac{16a^2 - 1}{1 - 2a + a^2}$

8. $\dfrac{x^4 - y^4}{x^2 + 2xy + y^2} \div \dfrac{x^2 + y^2}{y^2 - x^2}$

9. $\dfrac{3a^2 + 6ab + 3b^2}{4a^2 - 16b^2} \cdot \dfrac{2a - 4b}{9a + 9b}$

10. $\dfrac{x^2 + xy - 2y^2}{4x^2 - 4y^2} \cdot \dfrac{6x^2 + 2xy - 4y^2}{3x^2 + 4xy - 4y^2}$

11. $\dfrac{5x^2 + 10x - 120}{x^2 + 4x - 12} \div \dfrac{6x^2 - 14x - 40}{3x^2 - x - 10}$

12. $\dfrac{2x^2 - 11x - 6}{3x^2 + x - 2} \div \dfrac{x^2 - 3x - 18}{x^2 + 3x + 2}$

Complex Fractions

A complex fraction contains a fraction in its numerator and/or in its denominator. For example:

$$\frac{\dfrac{3}{4}}{x - 1} \cdot \qquad \frac{x + \dfrac{1}{x}}{3} \cdot \qquad \frac{5}{1 - \dfrac{1}{x}} \cdot$$

The dot at the end of a vinculum is used to identify the main numerator and denominator. For example,

$$\frac{\dfrac{2}{3}}{5}$$

could mean either $\frac{2}{3} \div 5$ or $2 \div \frac{3}{5}$. To prevent this ambiguity, the dot is used. Therefore,

$$\frac{\dfrac{2}{3}}{\overset{.}{5}}$$

means $\frac{2}{3} \div 5$, and

$$\frac{\overset{.}{\dfrac{2}{3}}}{5}$$

means $2 \div \frac{3}{5}$.

Complex fractions can be simplified by multiplying both the numerator and denominator by a factor called the lowest common multiple (LCM). The LCM is actually the LCD of the individual denominators of the fractions in the numerator and/or denominator. For example, in the complex fraction

$$\frac{3 - \dfrac{1}{a}}{\dfrac{2}{5} - 1}\cdot$$

the LCM is $5a$, which is the LCD for the fractions $\frac{2}{5}$ and $1/a$. It is called a common multiple rather than a common denominator because it is used as a multiplier rather than a denominator.

Now that we have the necessary LCM, let us see how we use it:

$$\frac{\left(3 - \dfrac{1}{a}\right)(5a)}{(\frac{2}{5} - 1)(5a)} = \frac{15a - 5}{2a - 5a} \quad \text{or} \quad \frac{15a - 5}{-3a}$$

Note that multiplying the numerator by $5a$ eliminates the denominator a, and multiplying the denominator by $5a$ eliminates the denominator 5. The resulting fraction is no longer complex.

example   Simplify the following fraction, using the LCM:

$$\frac{3 - \dfrac{1}{x - 1}}{\dfrac{2}{x + 1} - 1}\cdot$$

The LCM is $(x - 1)(x + 1)$. Multiply the numerator and denominator by this multiple:

$$\frac{\left(3 - \dfrac{1}{x - 1}\right)(x - 1)(x + 1)}{\left(\dfrac{2}{x + 1} - 1\right)(x - 1)(x + 1)} = \frac{3(x^2 - 1) - 1(x + 1)}{2(x - 1) - (x^2 - 1)}$$

$$= \frac{3x^2 - 3 - x - 1}{2x - 2 - x^2 + 1} = \frac{3x^2 - x - 4}{-x^2 + 2x - 1}$$

EXERCISE 5.3E

Simplify the complex fractions, using the LCM.

1. $\dfrac{\frac{3}{4}}{\frac{1}{3}}.$

2. $\dfrac{1 + \frac{3}{5}}{\frac{1}{4} - 1}.$

3. $\dfrac{x + \frac{1}{3}}{5 - \frac{1}{x}}.$

4. $\dfrac{1 + \dfrac{1}{x - 1}}{1 - \dfrac{1}{x + 1}}.$

5. $\dfrac{3x - \dfrac{1}{x}}{\dfrac{3}{x} - 5}.$

6. $\dfrac{x - \dfrac{1}{x^2}}{\dfrac{1}{x^2} - 1}.$

7. $\dfrac{\dfrac{x - 5}{x + 3}}{\dfrac{x + 1}{x - 2}}.$ *mult. only*

8. $\dfrac{\dfrac{1 - a}{1 + a}}{\dfrac{a + 2}{a - 2}}.$

Fractional Equations

The name "fractional equation" applies to an equation with terms that are fractions. For example, the equations

$$\frac{3}{x} - 5x + \frac{1}{x} = 6$$

and

$$\frac{1}{x - 1} + 3 = \frac{x - 4}{5}$$

are fractional equations. When solving fractional equations, the first practical step is to clear the equation of fractions. This is done by multiplying the equation through by the LCM of all the denominators in the equation.

example   Use the LCM to clear the following equation of fractions, solve for x, and check:

$$\frac{2x}{3} - \frac{3x}{4} = \frac{1}{2} \tag{1}$$

The LCM $= 12$.

$$12\left(\frac{2x}{3}\right) - 12\left(\frac{3x}{4}\right) = 12\left(\frac{1}{2}\right) \tag{2}$$

or

$$8x - 9x = 6$$

C.T., $-x = 6$ (3)

$\cdot(-1)$, $x = -6$ (4)

$\surd$, $\dfrac{2(-6)}{3} - \dfrac{3(-6)}{4} = \dfrac{1}{2}$ (5)

Simplify, $\dfrac{-12}{3} + \dfrac{18}{4} = \dfrac{1}{2}$ (6)

$$-4 + 4\tfrac{1}{2} = \tfrac{1}{2}$$

$$\tfrac{1}{2} = \tfrac{1}{2} \surd$$

example   Solve the following equation for x:

$$\frac{1}{x-1} + 5 = \frac{x}{2} \tag{1}$$

The LCM for this equation is $(x-1)(2)$. Multiplying each term in the equation by this multiple gives

$$\frac{1}{x-1}(x-1)(2) + 5(x-1)(2) = \frac{x}{2}(x-1)(2) \tag{2}$$

or

$$2 + 10x - 10 = x^2 - x$$

$-10x, +8$, $0 = x^2 - 11x + 8$ (3)

By the quadratic formula

$$x = \frac{11 \pm \sqrt{121 - 32}}{2}$$

The roots are

$$x_1 = \frac{11 + \sqrt{89}}{2} \qquad x_2 = \frac{11 - \sqrt{89}}{2} \tag{4}$$

example   Solve for R and check:

$$\frac{3R}{R-1} = 4 + \frac{3}{R-1} \tag{1}$$

$\cdot(R-1), \qquad (R\!-\!1)\dfrac{3R}{(R\!-\!1)} = 4(R-1) + \dfrac{3(R\!-\!1)}{(R\!-\!1)} \tag{2}$

$$3R = 4R - 4 + 3 \tag{3}$$

C.T., $\qquad\qquad 3R = 4R - 1 \tag{4}$

$-3R, +1, \qquad\qquad 1 = R \tag{5}$

$\checkmark, \qquad\qquad \dfrac{3 \cdot 1}{1-1} = 4 + \dfrac{3}{1-1} \tag{6}$

But $\frac{3}{0}$ is not a number. Therefore, there is no solution to this equation. Remember that the existence of an equality may depend on some property or value of the variable quantity. If no such property or value exists, there is no equality.

example   Solve for x and check:

$$\frac{2x+4}{a} = 6 + \frac{4}{a} \tag{1}$$

$\cdot a, \qquad \dfrac{2x+4}{a} \cdot a = 6a + \dfrac{4}{a} \cdot a \tag{2}$

$-4, \qquad\qquad 2x = 6a \tag{3}$

$\div 2, \qquad\qquad x = 3a \tag{4}$

$\checkmark, \qquad \dfrac{2(3a)+4}{a} = 6 + \dfrac{4}{a} \tag{5}$

C.T., $\qquad \dfrac{6a+4}{a} = \dfrac{6a+4}{a} \checkmark \tag{6}$

Remember that the literal numbers a, b, c, d, etc., are considered to be constant even though you may not know their numeric value.

EXERCISE 5.3F

Solve the following equations.

1. $\dfrac{3}{2y} + \dfrac{2}{y} = \dfrac{4}{y} + \dfrac{1}{4}$

2. $\dfrac{15}{x} = \dfrac{3}{2}$

3. $\dfrac{5s}{2} - \dfrac{2s}{3} = \dfrac{11}{8}$

4. $\dfrac{x}{2} + \dfrac{x}{3} = 5$

5. $\dfrac{V+8}{4} - \dfrac{V-4}{8} = \dfrac{7}{2}$

6. $\dfrac{3W}{4} + \dfrac{1}{6} = 2W - \dfrac{7}{3}$

7. $\dfrac{8}{s-4} = \dfrac{6}{s-3} + \dfrac{2}{s-6}$

8. $\dfrac{8}{x+4} = \dfrac{6}{x-4}$

9. $(y+5)(y+1) = (y+3)(y+2)$

10. $(x+1)^2 = x^2 + 9$

11. $\dfrac{x}{a} = 2$

12. $\dfrac{6a}{x} = 2$

13. $\dfrac{4x}{3} - 14 = \dfrac{x}{c}$

14. $\dfrac{R}{6} + \dfrac{R}{4} = -\dfrac{R}{3}$

15. $\dfrac{1}{s-2} + \dfrac{3}{s+2} = 0$

16. $\dfrac{5a}{2a-3} = a$

Challenge Problems

17. Check the roots of the equation in the second example, page 173.

5.4 EXPONENTS AND RADICALS

In this section, we shall study the relationships between exponents and indicated roots such as $\sqrt{}$ and $\sqrt[3]{}$. In solving the equation $y^2 = 25$, we are looking for the value or values of y that will satisfy the equation. In this case we apply the square root operation to both members and obtain $y = \pm 5$. Both $+5$ and -5 are roots of 25. When only the positive root of a number is indicated, we write $y = \sqrt{n}$; for the negative root, we write $y = -\sqrt{n}$. When both roots are indicated, we write $y = \pm\sqrt{n}$.

Consider the following equivalent equations: $y = \sqrt{64}$ and $y = \sqrt{2^6}$. The $\sqrt{}$ sign carries the understood index number 2; i.e., $\sqrt{n}$ means $\sqrt[2]{n}$. For higher order roots such as $\sqrt[3]{n}$ and $\sqrt[4]{n}$, the index numbers are always written. The number written under a radical sign is called the *radicand*. You know that $y = \sqrt{64} = 8$, which $= 2^3$. What can you do to $y = \sqrt{2^6}$ to end up with 2^3? You could remove the radical sign and divide the exponent 6 by 2 (2 being the index of the root). Now try this procedure on $\sqrt{64}$ written as $\sqrt{8^2}$. Dividing the exponent 2 by the index number 2 gives 8^1 which is the correct positive root of 64.

Try one more: $\sqrt{81} = \sqrt{3^4} = \sqrt{9^2}$. Dividing the exponents 4 and 2 by the index number 2 gives 3^2 and 9^1 for the positive root of 81. These examples illustrate a method of finding an indicated root of numbers written in power form.

To find the nth root of a quantity written with an exponent, divide the exponent by the index number n.

examples

$$\sqrt[3]{x^6} = x^{6/3} = x^2$$

$$\sqrt[4]{y^8} = y^{8/4} = y^2$$

$$\sqrt[5]{32} = \sqrt[5]{2^5} = 2^{5/5} = 2^1 = 2$$

4th root applies to both 16 and 2 [handwritten]

EXERCISE 5.4A

In Problems 1–10, solve the equations for x and check.

1. $x = \sqrt{36}$
2. $x = \sqrt[3]{27}$
3. $x^2 = \frac{1}{4}$
4. $x = \sqrt[4]{16}$
5. $x = \sqrt{5^6}$

6. $x = \sqrt[3]{3^9}$
7. $x = \sqrt[4]{16z^8}$
8. $x = -\sqrt{16}$
9. $x = -\sqrt[4]{2^{12}}$
10. $x^4 = \frac{1}{16}$

Simplify the expressions in Problems 11–16.

11. $\sqrt[3]{2^6}$
12. $\sqrt[4]{3^8}$
13. $\sqrt[5]{2^{20}}$

14. $-\sqrt{2^4}$
15. $\pm\sqrt{9}$
16. $\pm\sqrt{x^6}$

Use the slide rule to approximate the solutions for Problems 17–20.

17. $\sqrt{86.0}$
18. $\sqrt[3]{512}$

19. $x^2 = 44$
20. $x^3 = 52$

Consider the statement $y = \sqrt{5}$. Remembering that 5 has an exponent of 1, i.e., $5 = 5^1$, attempt to obtain the root by removing the radical sign and dividing the exponent by the index of the root. Then you have $y = \sqrt{5} = 5^{1/2}$. The $\frac{1}{2}$ exponent is another way of indicating the square root of a number. For example, $9^{1/2} = 3$, $4^{1/2} = 2$, $16^{1/2} = 4$, and $3^{1/2} = \sqrt{3}$.

A root of any positive quantity can be indicated by dividing its exponent by the index of the root. For example, $\sqrt[3]{4} = 4^{1/3}$, $\sqrt[5]{12} = 12^{1/5}$, $\sqrt[3]{x^2} = x^{2/3}$, and $\sqrt[4]{y^3} = y^{3/4}$.

To change a number with a fractional exponent to a number written with a radical, the denominator of the exponent becomes the index of the root and the numerator becomes the exponent on the radicand (the quantity under the radical):

examples

$$25^{2/3} = \sqrt[3]{25^2}$$

$$7^{3/4} = \sqrt[4]{7^3}$$

EXERCISE 5.4B

In Problems 1–10, remove the radical sign by using fractional exponents.

1. $\sqrt{x}$ $= x^{\frac{1}{2}}$
2. $\sqrt[3]{2}$ $= 2^{\frac{1}{3}}$
3. $\sqrt{3^3}$ $= 3^{\frac{3}{2}}$
4. $\sqrt[3]{2^2}$ $= 2^{\frac{2}{3}}$
5. $\sqrt[5]{x^3}$
6. $\sqrt[4]{x^2}$
7. $\sqrt[3]{x^1 y^2}$ (*Hint:* Both exponents must be divided by 3.)
8. $\sqrt[4]{x^3 y^2 z^1}$
9. $\sqrt{x^4}$ $= x^2$
10. $\sqrt[5]{3x^2 y^3}$ $= 3^{\frac{1}{5}} x^{\frac{2}{5}} y^{\frac{3}{5}}$

In Problems 11–16, write an equivalent expression using a radical sign.

11. $5^{1/2}$
12. $3^{1/3} x^{2/3}$
13. $(4y)^{2/3}$ $= \sqrt[3]{(4y)^2}$ or $\sqrt[3]{16 y^2}$

14. $3^{1/2} x^{1/2}$
15. $(28z)^{3/4}$ $= \sqrt[4]{2^{?} y^3 z^?}$ mult. out
16. $(12T)^{1/3}$

Multiplication of Radicals

Irrational quantities such as $\sqrt{a}$ and $\sqrt{b}$ can be multiplied together, giving $\sqrt{ab}$, if: (1) a and b are positive numbers and (2) the two roots are of the same order. Sometimes the product turns out to be a rational number. For example:

$$\sqrt{20}\sqrt{5} = \sqrt{100} = 10$$

$$\sqrt[3]{9}\sqrt[3]{3} = \sqrt[3]{27} = 3$$

$$\sqrt[4]{7}\sqrt[4]{5} = \sqrt[4]{35}$$

Frequently an irrational number can be simplified by factoring into a rational factor and an irrational factor. For example:

$$\sqrt{80} = \sqrt{16 \cdot 5} = (\sqrt{16})(\sqrt{5}) = (4)(\sqrt{5})$$

$$\sqrt[3]{54} = \sqrt[3]{27 \cdot 2} = \sqrt[3]{27}\sqrt[3]{2} = 3\sqrt[3]{2}$$

$$\sqrt{\tfrac{1}{8}} = \sqrt{\tfrac{1}{4} \cdot \tfrac{1}{2}} = \sqrt{\tfrac{1}{4}} \cdot \sqrt{\tfrac{1}{2}} = \tfrac{1}{2}\sqrt{\tfrac{1}{2}}$$

A radical containing a fraction may be simplified by multiplying the numerator and denominator of the fraction by a factor that will make the denominator a perfect square, cube, etc., depending on the order of the root.

example   Simplify $\sqrt{\tfrac{1}{2}}$. First multiply the numerator and the denominator by 2:

$$\sqrt{\tfrac{2}{4}}$$

Then factor and simplify:

$$\sqrt{\tfrac{1}{4} \cdot 2} = \sqrt{\tfrac{1}{4}} \cdot \sqrt{2} = \tfrac{1}{2}\sqrt{2} \quad \text{or} \quad \frac{\sqrt{2}}{2}$$

example   Simplify $\sqrt[3]{\tfrac{2}{3}}$. First multiply the numerator and the denominator by 9:

$$\sqrt[3]{\tfrac{18}{27}}$$

Then factor and simplify:

$$\sqrt[3]{\tfrac{1}{27} \cdot 18} = \sqrt[3]{\tfrac{1}{27}} \cdot \sqrt[3]{18} = \tfrac{1}{3}\sqrt[3]{18} \quad \text{or} \quad \frac{\sqrt[3]{18}}{3}$$

example   Simplify $\sqrt[4]{\tfrac{1}{8}}$. First multiply the numerator and the denominator by 2:

$$\sqrt[4]{\tfrac{2}{16}}$$

Then factor and simplify:

$$\sqrt[4]{\tfrac{1}{16} \cdot 2} = \sqrt[4]{\tfrac{1}{16}} \cdot \sqrt[4]{2} = \tfrac{1}{2}\sqrt[4]{2} \quad \text{or} \quad \frac{\sqrt[4]{2}}{2}$$

Negative Exponents

If you apply the rule of exponents to a fraction such as x^3/x^4, you end up with x^{-1}. What does this mean? Had you reduced the fraction in the regular manner, you would have obtained $1/x$; i.e.,

$$\frac{x^3}{x^4} = \frac{1}{x}$$

Therefore, x^{-1} equals $1/x$. Then, $5^{-1} = \frac{1}{5}$ and $3^{-1} = \frac{1}{3}$. Also:

$$x^{-2} = \frac{1}{x^2} \qquad y^{-3} = \frac{1}{y^3} \qquad z^{-5} = \frac{1}{z^5}$$

Thus, you can change *a factor* from the numerator to the denominator or vice versa by changing the sign of its exponent.

examples

$$\frac{1}{x^{-2}} = \frac{1}{\dfrac{1}{x^2}} \cdot = x^2$$

$$\frac{1}{5^{-2}} = \frac{1}{\dfrac{1}{5^2}} \cdot = 5^2$$

$$\left(\frac{2}{3}\right)^{-1} = \frac{1}{(\frac{2}{3})^1} = \frac{3}{2}$$

EXERCISE 5.4C

In Problems 1–10, multiply the given factors and simplify when possible.

1. $\sqrt{5} \cdot \sqrt{7}$
2. $\sqrt[3]{2} \cdot \sqrt[3]{10}$
3. $\sqrt[5]{3} \cdot \sqrt[5]{9}$
4. $\sqrt{2} \cdot \sqrt{3}$
5. $\sqrt{3} \cdot \sqrt{12}$

6. $\sqrt{27} \cdot \sqrt{3}$
7. $\sqrt[3]{25} \cdot \sqrt[3]{5}$
8. $\sqrt[3]{3} \cdot \sqrt[3]{3}$
9. $\sqrt{\frac{1}{2}} \cdot \sqrt{\frac{1}{3}}$
10. $\sqrt[3]{\frac{1}{4}} \cdot \sqrt[3]{\frac{1}{2}}$

In Problems 11–18 simplify each of the given quantities.

11. $\sqrt{\frac{1}{3}}$
12. $\sqrt[3]{\frac{1}{2}}$
13. $\sqrt{20}$
14. $\sqrt[3]{16}$

15. $\sqrt{0.25}$
16. $\sqrt[3]{0.027}$
17. $\sqrt{\frac{2}{3}}$
18. $\sqrt[3]{\frac{3}{4}}$

Determine the value of each expression in Problems 19–26 by rewriting with positive exponents.

19. 3^{-3} $\frac{1}{3^3} = \frac{1}{\sqrt[3]{9}}$ ... $= \frac{1}{2}$
20. $4^{-1/2}$
21. $(\frac{1}{8})^{-1/3}$
22. $(\frac{3}{4})^{-3}$ $= (\frac{4}{3})^3 = \frac{64}{27}$

23. $(\frac{4}{9})^{-1/2}$
24. $(0.001)^{-1/3}$
25. $(0.125)^{-1}$
26. 1.2×10^{-3}

In Problems 27–34, write each expression without zero or negative exponents and simplify.

27. $3^{1/3} \cdot 3^{2/3}$
28. $8^{1/2} \cdot 4^{1/2}$
29. $(16y^4)^{-1/2}$
30. $(9x^2)^{-1/2}$

31. $x^4 \cdot x^{-4}$
32. $3^3 \cdot 3^{-3}$
33. $8x^{-1}$
34. $(16y^2)^{-1/2}$

Equations Involving Radicals

example Solve for x and check:

$$\sqrt{x - 1} = 5 \tag{1}$$

Square both members:

$$x - 1 = 25 \tag{2}$$

$$+ 1, \qquad x = 26 \tag{3}$$

$$\checkmark, \qquad \sqrt{26 - 1} = 5 \tag{4}$$

$$\sqrt{25} = 5$$

$$5 = 5 \checkmark$$

example Solve for x and check:

$$2\sqrt[3]{x + 2} = 8 \tag{1}$$

Cube both members:

$$8(x + 2) = 512 \tag{2}$$

$$\div 8, \qquad x + 2 = 64 \tag{3}$$

$$- 2, \qquad x = 62 \tag{4}$$

$$\checkmark, \qquad 2\sqrt[3]{62 + 2} = 8 \tag{5}$$

$$2\sqrt[3]{64} = 8$$

$$2(4) = 8$$

$$8 = 8 \checkmark$$

example   Solve for x and check:

$$\sqrt{3x - 5} = \sqrt{x + 3} \qquad (1)$$

Square both members:

$$3x - 5 = x + 3 \qquad (2)$$

$$- x, + 5, \qquad\qquad 2x = 8 \qquad (3)$$

$$\div 2, \qquad\qquad x = 4 \qquad (4)$$

$$\checkmark, \qquad \sqrt{3(4) - 5} = \sqrt{4 + 3} \qquad (5)$$

$$\sqrt{12 - 5} = \sqrt{7}$$

$$\sqrt{7} = \sqrt{7} \; \checkmark$$

example   Solve for y and check:

$$y + 1 + \sqrt{y + 7} = 0 \qquad (1)$$

$$- \sqrt{y + 7}*, \qquad y + 1 = - \sqrt{y + 7} \qquad (2)$$

Square both members:

$$y^2 + 2y + 1 = y + 7 \qquad (3)$$

$$- y, - 7, \qquad y^2 + y - 6 = 0 \qquad (4)$$

$$\text{Factor}, \qquad (y + 3)(y - 2) = 0 \qquad (5)$$

The indicated roots are

$$y = - 3, \qquad y = 2$$

Check. If $y = - 3$:

$$- 3 + 1 + \sqrt{- 3 + 7} = 0$$

$$- 2 + \sqrt{4} = 0 \qquad (6)$$

$$- 2 + 2 = 0$$

---

* To see why you take this step, try squaring both members as they stand and examine the result.

Therefore, -3 is a root. If $y = 2$:

$$2 + 1 + \sqrt{2 + 7} = 0$$

$$3 + \sqrt{9} = 0$$

$$3 + 3 \neq 0$$

Therefore, 2 is not a root.* This example further illustrates the need for checking.

EXERCISE 5.4D

Solve the following equations and check.

1. $\sqrt{x - 4} = 12$
2. $\sqrt[3]{x + 1} = 2$
3. $\sqrt{2x - 3} = \sqrt{x + 7}$
4. $x + \sqrt{x^2 + 1} = 1$
5. $\sqrt{x - 1} = \dfrac{4}{\sqrt{x - 1}}$
6. $\dfrac{1}{\sqrt{x - 3}} + \dfrac{1}{2} = 1$

7. $\sqrt{2x - 5} = -3$ *N.G.*
8. $5 + \sqrt{3 - x} = 0$
9. $\sqrt{x - 2} + \sqrt{x + 2} = 1$
10. $\sqrt{x + 3} + 1 = \sqrt{5x + 4}$
11. $\sqrt{x - 1} + \sqrt{3x + 3} = 4$
12. $(x + 4)^{1/2} = 5$
13. $\dfrac{1}{(x - 1)^{1/2}} = (x - 1)^{3/2}$
14. $(x + 3)^{1/3} = 1$

5.5 WORD PROBLEMS

Technicians are frequently required to formulate word problems from experimental data, from known relationships, and from observations. Once the problem is written, it must be solved to see that it is properly written and that it does indeed have a solution which confirms the expressed relationship(s). It is therefore important to spend some time in a systematic attack on typical word problems.

Equivalent Statements

Sometimes just the writing of an equivalent statement removes some of the difficulty in a problem.

Given statement: Three-eighths of a new tire tread has worn away.

Equivalent statement: Five-eighths of the tire tread remains.

Given statement: A steel rod has lost 20% of its original tensile strength.

Equivalent statement: A steel rod has retained 80% of its original tensile strength.

---

* Extraneous roots are frequently introduced into a discussion when the members of an equation are squared or when the members are multiplied by a factor containing the variable.

Assignment of Letters

The assignment of a letter for each unknown quantity in the problem should be very explicit. For example, saying "let x = length" is poor. Instead say: "let x = length of the diagonal in feet." Frequently it is possible to reduce the number of letters needed to represent the unknown quantities. For example, if a wire of known length (100 ft) is cut into two pieces and you wish to represent the length of each piece, you could say: "Let x = length of the shorter piece in feet, and let y = length of the longer piece in feet." Or you could represent the longer piece as $(100 - x)$ ft, thus saving the use of a second variable and keeping the problem simpler.

EXERCISE 5.5A

For each statement in Problems 1–20, write as many equivalent statements as indicated by the number in parentheses.

1. Three of 8 windows were sold.   (1)
2. Four inches were cut from a 20-in. board.   (1)
3. $N = \frac{1}{2}W$ (1)
4. $N_1 - N_2 = 24$ (2)
5. The cylinder is one-fourth full.   (1)
6. He traveled five-eighths of the way home.   (1)
7. The lens filters out 20% of the light.   (1)
8. Twenty is four-fifths of 25.   (1)
9. A certain piston weighs one-half as much as its connecting rod.   (1)
10. Of the land area, 80% is covered by forest.   (1)
11. A water solution is 20% salt.   (1)
12. On a hot day, gasoline will expand to 105% of its volume at 50°F.   (1)
13. The tensile strength of copper wire is approximately twice that of aluminum wire.   (1)
14. The work is 80% complete.   (1)
15. $W = 2D$ (1)
16. $N_1 + N_2 + N_3 = 165$ (5)
17. $R = \frac{4}{5}T$ (2)
18. $c^2 = a^2 + b^2$ (2)
19. The denominator of a fraction exceeds the numerator by 5.   (1)
20. The length exceeds the width by 82 ft.   (2)

In Problems 21–30, assign appropriate letters to stand for the quantities involved and write at least two equivalent algebraic statements expressing the given or implied relationships between the quantities. Do not solve the problems but concentrate on establishing several equivalent relationships.

21. The number of quarters collected exceeded the number of dimes collected by 42.
22. The annexed land cost \$45/acre more than the cost of the original land.

23. A bar contains 3 times more tin than it does silver.
24. A 60-lb bar is made from silver and tin. The bar contains 300 % more tin than it does silver.
25. Two pipes are used to fill a tank at the rate of 80 gal/min. Water flows through the first pipe at four-fifths the rate of the second pipe.
26. Two tanks contained equal amounts of solvent. After 40 gal was transferred from tank 1 to tank 2, the first tank had three-fourths as much as the second.
27. A 36-ft long steel rod is to be cut into 3 pieces so that the longest piece is 5 times the length of the short piece and the shortest piece is one-third the length of the remaining piece. In addition to writing the necessary algebraic statements, make an appropriate sketch which helps illustrate these data.
28. A long horizontal beam is supported by 3 wires of different sizes. The medium-size wire has a safe load capacity which is 80 % that of the larger wire. The larger wire has a safe load capacity which is 1000 lb more than twice the capacity of the smaller wire. Together the wires will support 5100 lb.
29. When a tank is one-fourth full of fuel, the weight of the tank equals the weight of the fuel. When three-fourths full, the tank weighs 40 lb less than the contained fuel.
30. Three resistors in series have a total resistance of 6800 ohms. The resistance of the first is 300 ohms less than that of the second and the resistance of the third is 3 times the resistance of the first.

Problems Leading to Quadratic Equations

Many physical relationships give rise to quadratic equations. In this section we shall consider several different typical applications. Many other applications will be discussed in Chapter 7 where we shall combine our knowledge of the quadratic function with graphing techniques.

example   A strip of metal 16 in. wide is to be formed into a trough with a rectangular cross section (open top); the cross-sectional area is to be $31\frac{1}{2}$ sq in. Find the depth of the trough. Let $x = $ depth of trough in inches and $16 - 2x = $ width in inches. Then:

$$\text{area} = (x)(16 - 2x) = -2x^2 + 16x = 31\tfrac{1}{2} \text{ sq in.}$$

$$2x^2 - 16x + 31\tfrac{1}{2} = 0 \tag{1}$$

$$\cdot\, 2, \qquad 4x^2 - 32x + 63 = 0 \tag{2}$$

$$\text{Factor,} \qquad (2x - 7)(2x - 9) = 0 \tag{3}$$

Set factors $= 0$:

$$2x - 7 = 0 \quad \text{and} \quad 2x - 9 = 0 \tag{4}$$

$$2x = 7 \qquad\qquad 2x = 9 \tag{5}$$

$$x = \tfrac{7}{2} \qquad\qquad x = \tfrac{9}{2} \tag{6}$$

This tells us that two different troughs could be formed, one with a depth of $3\frac{1}{2}$ in. and the other with a depth of $4\frac{1}{2}$ in.

Check:

$$2 \text{ sides, each } 3\tfrac{1}{2} \text{ in. } = 7 \text{ in.}$$

$$\text{bottom} = 16 - 7 = 9 \text{ in.}$$

$$\text{area} = (3\tfrac{1}{2})(9) = 31\tfrac{1}{2} \text{ sq. in. } \checkmark$$

Also:

$$2 \text{ sides, each } 4\tfrac{1}{2} \text{ in. } = 9 \text{ in.}$$

$$\text{bottom} = 16 - 9 = 7 \text{ in.}$$

$$\text{area} = (7)(4\tfrac{1}{2}) = 31\tfrac{1}{2} \text{ sq in. } \checkmark$$

It is important to note that in the last example, after assigning the letter x to represent the depth, we were able to write the width in terms of x also.

example The cross-sectional area of the L-shaped beam shown in Figure 5-1 must be 6.0 sq in. What must be the thickness of the metal? If you visualize

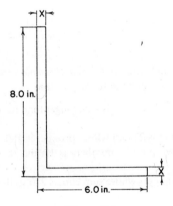

Figure 5-1

the beam as consisting of two rectangular parts, the areas of the two parts are $8x$ in.^{2} and $(6 - x)x$ in.^{2}, or $(8x + 6x - x^2)$ sq in., which must equal 6 sq in. Therefore,

$$- x^2 + 14x = 6$$

Rewrite:

$$x^2 - 14x + 6 = 0 \tag{1}$$

$$x = \frac{14 \pm \sqrt{196 - 24}}{2} \tag{2}$$

$$x = \frac{14 \pm \sqrt{172}}{2} \tag{3}$$

$$x_1 \approx 13.6 \text{ in.} \tag{4}$$

$$x_2 \approx 0.44 \text{ in.}$$

The only likely answer is 0.44 in. because an 8-in. × 6-in. angle iron could not possibly have a thickness of 13.6 in. So throw out this answer as being extraneous. To check the 0.44 in. answer, substitute 0.44 for x and compute the areas of the two rectangular sections:

$$A_1 = (8.0 \text{ in.})(0.44 \text{ in.}) = 3.52 \text{ sq in.}$$

$$A_2 = (6.0 \text{ in.} - 0.44 \text{ in.})(0.44 \text{ in.}) = 2.45 \text{ sq in.}$$

$$A_1 + A_2 = (3.52 + 2.45) \text{ sq in.} = 5.97 \approx 6.0 \text{ sq in.}$$

EXERCISE 5.5B

Represent the two variable quantities in terms of the same variable, express the given relationship as a quadratic function, and solve the problem.

1. The length of a rectangular-shaped die is 4 in. longer than its width. If its area is $26\frac{1}{4}$ sq in., what are its dimensions?
2. Find two factors whose sum is 27 and whose product is 180.
3. The difference between two positive numbers is 11 and their product is 242. Find the numbers.
4. Determine two positive numbers whose sum equals 240 so that one number is the square of the other.
5. Find two consecutive, positive, even integers whose product is 528.
6. Find two consecutive, positive, odd integers whose product is 1295.
7. The square of a certain positive number exceeds 10 by as much as 10 exceeds the number. Find the number.
8. The difference between the square of a certain positive number and 28 is equal to the difference between 28 and the number. Find the number.
9. A strip of metal 20 in. wide is to be bent into a trough (open top) with a rectangular cross section of 50 sq in. What should be its dimensions?

10. If the trough in Problem 9 is to have a closed top and a cross-sectional area of 24 sq in., what must be its dimensions?

11. The hypotenuse of a right triangle is 2 ft longer than the longer leg which is 14 ft longer than the shorter leg. Find the lengths of the three sides.

12. The area of a rectangular-shaped parking lot, 420 ft long and 360 ft wide, is to be increased by 128,800 sq ft. If the length and width are increased by the same amount, what will be the dimensions of the larger lot?

13. Telegraph poles are spaced at equal intervals along a railroad. If 2 more poles are used per mile, the distance between poles would decrease by 11 ft. How many poles are used per mile?

14. A triangular-shaped plate has an altitude of 12 in. and a base of 16 in. A strip is to be cut by a line parallel to the base so that the remaining triangle has an area of 54 sq in. How wide is the strip?

A Basic Procedure for Solving Problems

One big stumbling block in the path to solving both technical and "everyday" problems is the *failure to use the information at hand*. There is no magic formula by which all problems can be solved. Yet there are some basic steps which will help in arriving at a solution. In conjunction with these basic steps the successful technician pays careful attention to form. By *form* we mean the systematic organization and structure given to the interrelated parts of a problem.

Basic steps. Let us examine the basic steps first and then consider a basic form.

1. Study the given information about the situation and conditions from which the problem grows. Write down all important information and known data about the problem.

2. If the problem involves a diagram or figure make a neat and accurate drawing (to scale when appropriate), showing dimensions and labeling important parts.

3. If a problem exists, know what the problem is. A simple statement of the problem will help you to see what you are to find or determine. This may be a statement such as: "Determine the weight (in grams) of the material lost by oxidation."

4. Make a list of statements that express or establish relationships between known and/or unknown facts or parts of the problem. Examples of such statements are:

   (1) Mathematical formulas
   (2) Algebraic statements
   (3) Geometric properties or relationships
   (4) Definitions
   (5) Conversion factors

(6) Trigonometric relationships

(7) Simple word statements including hints or reminders

(8) Assignment of letter(s) for unknown factor(s)

(9) Equivalent statements about given or related data

5. Write the proposed steps that you plan to follow in solving the problem. In other words, prepare a simple statement outline of the solutions involved. Once you can identify the various parts that need a solution and can see the interdependence of these parts, you have the battle almost won. If you list an extra step, it can be deleted later. It is much easier to rearrange these steps into a logical order than to work your way up a blind alley of calculations only to discover that you should have worked some other part of the problem first.

The steps referred to here are not the steps performed in carrying out a mathematical operation, but are the broad sequential steps involved in the entire solution. Examples of such steps are:

(1) Compute the area of the ends of the tank.

(2) Determine the area of the inside walls of the tank.

(3) Calculate the volume of plastic required to cover the areas in Parts 1 and 2 to a thickness of $\frac{1}{8}$ in.

(4) Compute the cost of the plastic at $14.50/gal.

6. To solve the problem, follow the sequential order established in Step 5. The solution to each part can only grow out of one or more of the statements listed in Step 4. If a solution exists, there must be at least one relationship between known and/or unknown data which you have (or should have) listed in Step 4. Then solve each statement (there may be only one) in Step 4 for the unknown part or factor. If you seem to lack a suitable statement in Step 4, go back over Steps 1, 2, and 3 again. You are probably overlooking some important bit of information.

7. Check and review. Check all work, including calculations. Review your method and make notes on how you might improve the approach. Pay special attention to the parts that caused trouble.

In summary, the basic steps are:

1. Given information.
2. Figure.
3. Find.
4. Related information.
5. Proposed steps.
6. Solution.
7. Checking and review.

Basic general form (box diagram form). Some of the most highly paid technicians, engineers, and mathematicians use a problem-solving procedure called flow diagramming. Flow diagramming employs a basic box diagram form approach to problem solving. The box diagram method allows for the simplification of complicated problems; it puts related information into perspective and is effective. We shall follow a modified box diagram approach in solving the word problems in the next four examples. This form serves three important functions:

1. It is a helpful guide in following the seven basic steps.
2. It is valuable as a tool for identifying and studying all the important and *sometimes overlooked parts of a problem.*
3. It helps bring into focus interrelationships among parts of a problem.

Most problems need not be solved by the box diagram approach. However, you should become well enough acquainted with its uses so that, when confronted with a difficult problem, you can use it effectively.

example Find three unequal numbers whose sum is 165 if the difference between the smaller of the two numbers is 14 and the sum of the two larger numbers is 127. The solution is shown in Figure 5-2.

example A contractor has a large truck and a small truck. Using the large truck, it requires 32 hr hauling time to deliver a certain order. The smaller truck requires 48 hr of hauling time to deliver the same order. (Both trucks make several trips to deliver one order.) If the contractor uses both trucks to deliver the order, what is the hauling time required? The solution is shown in Figure 5-3.

example A bar of a certain metal alloy contains two ingredients, A and B. Find the weight of each ingredient if the ratio of their weights is $\frac{7}{9}$ and the difference in their weights is 80.0 g. The solution is shown in Figure 5-4.

example A steel tire for a railroad locomotive has an inside diameter of 59.82 in. at 70°F. The tire is to be heated and placed on a wheel with an outside diameter of 60.00 in. If the coefficient of expansion for steel is $6.36 \times 10^{-6}/$ degree Fahrenheit, determine the minimum temperature to which the tire must be heated to just slip on the wheel. Assume that the wheel is kept at a constant temperature of 70°F. The solution is shown in Figure 5-5.

EXERCISE 5.5C

Solve the following problems using a procedure similar to that used in the examples whenever appropriate. (Watch significant figures.)

1. A fraction whose denominator is 24 more than its numerator is equal to $\frac{5}{8}$. What is the fraction?
2. The sum of two numbers is 5 more than 5 times the smaller number. Find the two numbers if the sum is 205.

3. One number is 40 less than twice a second number. The sum of the two numbers is 110. What are the numbers?

4. Find three consecutive even integers whose sum is 72.

5. One leg of a right triangle is 5.0 in. and the hypotenuse is 1.0 in. longer than the other leg. Find the length of the hypotenuse.

6. The perimeter of an isosceles triangle is 34 in. The base is 2.0 in. shorter than one of the equal sides. What is the length of the base?

7. The perimeter of a rectangle is 100 in. and its area is 600 sq in. What are its dimensions?

8. The perimeter of a rectangle is 140 ft and its diagonal is 50 feet. What is the area of the rectangle?

9. The number of square inches in the area of a square equals 4 times the number of inches in its length. Find the length of an edge.

10. Find a positive number which, when increased by 15, is equal to 100 times the reciprocal of the number.

11. Find a positive number which, when diminished by 7, equals 60 times the reciprocal of the number.

12. What positive number equals 4 times its own reciprocal?

13. A man's age is now four-fifths of what it will be 12 yr from now. How old is he now?

14. A woman's age is 1 yr less than twice her daughter's age. If the sum of their ages is 59 yr, what are their ages?

1. Given Information	2. Figure
Sum of three unequal numbers is 165. Difference between two smaller numbers is 14. Sum of two larger numbers is 127.	Not needed.

3. Find	4. Related Information
Determine the three numbers satisfying the relationships shown in 4(b), 4(c), and 4(d).	(a) Let N_1 = smaller number. Let N_2 = middle number. Let N_3 = larger number.

5. Proposed Steps	4. Related Information (cont.)

(a) In equation 4(b), Substitute value for N_1 from equation 4(e) and for N_3 from equation 4(f).
 (b) Solve the equation from step 5(a) for N_2.
 (c) Using the value for N_2 and equation 4(e), solve for N_1.
 (d) Using the computed value for N_2 and equation 4(f), solve for N_3.
	(b) $N_1 + N_2 + N_3 = 165$
	 (c) $N_2 - N_1 = 14$
	 (d) $N_2 + N_3 = 127$
	 (e) $N_1 = N_2 - 14$
	     equivalent to (c)
	 (f) $N_3 = 127 - N_2$
	     equivalent to (d)

---

	6. Solution

Calculations:
(b)   Solve equation 6(a) for N_2:
  (1) $(N_2 - 14) + N_2 +$
              $(127 - N_2) = 165$
  (2)   Remove ( ):
      $N_2 - 14 + N_2 +$
            $127 - N_2 = 165$
  (3)   C.T., $N_2 + 113 = 165$
  (4) -113, $N_2 = 52$

Steps:
(a)   From 4(e), $N_1 = (N_2 - 14)$, and from 4(f), $N_3 = (127 - N_2)$. Substituting these values in equation 4(b), you have:

$$N_1 + N_2 + N_3 = 165$$
$$\downarrow \qquad \downarrow \qquad \downarrow$$
$$(N_2 - 14) + N_2 + (127 - N_2) = 165$$

(b)   Solve equation 6(a) for N_2:
$$N_2 = 52$$

(c)   From 4(e):
(With $N_2 = 52$)
$$N_1 = N_2 - 14$$
$$N_1 = 52 - 14$$
$$N_1 = 38$$

(d)   From 4(f):
(With $N_2 = 52$)
$$N_3 = 127 - N_2$$
$$N_3 = 127 - 52 = 75$$

---

7. Checking and Review

$$38 + 52 + 75 = 165 \checkmark$$
$$52 - 38 = 14 \checkmark$$
$$52 + 75 = 127 \checkmark$$

Figure 5-2

15.   A student has test scores of 70 and 83. To obtain an average of 81, what score must he make on the next test?
16.   A student has test scores of 64, 70, 78, 82, and three identical scores. If his average score is 75, what are the identical scores?
17.   The length of a desk top is 16 in. more than the width. If the top has an area of 720 sq in., what are the dimensions of the desk?
18.   A bottle and its stopper cost a dollar and a dime. If the bottle cost $1 more than the stopper, what did the stopper cost?
19.   Will 200 rods of fencing enclose a rectangular field whose area is 2600 sq rods?

1. Given Information	2. Figure
A contractor has two different size trucks. Using one truck, it requires 32 hr hauling time to deliver a certain order. Using the other truck, it takes 48 hr to deliver the same order.	Not needed.

	4. Related Information

3. Find

Determine how long it will take to deliver the order using both trucks.

4. Related Information

(a)   Let T_1 represent the truck requiring 32 hr.
(b)   Let T_2 represent the truck requiring 48 hr.
(c)   Let t represent time in hours.
(d) T_1 delivers $\frac{1}{32}$ of the order per hour.
(e) T_2 delivers $\frac{1}{48}$ of the order per hour.
(f) T_1 has rate $R_1 = \frac{1}{32}$ order/hr.
(g) T_2 has rate $R_2 = \frac{1}{48}$ order/hr.
(h)   Combined rate $= R_1 + R_2$
(i)   (Rate of hauling) (time) = amount hauled
(j)   Amount to be hauled is one order.

5. Proposed Steps

(a)   Determine the combined hauling rate.
(b)   Using 4(i) and combined hauling rate, determine total time for both trucks.

6. Solution

Calculations:

(a) $\frac{1}{32} + \frac{1}{48} = \frac{3}{96} + \frac{2}{96} = \frac{5}{96}$

Steps:

(a)   Combined hauling rate
$= R_1 + R_2$
$R_1 + R_2 = (\frac{1}{32} + \frac{1}{48})$ order/hr
$R_1 + R_2 = \frac{5}{96}$ order/hr

(b)   (Rate of hauling) (time) = amount hauled
$\frac{5}{96}(t) = 1$ (order)

Solve for t:

$t = \frac{96}{5}$ hr $= 19.2$ hr

(to two significant figures = 19 hr)

7. Checking and Review

$T_1, \quad (\frac{1}{32})(19.2) = 0.6$ (order),
$T_2, \quad (\frac{1}{48})(19.2) = 0.4$ (order),
together $= 1.0$ (order) $\checkmark$

Figure 5-3

1. Given Information	2. Figure

	Alloy bar
A metal alloy bar contains two in-gredients, *A* and *B*. The difference in the weights of *A* and *B* is 80.0 g. The ratio of their two weights is $\frac{7}{9}$.	contains metals *A* and *B*

	4. Related Information

3. Find	(a) Let *A* = weight (in grams) of metal *A*.
	(b) Let *B* = weight (in grams) of metal *B*.
What weight of each ingredient does the bar contain?	(c) $B - A = 80.0$ g
	(d) $A + 80.0 = B$ or $B = (A + 80.0$ g)

5. Proposed Steps	
(a) Determine the weight of *B* in terms of *A*.	(e) Ratio of weights = $\frac{7}{9}$:
(b) Write the ratio of *A*/*B* equal to $\frac{7}{9}$.	$$\frac{\text{weight of } A}{\text{weight of } B} = \frac{7}{9}$$
(c) Replace *B* in terms of *A*.	
(d) Solve for weight of *A*.	
(e) Solve for weight of *B*.	

6. Solution

Calculations:

(d) $\dfrac{A}{A + 80.0 \text{ g}} = \dfrac{7}{9}$ (1)

clear fractions,

$9A = (7A + 560 \text{ g})$ (2)

$- 7A,\quad 2A = 560 \text{ g}$ (3)

$\div 2,\quad A = 280 \text{ g}$ (4)

(e) $280 \text{ g} + 80.0 \text{ g} = 360 \text{ g}$

$B = 360 \text{ g}$

Steps:

(a) From 4(d), $B = (A + 80.0 \text{ g})$

(b) $A/B = \frac{7}{9}$

(c) Substitute for $B = A + 80.0$ g

$$\frac{A}{A + 80.0 \text{ g}} = \frac{7}{9}$$

(d) Solve for *A*,

$$\frac{A}{A + 80.0 \text{ g}} = \frac{7}{9}$$

$$A = 280 \text{ g}$$

(e) Solve for *B*, $B = (A + 80.0 \text{ g})$

$$B = 360 \text{ g}$$

7. Checking and Review

The solution is $28\bar{0}$ g of *A* and $36\bar{0}$ g of *B*.

$$\frac{280 \text{ g } (A)}{360 \text{ g } (B)} = \frac{28}{36} = \frac{7}{9} \quad \checkmark$$

$$360 - 280 = 80 \text{ g} \quad \checkmark$$

Figure 5-4

20. A circle whose radius is 5.0 in. is inscribed in a Rt △ *ABC* so that the circle is tangent to the hypotenuse at point *P*. If *AP* = 1õ in., what are the lengths of legs *AC* and *BC*?

21. A pipe is to fit into a square conduit so that the distance from the outside of the pipe to an inside corner of the conduit is 12 in. If the pipe is tangent to all four walls of the conduit, what is the outside diameter of the pipe?

22. A truck hauled a load up a mountain at 2õ mph and returned at 6õ mph. What was the average speed for the round trip?

23. A rectangular-shaped plate, 48 in. by 36 in., is to have a uniform width border painted on it so that the painted and unpainted areas are equal. What will be the width of the border?

1. Given Information	2. Figure
A steel tire with I.D. = 59.82 in. (at 70°F) is to be heated and shrunk onto a wheel with O.D. = 60.00 in. Coefficient of expansion for steel is 6.36×10^{-6} per degree Fahrenheit.	WHEEL TIRE O.D. = 60.00 in. I.D. = 59.82 in.
3. Find	
Determine the minimum temperature to which the tire must be heated to just slip on the wheel.	**4. Related Information**
5. Proposed Steps	(a) Coefficient of expansion for steel = $6.36 \times 10^{-6}/°F$. (b) Heating the tire causes an increase in the circumference of the tire and, hence, an increase in its diameter.
(a) Determine necessary increase in tire I.D. (b) Determine how much this increase (in tire I.D.) will increase the inside circumference. (c) Calculate increase in inside circumference of tire by $x°$ temperature change. (d) Equate needed increase, 5(b), with the increase produced by $x°$ temperature change, 5(c), and solve the equation for *x*.	(c) $C = \pi D$ and $D = \dfrac{C}{\pi}$ (d) The diameter of the tire must be increased to allow the tire to slip on. (e) Assume that the wheel will stay at 70°F. (f) Let *x* = the number of degrees above 70° required to produce the necessary expansion. (g) Required temperature = $(x + 70)°F$. (h) Circumference increase = $\pi \cdot$ (diameter increase). (i) Increase in circumference = $\pi(\text{I.D.}) (6.36 \times 10^{-6}) (x)$.

---

6. Solution

Calculations:
(a) 60.00 in. − 59.82 in. = 0.18 in.
(d) Solving for x:

$$\pi(0.18 \text{ in.}) = \pi(59.82 \text{ in.})$$
$$\times \frac{(6.36 \times 10^{-6})}{1\,°F} (x) \quad (1)$$

Dividing by π in.,

$$0.18 = (59.82)$$
$$\times \frac{(6.36 \times 10^{-6})}{1\,°F} (x) \quad (2)$$

Divide by

$$\frac{6.36 \times 10^{-6}}{1\,°F} \text{ and by } (59.82)$$

$$\frac{(0.18)\,(1\,°F)}{(59.82)\,(6.36 \times 10^{-6})} = x \quad (3)$$

$$x \approx 500°F \quad (4)$$

By slide rule, $x = 473°F \quad (5)$

Steps:
(a) Tire I.D. must increase from 59.82 in. to 60.00 in.

I.D. increase = 0.18 in.

(b) C (increase) = π (diameter increase)
C (increase) = $\pi(0.18 \text{ in.})$

(c) C increase due to $x°F$ temperature change:

$$C(\text{increase}) = \pi(59.82 \text{ in.})$$
$$\times \frac{(6.36 \times 10^{-6})\,(x)\,°F}{1°F}$$

(d) From (b) and (c):

$$\pi(0.18 \text{ in.}) = \pi(59.82 \text{ in.})$$
$$\times (6.36 \times 10^{-6})\,(x)$$

Solve for x:

$$x = 473°$$

(e) Required temperature

$$(x + 70)°$$

Required temperature =

$$473° + 70° = 543°F$$

---

7. Checking and Review

Proposed check: Determine the expansion in the diameter due to a temperature increase of 473°F.

$$\text{I.D. increase} = \frac{C \text{ (increase)}}{\pi} = \frac{(\pi)\,(59.82 \text{ in.})\,(6.36 \times 10^{-6})\,(473°)}{\pi}$$

$$\text{I.D. increase} = (5.982 \times 10 \text{ in.}) \frac{6.36 \times 10^{-6}}{1°F} (4.73 \times 10^2)°F$$

$$= (5.982)\,(6.36)\,(4.73) \times 10^{-3} \text{ in.}$$

$$= 0.18 \text{ in.} \quad \checkmark$$

Figure 5-5

24. An open box is to be formed from a square piece of sheet iron by cutting a 4.00 in. square from each corner and folding up the four sides. The volume of the box must be 484 cu in. What must be the size of the sheet iron?

25. A wire is bent into a rectangle whose length is 3 times its width. If the length is shortened by 4.0 in. and the 4.0 in. are added to the width, the original area will be increased by 144 sq in. What is the length of the wire?

26. A 48-in. wire is cut into two pieces and a square is formed from each piece. If the areas of the squares differ by 48 sq in., what are the lengths of the two pieces of wire?

27. A water tank is filled by two inlet pipes and drained by one outlet pipe. One input pipe can fill the tank in 8.0 hr and the other can fill the tank in 10 hr.
    (a) How long will it take to fill the tank with both pipes (drain closed)?
    (b) Starting with the tank empty and using both input pipes, how long will it take to fill the tank with the drain open if a full tank will drain in 20.0 hr?

28. A certain mixing tank is fed by three input pipes and drained by one outlet pipe. The first pipe can fill the tank in 4.0 hr, the second can fill the tank in 6.0 hr, and the third can fill the tank in the time required for the first two running together.
    (a) How long will it take to fill the tank with all three pipes (drain closed)?
    (b) Starting with the tank empty and the drain open, how long will it take to fill the tank with all three pipes if a full tank will drain in 12 hr?

29. A road crosses a second road, which is twice as wide, forming angles of 30° and 150°. If the area of the intersection is 7056 ft^2, what is the width of each road?

30. Two conveyors (equal in length) move ore from a stockpile to a hopper. If the first conveyor has twice the load per foot as the second and moves $\frac{1}{10}$ ft/sec while the second has a load of 88 lb/ft and moves $\frac{1}{6}$ ft/sec, what is the average rate at which the ore enters the hopper?

Solution Problems

Almost all solution problems* can be solved by the aid of the simple diagram shown in Figure 5-6. This diagram suggests that two solutions are to be mixed together to form a third solution. Even when one solution is being poured into the other, it is best to think of the two as being poured into a third container. This approach helps to formulate the resulting algebraic equation.

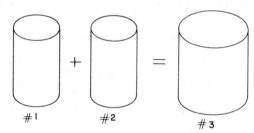

Figure 5-6

* Excluding those liquids that lose volume when mixed together.

Start with all three containers empty and place 1 lb of sugar in the first container (#1) and 2 lb of sugar in the second container (#2). Then add 1 lb of water to #1. The first solution now weighs 2 lb. Once all the sugar dissolves, the solution is one-half or 50% sugar. You know three things about this solution:

1. Total amount = 2 lb.
2. Amount of pure sugar = 1 lb.
3. Amount of pure sugar as a percent of total = 50%.

Now add 8 lb of water to #1, giving 10 lb of solution. You now have the following:

1. Total amount = 10 lb.
2. Amount of pure sugar = 1 lb (no change).
3. Amount of pure sugar as a percent of total = 10%.

Turning to container #2 with its 2 lb of sugar, add 3 lb of water. You then have:

1. Total amount = 5 lb.
2. Amount of pure sugar = 2 lb.
3. Amount of pure sugar as a percent of total = 40%.

Now pour both solutions together into container #3 and see what you can conclude.

1. *Amount of first solution plus amount of second solution equals amount of final solution:*

$$10 \text{ lb } (\#1) + 5 \text{ lb } (\#2) = 15 \text{ lb } (\#3)$$

2. *Amount of pure solute* in first solution plus amount of pure solute in second solution equals amount of pure solute in final solution:*

$$1 \text{ lb pure sugar} + 2 \text{ lb pure sugar} = 3 \text{ lb pure sugar}$$

These two basic equations take the headache out of solution problems. To determine the percent of sugar in the final solution, you only need to compare the 3 lb with the 15 lb and you have 20% for its strength.

example　How many gallons of a 20% acid solution must be added to 25 gal of a 10% acid solution to obtain a solution that is 12% acid? First, put the given data on the container diagram and see what else you might need (Figure 5-7). You want to know how many gallons of #2 must be added to #1. Assign a letter for this quantity and see how this changes the picture. Let x = number of gallons of 20% solution needed for the required dilution.

---

* Solute is the name given to the material dissolved in a liquid. The solute may be sugar, salt, acid, alcohol, butterfat, etc.

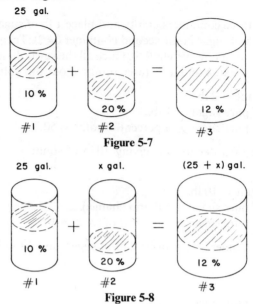

25 gal.

10 %   20 %   12 %

#1   #2   #3

Figure 5-7

25 gal.   x gal.   (25 + x) gal.

10 %   20 %   12 %

#1   #2   #3

Figure 5-8

1. Given Information	2. Figure
Starting with 25.0 gal of solution testing 10.0% acid, add a 20.0% acid solution to obtain a 12.0% acid solution.	25.0 gal. x gal. (25.0 + x) gal. 10.0% + 20.0% = 12.0% #1 #2 #3
3. Find	4. Related Information
How many gallons of the 20% acid solution are needed?	(a) Let x = number of gallons of the 20.0% acid.
	(b) Volume of #3 = volume of #1 + volume of #2.
5. Proposed Steps	(c) Pure acid in a solution = (volume) (percent acid)
(a) Compute volume of final mixture.	(d) Amount pure acid #1 + amount pure acid #2 = pure acid #3.
(b) Compute amount of pure acid in #1.	
(c) Compute amount of pure acid in #2.	
(d) Compute amount of pure acid in #3.	
(e) Using 4(d), write an equation and solve.	

---

6. Solution

Calculations:
(d) Acid in #3,
    (25.0 gal) (0.120) = 3.00 gal
    (x gal) (0.120) = 0.120x gal
(e) Using statement 4(d),
    (2.50 + 0.200x) gal =
    (3.00 + 0.120x) gal   (1)

Divide by 1 gal

2.50 + 0.200x =
3.00 + 0.120x (2)

Subtract 2.50 and 0.120x,

0.080x = 0.50   (3)

÷0.080, x = 6.25   (4)

Steps:
(a) Volume #3:
    Volume #3
      = 25.0 gal + x gal
      = (25.0 + x) gal
(b) Amount pure acid #1:
    Amount acid #1
      = (25.0 gal) (10.0 %)
      = 2.5 gal
(c) Amount pure acid #2:
    Amount acid #2
      = (x gal) (20.0 %)
      = 0.200x gal
(d) Amount pure acid #3:
      = (25.0 + x) gal (12.0 %)
      = 3.00 gal + 0.120x gal
(e) Using 4(d):
    2.50 gal + 0.200x gal =
    3.00 gal + 0.120x gal

    x = 6.25 gal of 20.0 % Acid

---

7. Checking and Review

The solution to the problem is 6.25 gal of 20.0 % acid.

25.0 gal	x = 6.25 gal	(25.0 + x) gal
10.0 %	20.0 %	25.0 + 6.25 = 31.25 gal
contain	contain	31.25 gal
2.50 gal pure acid	1.25 gal pure acid	12.0 % acid
		contain 3.75 gal pure acid
2.50 gal +	1.25 gal =	3.75 gal √

Figure 5-9

Note that you can now obtain an expression for the volume of the resulting solution; i.e., 25 gal + x gal = (25 + x) gal. See Figure 5-8. (Assume three digit accuracy.)

1. How much pure acid is contained in #1?

    (10 %)(25 gal) = 2.5 gal

2. How much pure acid is contained in #2?

    (20 %)(x gal) = 0.20x gal

3. How much pure acid is contained in #3?

$$(12\%)(25 + x)\,\text{gal} = (0.12)(25 + x)\,\text{gal}$$

The only places the acid in #3 comes from are #1 and #2. Therefore, pure acid in #1 plus pure acid in #2 equals pure acid in #3:

$$2.5\,\text{gal} + 0.20x\,\text{gal} = 0.12(25 + x)\,\text{gal}$$

You now have an equation containing the variable x whose value you seek. Further, note that you did not begin with an attempt to write an equation involving x. The only effort made was to determine the three basic facts about each solution: (1) its volume, (2) its amount of pure acid, and (3) its strength as a percent. Figure 5-9 shows the complete solution of this example using the box diagram procedure.

example How much pure lead must be added to 200 lb of an alloy that is 60% tin and 40% lead to produce an alloy that is 44% lead? (An alloy is simply a solid solution. Therefore the same solution diagram can be used.) Let x

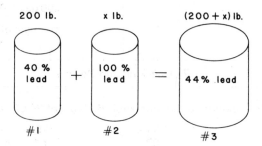

Figure 5-10

equal the number of pounds of pure lead which must be added. Putting this data on the containers, you have Figure 5-10. Now determine the pure lead in each container. In:

#1, pure lead $= (40\%)(200\,\text{lb}) = 80\,\text{lb}$

#2, pure lead $= (100\%)(x\,\text{lb}) = x\,\text{lb}$

#3, pure lead $= (44\%)(200 + x)\,\text{lb} = (0.44)(200 + x)\,\text{lb}$

Lead from #1 plus lead from #2 equals lead in #3:

$$80\,\text{lb} + x\,\text{lb} = (0.44)(200 + x)\,\text{lb}$$

Solve this equation for x:

$$x = \frac{100}{7} \text{ lb}$$

The checking is left as an exercise.

EXERCISE 5.5D

Solve the following problems using the procedure shown in Figure 5-9. (Watch significant figures.)

1. If 40.0 gal of a solution testing 30.0% acid were mixed with 100.0 gal of a solution testing 60.0% acid, what strength was the mixture?
2. What quantities of syrups testing 20.0% sugar (by weight) and 58.0% sugar must be mixed to produce 45Õ lb of syrup testing 30.0% sugar?
3. What quantities of two different batches of an insecticide, one containing 0.010% active ingredient and the other containing 0.040% active ingredient, must be mixed to produce 1280 lb that test 0.015% active ingredient?
4. One bar of metal is 40.0% silver and another is 80.0% silver. How many grams of each must be melted together to produce a 64 g bar testing 75% silver?
5. How much distilled water must be added to 80.0 ml of a solution testing 20.0% alcohol by volume to produce a solution testing 2.00% alcohol?
6. A shop foreman wishes to prepare 3Õ gal of cutting fluid consisting of 3Õ% lard oil and 7Õ% petroleum machine oil. He has in stock a supply of cutting fluid which is 5Õ% lard oil and 5Õ% machine oil. He also has an adequate supply of pure machine oil. How much of the 50–50 mixture and how much of the pure machine oil must he use to prepare the desired amount of cutting oil?
7. What quantities of two liquids, one testing 95% alcohol (by volume) and the other testing 15% alcohol, must be mixed to prepare 25 ℓ of 35% alcohol?
8. A certain company had in stock 2800 lb of solder testing 55% tin (45% lead) and 4400 lb of solder testing 5Õ% tin (5Õ% lead). The company wished to melt these two solders together and add enough more pure tin and lead to fill an order for 8800 lb of solder testing 52% tin and 48% lead. How much pure tin and how much pure lead were needed?
9. How much milk testing 6.4% butterfat must be added to 4200 gal of cream testing 2Õ% butterfat to produce cream testing 12% butterfat?
10. A certain power company purchases coal from two different mines, A and B. The coals are ground and mixed before being blown into the furnaces. Due to the differences in transportation costs, the most economic mixture is 7 lb from mine A to 4 lb from mine B. If the coal from mine A contains 8200 B.t.u./lb and the coal from mine B contains 12,400 B.t.u./lb, what is the B.t.u. rating of the fuel mixture? (Assume two-digit accuracy.)

Challenge Problems

11. A hot water-heating system in a mountain ski resort contains 450 gal of 25% antifreeze. How many gallons must be drawn off and replaced with pure antifreeze to raise its strength to 45%?
12. Work Problem 11 replacing the drawn off antifreeze with a 9Õ% solution.

⑥ Mensuration

Mensuration is the area of applied mathematics that deals with the measures of lengths of lines, areas of surfaces, and volumes of solids. Most measurements of areas and volumes are ultimately based on measurements of lengths.

Some measurements are determined by direct means and others by indirect means. For example, the length and width of this page may be measured directly with a ruler or scale; its area may be determined indirectly by calculation. Almost all indirect measurements are determined by the use of a formula. The unit of area is the square whose side is 1 unit of length. For example, if the chosen unit of length is the centimeter, the inch, the kilometer, or the mile, the corresponding unit of area is 1 sq cm, 1 sq in., 1 sq km, or 1 sq mile, respectively.

6.1 MEASURE OF AREAS

In contrast to lengths, which are one dimensional, areas are two dimensional. The calculation of any area must include two basic measurements (direct or indirect) made at right angles to each other. This requirement becomes fairly clear when we realize that if a surface had only length and no width, its area would be zero.

Areas of surfaces are usually calculated by formulas if the surfaces are bounded in such a way that regular plane figures result. The area of a surface formed by irregular boundary lines may be approximated by various methods. We shall discuss such methods in Section 6.3.

Figures 6-1 through 6-15 show many basic plane geometric figures and list their corresponding formulas* relative to their dimensions and properties. The abbreviations used in these formulas are given here for reference:

$$l = \text{length} \qquad\qquad d = \text{diameter}$$
$$w = \text{width} \qquad\qquad C = \text{circumference}$$
$$h = \text{height or altitude} \qquad s = \text{length of an edge}$$
$$b = \text{base} \qquad\qquad s = \text{slant height}$$
$$a = \text{altitude} \qquad\qquad S = \text{lateral surface area}$$

* Many of the formulas listed are not covered in this chapter. However, they are included for future reference purposes.

d = diagonal (polygons) T = total surface area
P = perimeter V = volume
A = area D = outside diameter
B = area of base l = arc length
r = radius c = chord length

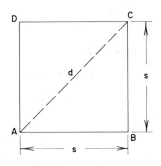

Square:
$$A = s^2$$
$$A = \tfrac{1}{2}d^2$$
$$d = \sqrt{2}\,s$$
$$s = 0.7071d$$

Figure 6-1

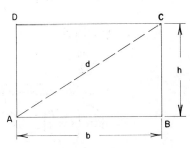

Rectangle:
$$A = bh$$
$$d = \sqrt{b^2 + h^2}$$

Figure 6-2

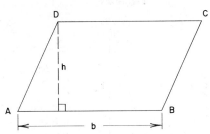

Parallelogram:
$$A = bh$$

Figure 6-3

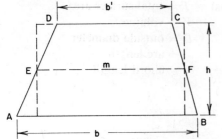

Figure 6-4

Trapezoid:

$$A = \tfrac{1}{2}(b + b')h \qquad \text{or} \qquad \frac{(b + b')h}{2}$$

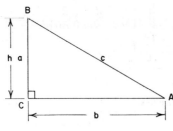

Figure 6-5

Right Triangle:

$$A = \frac{bh}{2}$$

$$A = \sqrt{s(s - a)(s - b)(s - c)},$$

$$\text{where} \left(s = \frac{P}{2} \right)$$

$$A = \tfrac{1}{2}bc \sin A$$
$$A = \tfrac{1}{2}hc \cos A$$

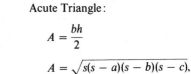

Figure 6-6

Acute Triangle:

$$A = \frac{bh}{2}$$

$$A = \sqrt{s(s - a)(s - b)(s - c)},$$

$$\text{where} \left(s = \frac{P}{2} \right)$$

$$A = \tfrac{1}{2}bc \sin A$$
$$A = \tfrac{1}{2}b(c \cos A + a \cos c)$$

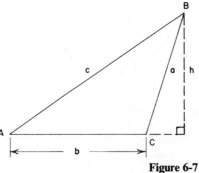

Figure 6-7

Obtuse Triangle:

$$A = \frac{bh}{2}$$

$$A = \sqrt{s(s - a)(s - b)(s - c)}$$
$$A = \tfrac{1}{2}bc \sin A$$
$$A = \tfrac{1}{2}ac \sin B$$

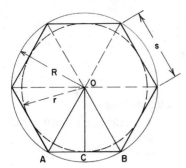

Regular Hexagon:

R = radius of circumscribed circle
r = radius of inscribed circle
$A = 2.598s^2 = 2.598R^2 = 3.464r^2$
$R = s = 1.155r$
$r = 0.866s = 0.866R$

$$A = \frac{6sr}{2} = 3sr = 3Rr$$

Figure 6-8

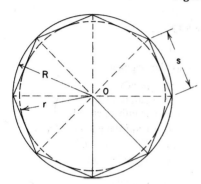

Regular Octagon:

R = radius of circumscribed circle
r = radius of inscribed circle

$$A = \frac{8rs}{2} = 4rs$$

$A = 4.828s^2 = 2.828R^2 = 3.314r^2$
$r = 1.207s = 0.924R$
$s = 0.765R = 0.828r$

Figure 6-9

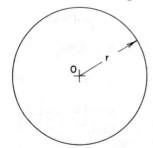

Circle:

$A = \pi r^2 = \frac{1}{4}\pi d^2$
$d = 2r$
$C = \pi d = 2\pi r$

$$r = \sqrt{\frac{A}{\pi}} = 0.564\sqrt{A}$$

Figure 6-10

Circular Sector:

θ = angle in degrees

$$A = \frac{\theta}{360°}\pi r^2$$

$$A = \frac{1}{2}rl$$

$$l = \text{arc length} = \frac{\theta}{360°} \cdot 2\pi r$$

$$\theta = \frac{57.296l}{r}, \qquad r = \frac{2A}{l}$$

Figure 6-11

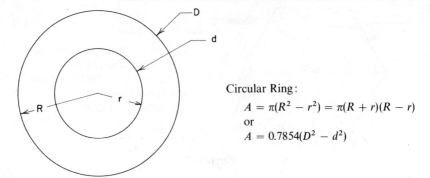

Circular Ring:

$$A = \pi(R^2 - r^2) = \pi(R + r)(R - r)$$
or
$$A = 0.7854(D^2 - d^2)$$

Figure 6-12

Circular Segment:

$A = $ area
$l = $ length of arc
$c = $ length of chord
$\theta = $ central angle in degrees
$A = \frac{1}{2}[rl - c(r - h)]$
$c = 2\sqrt{h(2r - h)}$

$$r = \frac{c^2 + 4h^2}{8h}$$

$$l = 0.01745r\theta$$

$$\theta = 57.296\frac{l}{r}$$

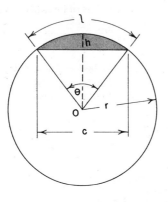

Figure 6-13

$$h = r - \frac{1}{2}\sqrt{4r^2 - c^2}$$

$$h = r\left[1 - \cos\left(\frac{\theta}{2}\right)\right]$$

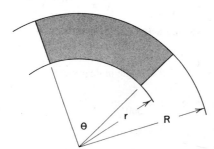

Circular Ring Sector:

$$A = \frac{\theta\pi}{360°}(R^2 - r^2)$$

$$A = 0.00873\theta(R^2 - r^2)$$

Figure 6-14

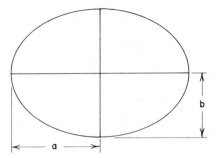

Ellipse:

$$A = \pi ab$$
$$P \approx \pi\sqrt{2(a^2 + b^2)}$$

A closer approximation for P is:

$$P \approx \pi\sqrt{2(a^2 + b^2)} - (a - b)^2/2.2$$

Figure 6-15

EXERCISE 6.1

Use information from Figures 6-1 through 6-15 as needed in solving the following problems.

1. Develop formulas for the perimeters of the following:

    (a) Square.
    (b) Rectangle.
    (c) Equilateral triangle.
    (d) Regular hexagon.

2. Using the results in Problem 1, compute the perimeters of the following:

    (a) Square with $s = 4.202$ in.
    (b) Rectangle with $b = 14.5$ cm and $h = 2.21$ cm.
    (c) Equilateral triangle with sides of 10.25 mm.
    (d) Regular hexagon with side $s = 8.05$ m.

3. How many rods of fencing are required to enclose a rectangular field measuring 48 rods by 92 rods? What is the area of the field?

4. A concrete bridge support has a square cross section measuring 4.20 ft on a side. Determine the length of the diagonal.

5. A building contractor staked out the corners of an 80.0 ft by 50.0-ft foundation for a building. What should the diagonals measure?

6. Two 80.0-ft long diagonals of a rectangle intersect, forming two acute angles of 60.0°. Determine the dimensions and area of the rectangle.

7. Find the area of a square whose diagonal is $4\bar{0}$ in.

8. Determine the tensile strength of a bronze rod whose square cross section measures 0.850 in. on an edge. Tensile strength of bronze is 6.0×10^4 lb/sq in.

9. Determine the cross-sectional area of a triangular-shaped steel bar with sides of 2.50, 1.50, and 2.00 in.

10. A triangle has sides of 8.00, 9.00, and 10.0 in. Find the perimeter and area of the triangle formed by joining the midpoints of the sides of the original triangle.

11. Determine the area of an equilateral triangle whose sides measure 4.00 in.

12. If a trapezoid has bases of 12 and 18 in. and an altitude of 4.0 in., what are the lengths of the nonparallel sides if they are equal?

13. A trapezoid with bases of 48.2 and 22.6 cm has an altitude of 20.0 cm. Determine the length of the line segment joining the midpoints of the nonparallel sides.

14. What is the cross-sectional area of a heating duct whose cross section is a parallelogram with a base of 24.0 in. and an altitude of 18.0 in.?

15. Two sides of a parallelogram-shaped die are 14.0 and 18.0 in.; the included angle is 30.0°. Determine the area of the parallelogram and the length of the shorter diagonal.

16. A parallelogram has an area of 85 cm². If the obtuse angles measure 150° each and one side measures 17 cm, determine the length of the other side.

17. What must be the diameter of a circular heating duct with the same cross-sectional area as a rectangular duct measuring 28.0 in. by 22.0 in.?

18. How much less material (as a percent) does the circular duct require than the rectangular duct in Problem 17?

19. If a piece of sheet metal 4.00 ft wide is to be formed into a duct with either a circular cross section ($C = 4.00$ ft) or a square cross section ($P = 4.00$ ft), how would the cross-sectional area of the circular duct compare (as a percent) with the cross-sectional area of the square duct?

20. Determine the diameter of a circular duct having the same cross-sectional area as four smaller ducts whose diameters are 3.0, 4.0, 5.0, and 6.0 in.

21. Three circles, each having a radius of 8.00 in. are mutually tangent to each other. Determine the area of the section bounded by the circles.

22. Develop a general formula for computing the area in Problem 21. (*Hint*: Join the centers forming an equilateral triangle and use "r" for the radius of each circle.)

23. Using the formula from Problem 22, compute the radii of the three mutually tangent circles (equal radii) which bound an area of 2.560 sq in.

24. A steel washer measuring 1.00 in. in diameter has a 0.500 in. diameter hole drilled through its center. What is the area of one face of the washer?

25. How large should the hole in a washer be so that the cross-sectional area of the hole equals the cross-sectional area of one face of the washer?

26. The counterweight on a 60.0 in. diameter driving wheel subtends a central angle of 60.0° (Figure 6-16). Determine the distance h and the area of the segment.

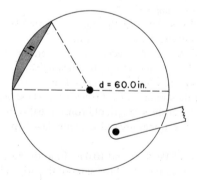

Figure 6-16

27. Work Problem 26 for a driving wheel with a diameter of 72.0 in.
28. Determine the area of a circular ring whose inside diameter $d = 4.00$ in. and whose outside diameter $D = 8.00$ in.
29. Compute the tensile strength of a hollow bronze rod having I.D. $= 0.50$ in. and O.D. $= 1.00$ in. if the tensile strength of bronze is 6.0×10^4 lb/sq in.
30. Determine the surface area of the water in the elliptical-shaped swimming pool shown in Figure 6-17 if the major axis measures 32 ft and the minor axis measures 18 ft.

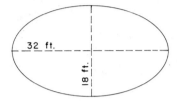

Figure 6-17

31. A concrete pillar for a modern building has an elliptical cross section with a major axis measuring 60.0 in. and a minor axis measuring 40.0 in. If the safe load for this pillar is 1.20×10^3 lb/sq in., what is the maximum safe load for the pillar?
32. A sheet metal pattern called for a circular sector having a vertex angle of 36° to be cut from a circle having a diameter of $2\hat{0}$ in. Determine the area of the sector.
33. Determine the area of the circular ring sector shown in Figure 6-18.

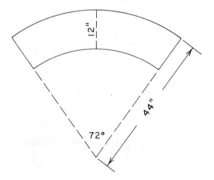

Figure 6-18

6.2 MEASURE OF VOLUMES

The unit of volume is the cube whose edges are 1 unit of length. Volumes are three-dimensional quantities; they involve three linear measurements in mutually perpendicular directions. These three basic measurements are referred to as length, width, and height (or depth).

Volumes of regular geometric solids are calculated from formulas. Figures 6-19 through 6-29 show the most commonly used figures along with their corresponding properties and formulas.

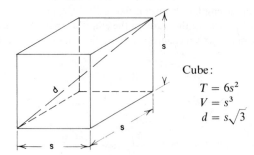

Cube:

$$T = 6s^2$$
$$V = s^3$$
$$d = s\sqrt{3}$$

Figure 6-19

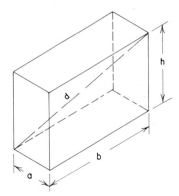

Rectangular Prism:

$$T = 2(ab + bh + ah)$$
$$V = abh \qquad \text{or} \qquad V = lwh$$
$$d = \sqrt{a^2 + b^2 + h^2}$$

or

$$d = \sqrt{l^2 + w^2 + h^2}$$

Figure 6-20

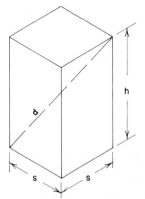

Square Base Prism:

$$T = 2s^2 + 4sh$$
$$V = s^2h$$
$$d = \sqrt{h^2 + 2s^2}$$

For any prism having a base in the form of a triangle, square, rectangle, hexagon, etc.;

$$V = Bh \qquad (B = \text{area of base})$$

Figure 6-21

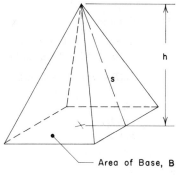

Pyramid:

$$V = \tfrac{1}{3}Bh$$
$$S = \tfrac{1}{2}Ps \qquad (P = \text{perimeter of base})$$
$$T = \tfrac{1}{2}Ps + B$$

Area of Base, B

Figure 6-22

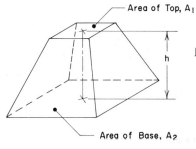

Area of Top, A_1

Frustum of Pyramid:

$$V = \frac{h}{3}(A_1 + A_2 + \sqrt{A_1 \times A_2})$$

Area of Base, A_2

Figure 6-23

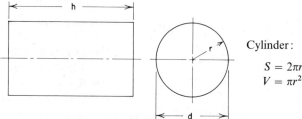

Cylinder:

$$S = 2\pi rh$$
$$V = \pi r^2 h$$

Figure 6-24

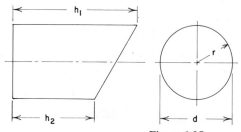

Portion of a Cylinder:

$$S = \pi r(h_1 + h_2)$$
$$V = \frac{\pi r^2}{2}(h_1 + h_2)$$

Figure 6-25

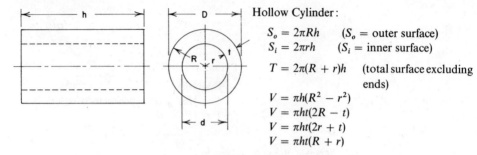

Hollow Cylinder:

$S_o = 2\pi Rh$ $(S_o = $ outer surface)
$S_i = 2\pi rh$ $(S_i = $ inner surface)

$T = 2\pi(R + r)h$ (total surface excluding ends)

$V = \pi h(R^2 - r^2)$
$V = \pi ht(2R - t)$
$V = \pi ht(2r + t)$
$V = \pi ht(R + r)$

Figure 6-26

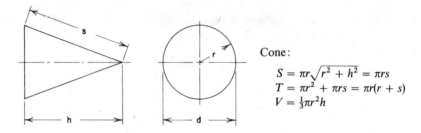

Cone:

$S = \pi r\sqrt{r^2 + h^2} = \pi rs$
$T = \pi r^2 + \pi rs = \pi r(r + s)$
$V = \frac{1}{3}\pi r^2 h$

Figure 6-27

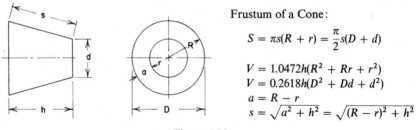

Frustum of a Cone:

$$S = \pi s(R + r) = \frac{\pi}{2}s(D + d)$$

$V = 1.0472h(R^2 + Rr + r^2)$
$V = 0.2618h(D^2 + Dd + d^2)$
$a = R - r$
$s = \sqrt{a^2 + h^2} = \sqrt{(R - r)^2 + h^2}$

Figure 6-28

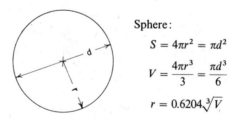

Sphere:

$S = 4\pi r^2 = \pi d^2$

$$V = \frac{4\pi r^3}{3} = \frac{\pi d^3}{6}$$

$r = 0.6204\sqrt[3]{V}$

Figure 6-29

EXERCISE 6.2

Refer to the figures and formulas in Figures 6-19 through 6-29 as needed.

1. A well-proportioned pedestal for displaying objects in museums and galleries can be made from two equal cubes placed one on top of the other. If such a pedestal was made from two 18.0-in. cubes of polished marble, what would be the weight of the pedestal and the length of its diagonal? (Density of marble is 168 lb/cu ft.)
2. The base of a lamp is to be made from two equal size cubes (edge = 12.0 in.) by standing one cube on a corner (so that its diagonal is vertical) in the center of the top face of the other. Determine the height of the base.
3. Determine the volume and weight of air in a room which measures 32.0 ft long, 28.0 ft wide, and 10.5 ft high.
4. A wire heating element is to run between opposite vertices of a drying oven which measures 22 in. long, 16 in. wide, and 14 in. high. What must be the length of the heating element?
5. A geometric model of a rectangular parallelepiped, showing all edges and the four diagonals, is to be made from wire. If it is to measure 12.0 in. long, 8.00 in. wide, and 6.00 in. high, how long a piece of wire is needed? (Neglect any waste from cuts and welds.)
6. How large a cube can be machined from a cylindrical shaft which measures 3.000 in. in diameter?
7. How many cubic yards of asphaltic concrete are needed to pave 12 miles of a road if the rectangular cross section of the paving is 24 ft by 6.0 in.
8. A certain building code requires that soil-compaction tests be made each 4 ft in direction of length and each 2 ft in direction of width on earth fills which are to support a building. How many tests are required for a fill measuring 87 ft long and 36 ft wide? (No tests are made on the edges.)
9. Calculate the weight of gypsum mined from a vein measuring 18 ft thick, 44 ft wide, and 370 ft long.
10. How many gallons of oil can be stored in a pit tank which measures 240 ft long, 186 ft wide, and 16 ft deep?
11. A hexagonal prism is to be cut from a 28.0-ft marble column which measures 44.0 in. in diameter. What would be the volume and weight of the largest such prism that can be made from the column?
12. What would be the total surface area of the six rectangular faces for the prism in Problem 11?
13. How many 9-in. by 9-in. tiles are required to cover a floor which measures 27 ft by 39 ft?
14. A rectangular swimming pool is 75 ft long and 25 ft wide. If it is 12 ft deep at one end and 3.0 ft deep at the other end, and if the bottom has a uniform slope, how many gallons of water does the pool hold? What is the weight of the water?
15. The Great Pyramid in Egypt measures 764 ft on each edge of its square base. Its height is 451 ft. What is its volume?
16. If the stone from which the Great Pyramid (Problem 15) was made weighs 150 lb/cu ft, determine the total weight of the pyramid. (Neglect its foundation and spaces occupied by internal rooms.)

17. A concrete support for a railroad bridge has the shape of a pyramid with a square base measuring 14.00 ft on a side and having a height of 18.00 ft. How many cubic yards of concrete does the support contain?

18. Determine the surface area of the pyramid in Problem 17.

19. A grain hopper in a flour mill has the shape of an inverted pyramid. If the base is a regular hexagon with an 18.0-ft diagonal and the height of the hopper is 24.0 ft, how many bushels of grain will the hopper hold?

20. Determine the outside surface area for the hopper in Problem 19.

21. An overpass on a modern freeway is supported at its center point by a concrete column having the shape of a frustum of a pyramid. If the column is 24 ft high and has square bases with sides measuring 12 ft and 4.0 ft, respectively, determine the volume and weight of the support (1 cu ft of concrete weighs 150 lb).

22. Determine the surface area for the support in Problem 21.

23. If a manufacturer produces plastic dice which measure 0.625 in. on each edge, how many dice can be produced from 12.0 cu ft of plastic? (Make no allowance for waste.)

24. The weight of ice is approximately 92% as much as the weight of an equal volume of water. Water weighs 62.4 lb/cu ft. What would be the weight of a block of ice measuring 18.0 in. by 16.0 in. by 30.0 in.

25. What would be the weight of ice having an average thickness of 14 in. and covering a lake with a surface area of 3.5 acres?

26. What weight of concrete (concrete weighs 150 lb/cu ft) is needed to cover a rectangular parking lot measuring 280 ft by 120 ft if the thickness of the concrete is to be 4.0 in.?

27. How many cubic yards of earth must be removed from a trapezoidal cut 24.0 ft deep, 120.0 ft wide at the top, and 80.0 ft wide at the bottom if the cut is 3600 ft long?

28. A certain grade of stainless steel weighs 494 lb/cu ft. A 10.0-in. prism made from this steel has a uniform trapezoidal cross section which measures 2.50 in. and 3.70 in. on the parallel sides and 1.75 in. between the parallel sides. Determine the weight of the prism.

29. A metal duct to transport corrosive gases is to be coated inside with plastic measuring 0.0850 in. in thickness. How much plastic is required to coat 480 ft of this duct if its rectangular cross section measures 2.50 ft by 4.00 ft? (Neglect the material saved at the corners.)

30. How much plastic would be saved if the duct in Problem 29 was replaced by a cylindrical duct having a circular cross section with the same area?

31. A company has a contract to manufacture a large number of 80.0-gal cylindrical steel tanks (circular cross section). If the diameter of each tank must be 20.0 in. what must be the length of the tank?

32. If the tank in Problem 31 was specified to have a length of 6.00 ft, what would be the necessary diameter for the same volume?

33. A water tank is in the shape of an inverted cone having a base radius of 18.0 ft and an altitude of 16.0 ft.

    (a) How many gallons will the tank hold?

    (b) How much plastic is required to coat the outside surface (excluding the base) if a gallon of plastic covers 350 sq ft?

34. How many tons of cement may be stored in a silo (hollow cylinder) 24 ft in diameter and 110 ft tall? (Density of dry cement is 94 lb/cu ft.)

35.   A concrete form used in making railroad-bridge supports has the shape of a frustum of
      a cone with base radii of 12 ft and 8.0 ft. If the frustrum is 36 ft tall and has a 3.0 ft diameter
      hollow cylindrical core, determine the volume of the concrete needed for pouring one
      support.
36.   Determine the slant height and outside surface area of the frustrum in Problem 35.
37.   What would be the volume of the largest sphere that could be cut from a cube measuring
      12.00 in. on an edge?
38.   What would be the dimension of the largest cube that could be cut from a sphere having
      a 12.00-in. diameter?
39.   If the density of a certain grade of cast iron is 0.25 lb/cu in., what diameter hole (nearest
      64th in.) should be drilled through a $\frac{3}{4}$-in. plate to remove 4.0 oz of material?
40.   If the counterweight in Problem 26, Exercise 6.1, is 4.0 in. thick and has a density of
      0.26 lb/cu in., determine its weight.

6.3   MEASURE OF AREAS AND VOLUMES OF IRREGULAR FIGURES

There are many mathematical techniques for determining areas and volumes
of irregular figures. Most of these methods are covered in advanced mathematics
courses. In this section, we shall discuss two basic methods which will frequently
allow either the calculation or close approximation of an area or volume.

method I   Division of the irregular figure into parts having regular geometric shapes.

example   Figure 6-30 shows a figure divided by a dotted line into a trapezoid
            and a right triangle. The total area can be obtained by determining the
            areas of the two parts.

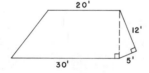

Figure 6-30

example   Figure 6-31 shows a machined part whose volume is needed. The total
            volume can be obtained by calculating the volumes of the two cylindrical
            parts and the conical part.

method II   Construction of equally spaced grid lines and counting of spaces.

example   Figure 6-32(a) shows an irregular shaped figure whose area is needed.
            Figure 6-32(b) shows the same figure with a system of grid lines constructed

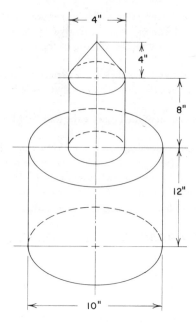

Figure 6-31

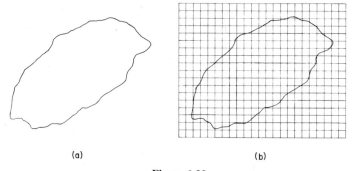

(a)   (b)

Figure 6-32

(to scale) on it. By counting the small squares and fractional parts close approximation of the area may be obtained.

example   Figure 6-33(a) shows an irregular shaped opening broached in a steel die. The volume of the opening can be determined by constructing grid lines Figure 6-33(b), counting the squares to determine surface area, and multiplying by the thickness of the die.

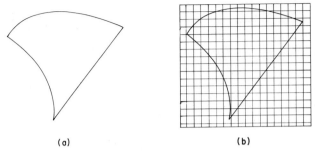

(a)   (b)

Figure 6-33

EXERCISE 6.3

1. Calculate the area and perimeter of the figure in Figure 6-30.
2. Calculate the area and perimeter of the quadrilateral shown in Figure 6-34.

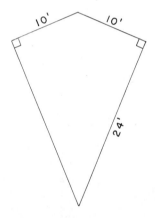

Figure 6-34

3. Determine the area and perimeter of the region shown in Figure 6-35.
4. Determine the area and perimeter of the region shown in Figure 6-36.
5. Approximate the volume of metal removed from the die in Figure 6-33(b) if the die is $\frac{3}{4}$ in. thick and the grid lines are drawn 5/in.
6. Approximate the area of the irregular region in Figure 6-32(b).
7. Approximate the volume of water (in acre-feet) in a lake (Figure 6-37) whose average depth is 24 ft.
8. Calculate the area of the template shown in Figure 6-38.
9. Calculate the outside surface area of the conical metal vent cap shown in Figure 6-39.

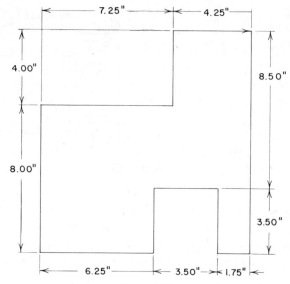

Figure 6-35

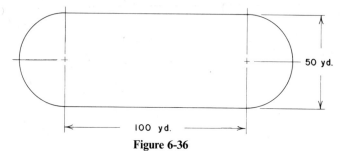

Figure 6-36

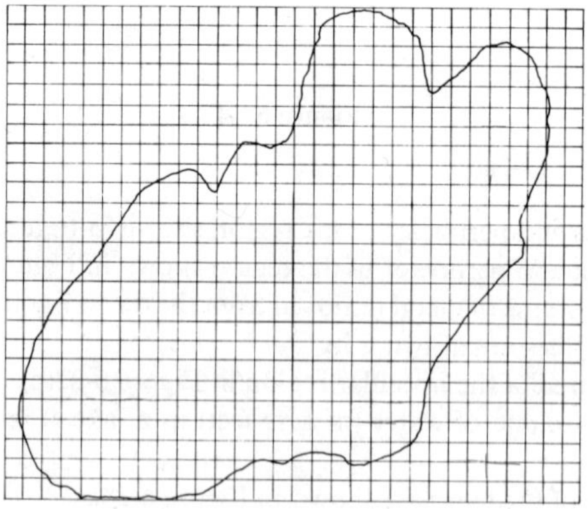

Scale — 1 space = 10 ft

Figure 6-37

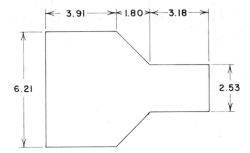

Figure 6-38

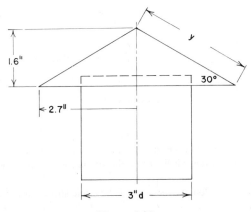

Figure 6-39

10. Assume that the elements drawn in Figure 6-40 form trapezoids. Measure the parallel sides with a steel rule and calculate the approximate area by summing the areas of the six trapezoids.

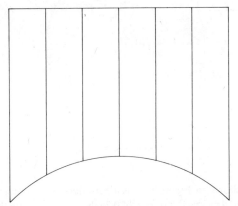

Figure 6-40

11. Calculate the volume of concrete required for 450 ft of curb and gutter having the cross section shown in Figure 6-41.

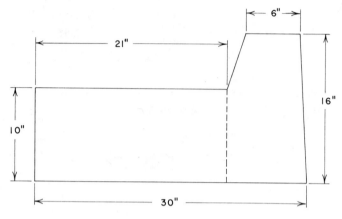

Figure 6-41

12. An underground silo is built in the shape of a right circular cylinder. Its inside diameter is 56.0 ft and its outside diameter is 64.0 ft. If the silo is 160.0 ft deep, how much concrete is contained in the walls?

13. Approximate and then calculate the area of the circle shown in Figure 6-42.

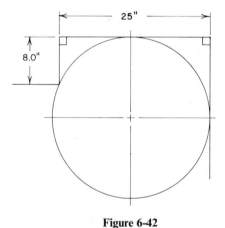

Figure 6-42

14. A cast iron ring, shown in Figure 6-43 is 1.5 in. thick and contains 24 holes, 1.0 in. in diameter. Using the density of cast iron of 0.26 lb/cu in., calculate the weight of the ring.

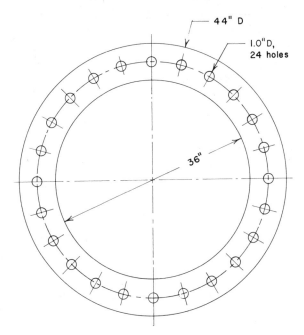

Figure 6-43

Challenge Problems

15. Calculate the volume of ore in the conical pile shown in Figure 6-44.
16. Calculate the area of one face of the snap ring shown in Figure 6-45.
17. Calculate the volume of the casting shown in Figure 6-46.

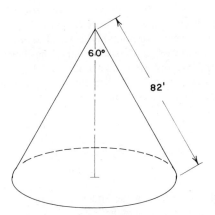

Figure 6-44

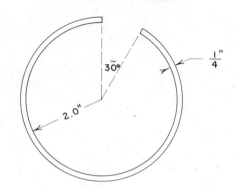

Figure 6-45

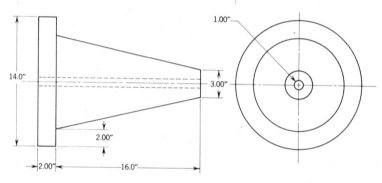

Figure 6-46

18. A geometric solid has a total surface area of 136 sq in. What is its volume if the figure is a cube? If the figure is a rectangular solid with a base 8.0 in. long and 4.0 in. wide, what is its volume? If the solid is a sphere, what is its volume?

19. An Egyptian king once gave a servant his choice of three gifts, providing that he conceal the gift in his pack and carry it home after dark. The choices were: (1) hollow sphere of gold 1 in. thick and 7 in. outside diameter, (2) a solid sphere of gold 10 in. in diameter, and (3) a solid sphere of gold 5 in. in diameter. Which was the best choice? Why?

20. If a cube of copper, 1.0 ft on each edge, is drawn into a wire 0.125 in. in diameter, what length wire results? By what percent is the surface area of the copper increased?

21. If 440 lb of gold are to be melted and poured into a hollow inverted cone with a radius of 6.0 in. and an altitude of 18 in., what will be the altitude of the gold cone?

22. A metal plate, 3.00 in. thick, has an elliptical cross section measuring 8.00 in. by 6.00 in. A tapered hole is machined through the center of the plate with diameter openings of 2.00 in. and 1.00 in. Calculate the volume of metal in the plate.

23. A cylindrical tank, 24.0 ft in diameter and 30.0 ft long, is capped on each end with cones having altitudes of 6.00 ft. Compute the volume of the tank and the weight of its contents when filled with a solvent which weighs 48.8 lb/cu ft.

7 Graphs and Their Uses

Graphs serve many purposes. A well-constructed graph is a valuable aid in the formulation and solution of a problem. Information presented in graphic form frequently allows for insight, interpretation, and planning that otherwise might not be possible. Technicians and engineers often use graphs to determine whether a certain problem has a solution or whether the solution is practical. They also use graphs to obtain solutions to problems which otherwise would require advanced mathematics beyond their present knowledge or numerous hours of calculations.

7.1 BAR, CIRCLE, AND LINE GRAPHS

The Bar Graph

The bar graph is usually used to show both the magnitude of the quantities involved and any variation in the magnitudes. Figure 7-1 shows a bar graph constructed for the following data.

Metal	Electric Conductivity
Silver	100.00
Copper	97.61
Gold	76.61
Aluminum	63.00
Tantalum	54.63
Magnesium	39.44
Sodium	31.98
Cadmium	24.38

A bar graph consists of heavy straight lines or long narrow rectangles whose lengths are proportional to the magnitudes of the quantities being presented. A well-constructed bar graph should display adequate information to allow interpretation by those for whom the graph is constructed, even after a long time. The scale should be clearly shown, the actual quantity represented by 1 unit of this scale should be precisely identified, and the graph should have an informative title. Some thought must be given to selection of the scale in order to present the desired

223

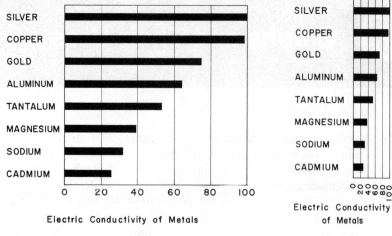

Figure 7-1	Figure 7-2

information clearly and to avoid distortion of relative differences. Figure 7-2 shows the same information as Figure 7-1, but the scale was poorly chosen; the person viewing the graph does not get as clear a picture of either the actual magnitudes of the quantities or their relative differences.

The Circle Graph

The circle graph (frequently called a pie chart) may also be used to show the relative sizes or magnitudes of items from a complete or closed set of data. One advantage over the bar graph is that the circle graph does not distort the relative

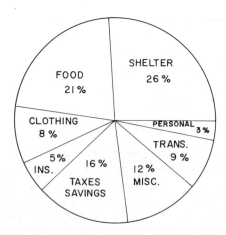

Figure 7-3

sizes of the quantities represented. Figure 7-3 shows a circle graph constructed for the following data.

Distribution of Income for a Family of Four
with Income of $15,000/Year

Item	Per-cent	Central Angle	Amount
Food	21	75.6°	$3,150.00
Shelter	26	93.6°	3,900.00
Personal	3	10.8°	450.00
Transportation	9	32.4°	1,350.00
Miscellaneous	12	43.2°	1,800.00
Taxes and savings	16	57.6°	2,400.00
Insurance	5	18.0°	750.00
Clothing	8	28.8°	1,200.00
	100	360.0°	$15,000.00

Each circular sector (wedge shaped) of the circle has a central angle which is the same percentage of 360° as the percentage of the quantity it represents. For example, if a particular item is 50% of some total, its sector has a central angle of $(50\%)(360°) = 180°$ or half the circle. To determine the size of each sector, compute each specific percentage of 360° and use a protractor to lay off the central angles. If the data are not given in percents, then the percent for each item must first be computed. When the total for the items is known, the slide rule makes short work of determining the corresponding central angle for each item.

For example, suppose the budget expenditures for a company were:

Postage	$72,500	→ 67.8°
Advertising	155,000	→
Insurance	32,000	→
Wages	93,500	→
Office rent	12,000	→
Transportation	16,000	→
Miscellaneous	4,000	→
	$385,000	→ 360°

You could compute the central angles for the various sectors as follows. Set 360 on *B* under 385 on *A*. Then move the hairline (H) to the first item, 72,500, on *A*. Read "678" under *H* on *B*; 72,500 is about $\frac{1}{5}$ of 385,000, so the angle must be about $\frac{1}{5}$ of 360° or about 70°. Therefore, the angle must be 67.8°.

With H over 155 on A read 145°   on B
With H over 320 on A read 30.0°   on B
With H over 935 on A read 87.4°   on B
With H over 120 on A read 11.2°   on B
With H over 160 on A read 15.0°   on B
With H over 400 on A read   3.7°   on B

The construction of this graph is left as an exercise. Sometimes it is desirable to express the percent for each sector of the graph as well as giving a numeric amount. Each sector should clearly state the name of the particular item represented and any numerical values that you wish to show. Each circle graph should have an informative title and sufficient additional information to enable the intended viewer

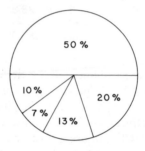

Figure 7-4

to understand the graph. Figure 7-4 shows a circle graph which is worthless if it is to show anything other than the percent of the total circle occupied by each sector.

Line Graphs and Broken Line Graphs

Figures 7-5 and 7-6 show a line graph and a broken line graph, respectively. When there is a continuous change from one entry to the next, a continuous smooth, curved line is drawn between the points. When there is not a continuous change from one entry to another, a series of straight-line segments join the points (hence the name broken line graph). The first example explains the construction of the graph in Figure 7-5, and the second example explains the graph in Figure 7-6.

example   The amount of light falling on a fixed area is inversely proportional to the square of the distance from the fixed area to the light source. In trying to calibrate an exposure meter, the meter was held 1 ft from a frosted-glass

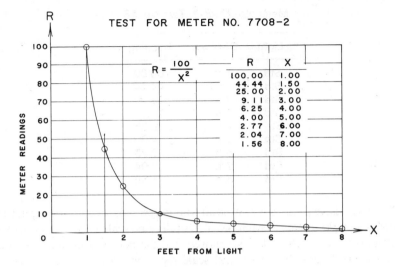

Figure 7-5

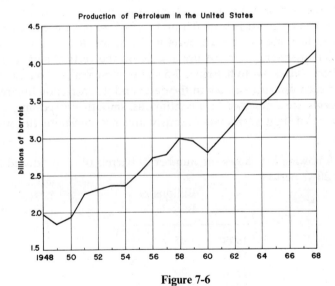

Figure 7-6

light source and the dial on the meter was set at 100. The meter was then moved back, making the distance 1.5 ft, and the meter reading was taken. This procedure was continued with the results shown in the following table.

Meter No. 7708-2, Test No. 18

Light	Distance	Reading
Std	1 ft	100.00
Std	1.5 ft	44.44
Std	2 ft	25.00
Std	3 ft	11.11
Std	4 ft	6.25
Std	5 ft	4.00
Std	6 ft	2.78
Std	7 ft	2.04
Std	8 ft	1.56

To show these data as a continual line graph, you need to construct two scales at right angles to each other. Usually one scale is horizontal and the other vertical. Each scale is so chosen that the magnitudes of all items can be shown. Normally the vertical scale represents the dependent quantity (when such dependency exists) and the horizontal scale the independent quantity. In this example the meter reading depends upon the distance; therefore the horizontal scale represents distance from the light source. The range of this factor is from 0 ft up to 8 ft. You must use a scale which allows you to represent 8 such units in the space desired for the graph. In a similar manner, you must use a scale on the vertical axis to allow a change from 100 down to 0. Figure 7-5 shows these scales. You then plot a point for each distance (shown in the data) and its corresponding reading. These points are connected by a continuous, smooth, curved line, because the amount of light varies continuously and not intermittently with the distance.

example   The following list shows the number of barrels of oil produced during a recent 20-yr period.

Year	Billions of Barrels
1948	1.95
1949	1.80
1950	1.90
1951	2.25
1952	2.31
1953	2.33
1954	2.33
1955	2.52
1956	2.71

1957	2.75
1958	3.00
1959	2.95
1960	2.79
1961	3.00
1962	3.20
1963	3.45
1964	3.44
1965	3.60
1966	3.90
1967	3.97
1968	4.20

In constructing the graph (Figure 7-6) to represent these data, either the vertical or horizontal scale could be used to represent the year or the production figure. In this figure the horizontal scale shows the 20-yr period and the vertical scale shows the number of barrels in units of 1 billion barrels. Because the change from one entry to the next is not continuous, the points are joined with straight-line segments.

EXERCISE 7.1

1. The following list shows the rate of flow in cubic feet per second for a river during part of a month. Construct a bar graph illustrating the given data.

December	Rate of Flow in 1000 cfs
1	19
3	97
5	121
7	80
9	72
11	68
13	67
15	78

2. Obtain data from Table 9, Appendix D, and prepare a bar graph for eight metals showing comparative weights in pounds per cubic foot.
3. Obtain data from Table 9, Appendix D, and prepare a bar graph for eight metals to compare their specific gravities.
4. Obtain data from Table 9, Appendix D, and prepare a bar graph for eight metals showing comparative melting points.

5. Prepare a bar graph for the data in Problem 2 to magnify the differences in the magnitudes of the quantities.

6. Prepare a bar graph for the data used in Problem 3 to minimize the differences in magnitudes of the quantities.

7. Construct a circle graph to illustrate the following land use data.

Land use for Hannensville

Use	Percent
Streets	16.2
Parks	8.3
Industrial	9.7
Commercial	3.1
Residential	51.3
Vacant	7.0
Streams	4.4

8. Construct a pie-chart (circle graph) for the following die-casting alloy.

Metal	Percent
Copper	6.9
Aluminum	18.3
Zinc	32.4
Tin	17.2
Lead	23.9
Antimony	1.3

9. The total cost of manufacturing a single part for a turret lathe is as follows: $7.00 for labor, $2.00 for overhead, $5.00 for materials, $4.50 for testing, and $2.75 for royalty. Construct a pie-chart to represent the production cost.

10. Construct a circle graph for the following data showing the number of cars owned by various age groups.

Age of Household Head	Number of Cars (in millions)
Under 25	2.7
25–34	9.6
35–44	11.9
45–54	10.8
55–64	8.5
65 and over	9.3

11. The following data were obtained in a tension test of a machine steel bar. Construct a continuous line graph, plotting elongation as the dependent variable on the vertical scale.

Applied Load in Pounds per Square Inch	Elongation per Inch of Length
1,500	0.00005
2,500	0.00008
5,000	0.00016
7,500	0.00025
10,000	0.00033
12,500	0.00042
15,000	0.00050
17,500	0.00057
20,000	0.00067
22,500	0.00077

12. Construct a broken line graph for the following data on factory sales of special motor vehicle types.

Year	Number of Vehicles
1930	250,000
1935	280,000
1940	500,000
1945	200,000
1950	800,000
1955	1,000,000
1960	1,600,000
1965	2,250,000

13. Construct a broken line graph for the following data.

Time	Temperature Reading
8:00 a.m.	60°F.
9:00 a.m.	72°F.
10:00 a.m.	76°F.
11:00 a.m.	81°F.
12:00 noon	83°F.
1:00 p.m.	87°F.
2:00 p.m.	86°F.
3:00 p.m.	84°F.
4:00 p.m.	79°F.
5:00 p.m.	73°F.

14. Construct a continuous line graph for the following data giving the height of a projectile above the ground as a function of time (after firing).

Height (feet)	Time (seconds)
0	0
112	1
192	2
240	3
256	4
240	5
192	6
112	7
0	8

15. Construct a line graph for the data in Problem 12 which will magnify the differences in the magnitudes of the quantities.

16. Construct a line graph for the data in Problem 12 which will minimize the differences in the magnitudes of the quantities.

17. Is it easier to read values from the graph in Problem 2 or in Problem 5? Why?

18. Is it easier to read values from the graph in Problem 3 or in Problem 6? Why?

19. Could a circle graph be used for Problem 1? Explain.

20. Could a circle graph be used for Problem 11? Explain.

21. Figure 7-7 is a graph from a recording barometer. The graph shows readings for a 24-hr period.

 (a) What was the reading at 11 p.m.?

 (b) What was the reading at 3 a.m.?

 (c) At what time did the pressure reach a maximum?

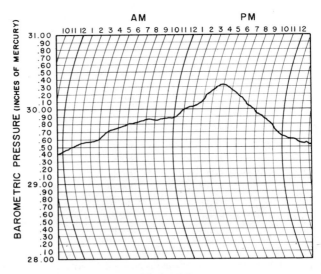

Figure 7-7

22. Figure 7-8 shows a graph from a temperature-recording device used on an industrial furnace.

(a) At what time was the maximum temperature reached?
(b) What was the maximum temperature?
(c) What was the average hourly reading for the 24 hr?
(d) What was the temperature at 11:30 a.m.?

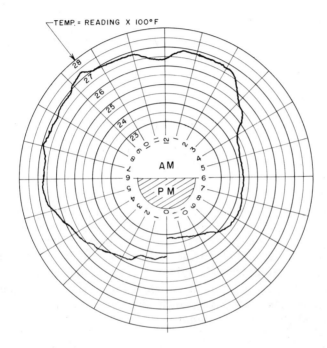

Figure 7-8

7.2 THE CARTESIAN (RECTANGULAR) COORDINATE SYSTEM

The rectangular coordinate system of graphing is an important tool for the technician, the engineer, and the mathematician. This system of graphing was developed by the French mathematician, Rene Descartes (1596–1650), and bears his Latinized name, Renatus Cartesius. He called it the right-angled coordinate system because the two basic lines (called axes) form four right angles. Figure 7-9 shows two perpendicular axes forming the four right angles which divide the plane into four quadrants. The horizontal line is usually designated as the x axis and the vertical line as the y axis. The point of intersection is called the origin (abbreviated

O in the figure) and serves as one of two important reference points. The positive direction may be shown in either of two ways:

    1. An arrowhead at the end of the line denotes the increasing or positive direction. The opposite direction is the negative direction.

    2. A plus sign (+) written with the letter (naming the axis) at the end of the

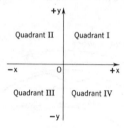

Figure 7-9

line shows the positive direction, and a minus sign with the same letter written at the opposite end of the line shows the negative direction. Figure 7-9 shows both methods.

    To make use of the two axes, a second reference point is needed. This point is chosen arbitrarily at some position to the right of the origin on the *x* axis. This second point establishes a "*unit distance*" for the system (Figure 7-10(a)). The unit distance is used to measure off any desired number of unit distances along both axes in all four directions from the origin. A different unit distance may be established for the *y* axis, but to begin with we shall use the same unit distance for both axes. The points along both axes which represent an integral number of unit distances are usually numbered as shown in Figure 7-10(b).

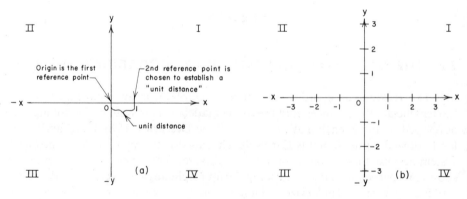

Figure 7-10

We can now designate any point on either axis by writing, for example, $y = -5$ to mean the point that is 5 units below the origin on the y axis, or $x = 4$ to mean the point that is 4 units to the right of the origin on the x axis. Also $x = 5.5$ means the point half way (in the positive direction) along the segment between points 5 and 6 on the x axis; $y = -3.2$ means the point two-tenths of the way (in the negative direction) along the segment between points -3 and -4 on the y axis.

We now make an assumption that any point in the entire plane can be identified by an ordered pair of signed numbers. An example of an ordered pair of numbers for a point is $(4, -5)$, meaning the point that is 4 units to the right of the y axis and 5 units below the x axis. Figure 7-11 shows the location of this point.

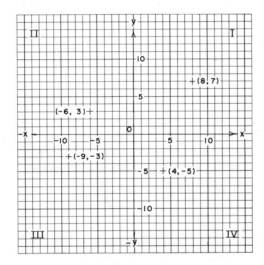

Figure 7-11

An ordered pair is always written with the x value first and the y value second. The two numbers are separated by a comma and enclosed with parentheses. The two numbers are called the *coordinates* of a point. Every distinct point has one pair of coordinate numbers. The x value (or number) is called the *abscissa* value or simply the abscissa; the y value (or number) is called the *ordinate* value or simply the ordinate. Whenever a specific point is located in the coordinate plane, the pair of numbers which designate the location of the point should be written near the point. When this is done, the point is said to be "plotted." A precise way of locating a point is to use two short intersecting lines: $+$.* See Figure 7-11.

---

* Another technique for plotting a point calls for locating a small point with a sharp pencil and drawing a small circle around the point: ⊙. See Figure 7-12.

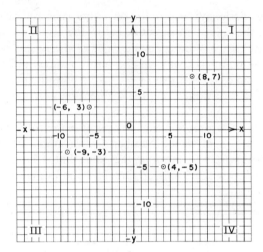

Figure 7-12

The lined side of engineering work paper, with 5 lines/in. and every fifth line darker, is very satisfactory for the construction of graphs using this coordinate system. The axes should be drawn along the darker lines. Label the origin *O* and number the integral points along both axes. Positive directions should be shown with either arrowtips or plus signs. See Figure 7-13.

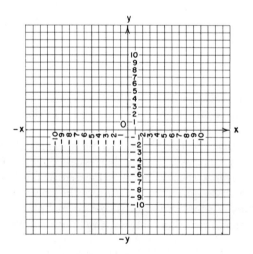

Figure 7-13

EXERCISE 7.2

1. Plot points whose coordinates are: $(0, 0)$, $(1, 1)$, $(5, 5)$, $(5, -5)$, $(-5, 5)$, $(-5, -5)$, $(10, -5)$, $(-10, 5)$, $(5, 10)$, $(-5, 10)$, $(5, -10)$, and $(-5, -10)$.
2. Plot points whose coordinates are: $(1, 1)$, $(2, 2)$, $(3, 3)$, $(4, 4)$, $(5, 5)$, $(10, 10)$, $(0, 0)$, $(-1, -1)$, $(-2, -2)$, $(-3, -3)$, $(-4, -4)$, $(-5, -5)$, and $(-10, -10)$.
3. Join all the points plotted in Problem 2 with a straight line. What can you say about the coordinates of any point on this line?
4. Could there be any point in the plane whose coordinates are equal (i.e., its abscissa value equals its ordinate value) that would not be on the line drawn in Problem 3? Explain.
5. Plot these points: $(0, 1)$, $(0, 2.5)$, $(0, 3.2)$, $(0, 5)$ $(0, -1)$, $(0, -2.9)$, $(0, -10)$, and $(0, 0)$. What do these points have in common?
6. Plot these points: $(0, 0)$, $(3, 0)$, $(5.1, 0)$, $(7, 0)$, $(0.8, 0)$, $(-2, 0)$, $(-5, 0)$, $(-7.5, 0)$, and $(-10, 0)$. What do these points have in common?
7. What are the coordinate values for each of the points shown in Figure 7-14? (Estimate values when needed.)

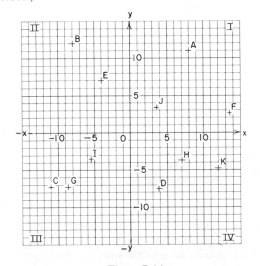

Figure 7-14

8. What are the coordinate values for each of the points shown in Figure 7-15? (Estimate when needed. Watch the scale.)
9. Plot 8 to 10 points, each having an abscissa value of 3. Then connect all plotted points with a straight line. What do all the points on this line have in common?
10. Plot 8 to 10 points, each having an ordinate value of 5. Then connect all plotted points with a straight line. What do all the points on this line have in common?
11. Plot 8 to 10 points so that $x = -8$ for each point, and connect the points with a straight line.
12. Plot 8 to 10 points so that $y = -3$ for each point, and connect the points with a straight line.

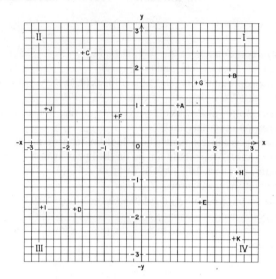

Figure 7-15

13. Locate all points whose abscissa values are 8.
14. Locate all points whose ordinate values are −7.
15. Locate all points whose ordinate values equal their abscissa values (i.e., $y = x$).
16. Plot 8 to 10 points, each having an ordinate value equal to twice its abscissa value (i.e., $y = 2x$), and connect the points with a straight line.
17. Plot 8 to 10 points, each having an ordinate value equal to one-half its abscissa value (i.e., $y = \frac{1}{2}x$), and connect the points with a straight line.
18. Plot 8 to 10 points whose ordinate values are the negative of their abscissa values (i.e., $y = -x$).
19. In what quadrants are the abscissa values positive? Negative?
20. In what quadrants are the ordinate values negative? Positive?
21. In what quadrants do the coordinates of a point have the same sign? Opposite signs?
22. In what quadrant is the abscissa negative and the ordinate positive? Abscissa positive and ordinate negative?
23. Two points are said to be symmetric to a line if the line is a perpendicular bisector of the segment joining the points. Are the points with coordinates (5, 6) and (− 5, 6) symmetric with respect to the *y* axis?
24. Are the points with coordinates (3, 7) and (3, − 7) symmetric to the *x* axis?

7.3 GRAPHING LINEAR EQUATIONS

In the last set of exercises you plotted several groups of points which fell on the same straight line. Each of these straight lines is actually the graph of a linear equation. Except for constant quantities such as $x = 4$ or $y = 3$, each linear equation involving two variable quantities expresses a direct variation between the two quantities. When a specific relationship is applied to the ordinate and abscissa values of a set of points, those points whose coordinates satisfy the specified relationship constitute the graph of the equation. For example, consider the equation $s = t$. To obtain the graph of this equation, plot several points whose ordinate values equal their abscissa values and connect these points with a straight line. In a case such as this, you would usually change the x axis to a t axis and the y axis to an s axis. The graph is shown in Figure 7-16.

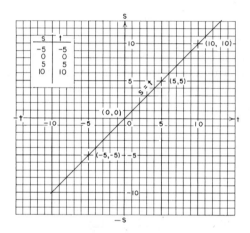

s	t
-5	-5
0	0
5	5
10	10

Figure 7-16   Graph of $s = t$

If weights are hung on a spring, the spring stretches longer and longer as each weight is added. If the same weights are transferred to a spring of a different size, they again produce a longer and longer stretch as they are added one by one. The chances are that the amount of stretch produced by the same total weight will be different for the two springs. The amount of stretch produced is dependent upon the amount of weight, while the weight is not dependent upon the stretch. The weight is called the independent variable and the stretch is called the dependent variable. When graphing relationships between two quantities, it is customary to plot the independent variable along the horizontal or x axis and the dependent variable along the vertical or y axis.

Some graphs can be drawn from an inspection of the equation. Equations such as $x = 5$ can be drawn simply by locating a few points whose abscissa values are 5 and then drawing a line through the points. Here is a basic procedure which will work for all linear equations involving two variables.

1. Solve the equation specifically for the dependent variable (usually y).

2. Choose a value for the independent variable (usually x) and substitute this value in the equation obtained in Step 1. Repeat this step several times.

3. Record the values chosen for x and the calculated values for y in a table as shown below for $y = 3x$.

x	y
-3	-9
-1	-3
0	0
2	6
4	12

Some people prefer the construction of this table in a horizontal form, but it is easier to associate the coordinates for each point when the vertical form is used. To illustrate this fact, we have added parentheses to the ordered pair of numbers forming the coordinates of the points to be plotted.

$$y = 3x$$

x	y
$(-3$	$-9)$
$(-1$	$-3)$
$(0$	$0)$
$(2$	$6)$
$(4$	$12)$

If you think of the line separating the coordinates as playing the same role as a comma, you then have the coordinates of each point to be plotted.

After determining the coordinates of several points, you are ready to construct the coordinate system. Experience will dictate how many points to plot and the scale for each axis. Use the same scale for both axes and let 1 small space (on the 5 lines/in. engineering paper) represent 1 unit. This procedure is not mandatory but it is practical.

Plot the points whose coordinates you have obtained and connect the points with a line. If the equation is linear, the points will fall on a straight line.

Figure 7-17 shows a graph of the equation $y = 3x$.

EXERCISE 7.3

Graph each of the following equations on a separate coordinate system. Plot at least 6 points for each.

1. $x = -4$
2. $y = 5$

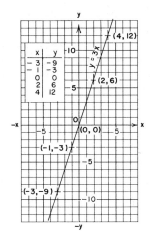

Figure 7-17 Graph of
$y = 3x$

3. $y = x$
4. $y = -x$ (Watch the "understood" coefficient.)
5. $y = 2x$
6. $y = \frac{1}{2}x$
7. $2y = 8x$
8. $y = x + 2$ (Each ordinate value will be 2 more than the abscissa value.)
9. $y = x - 4$
10. At what point does the line in Problem 8 cross the y axis? At what point does the line in Problem 9 cross the y axis?
11. Graph $3y + 6x = 12$
12. Graph $4x - 2y = 8$
13. Graph $y = x + 2$ and $y = x - 2$ on the same coordinate axes. What can be said about the lines?
14. Graph the lines $y = x + 2$ and $y = -x + 2$ on the same coordinate axes. What can be said about the lines?
15. Examine the graphs for Problems 8, 9, 13, and 14 to see what connection there might be between the constant term in each equation and the ordinate value at the point where each line crosses the y axis.
16. Does this relationship (Problem 15) hold for the constants in the original equations in Problems 11 and 12? Solve the equations in Problems 11 and 12 for y and then compare the constant terms with the ordinate values at the y-intercept points.
17. Graph $s = 2t$. Change axes to t and s so that coordinates are (t, s).
18. Graph $h = 4r - 5$. Use r and h axes.
19. The weight of a spring causes it to stretch 4 in. when it is suspended vertically. Then 5-lb weights are hung on the spring so that the stretch S (in inches) is given by $S = \frac{4}{5}W + 4$,

where W = weight in pounds. Plot the graph showing the stretch covering weights from 0 to 40 lb.

20. By extending the line in Problem 19, what would be the expected stretch when 50 lb is suspended from the spring? (Extending a graph to obtain data not previously measured is called *extrapolation*.) How far could you extend the line and extrapolate (with any degree of accuracy) what the stretch would be?

7.4 SLOPE AND THE PROPORTIONALITY CONSTANT

Figure 7-18 shows the graphs of several equations, each expressing that y varies directly as x but each having a different proportionality constant. Note that as the proportionality constant increases, the line becomes steeper (i.e., slopes

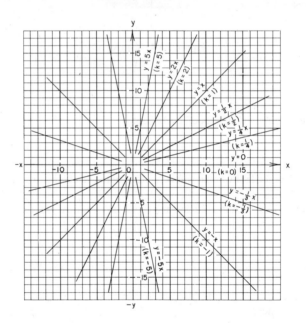

Figure 7-18 Graph Showing Relationship Between Proportionality Constant and Slope of the Line

upward). Also note that the smaller the constant becomes (increasing in the negative direction), the steeper the line slopes downward. To describe the steepness of a line mathematically, we must define the term *slope*. A simple and correct way of defining the slope of the line representing a linear equation is to say that the slope of the line is the constant of proportionality. However, in more advanced work you may have to talk about the slopes of lines which are not straight. So we shall start with a definition that will serve both our present and future needs.

Definition of Slope

The slope of a straight line is defined as the change in the ordinate (between any two points) divided by the corresponding change in the abscissa.

By examining the graph of $y = 2x$ in Figure 7-18, you see that for each unit change in x, y changes 2 units. The Greek letter delta (Δ) is commonly used to represent a change in a quantity. Therefore, Δx means a change in x and Δy means a change in y. The ratio $\Delta y/\Delta x$ is the algebraic definition of the slope of a straight line; Δy must be the change that corresponds to Δx. The letter m is commonly used to stand for slope:

$$m = \frac{\Delta y}{\Delta x} = \text{slope}$$

When a linear equation is written in the form of $y = kx + b$, the proportionality constant k is the slope of the line. Because $k = m$, the form $y = mx + b$ is called the *slope-intercept* form for a linear equation. The constant b is always the ordinate value for the point where the graph crosses the y axis; hence the name slope-intercept form.

example   Write the equation $3y - 6x = 12$ in the slope-intercept form, specify the slope, and give the coordinates of the y-intercept point. (Consider y the dependent variable.) To obtain the slope-intercept form of a linear equation, solve the equation explicitly for the dependent variable:

$$3y - 6x = 12 \tag{1}$$
$$+6x, \qquad 3y = 6x + 12 \tag{2}$$
$$\div 3, \qquad y = 2x + 4 \tag{3}$$

The slope $= 2$ and the y-intercept point is $(0, 4)$.

example   Write the given equation in the slope-intercept form, specify the slope, and give the coordinates of the intercept point. (Consider H the dependent variable.)

$$6L - 2H = 24 \tag{1}$$
$$-6L, \qquad -2H = -6L + 24 \tag{2}$$
$$\div (-2), \qquad H = 3L - 12 \tag{3}$$

The slope $= 3$ and the H-intercept point is $(0, -12)$.

To see how useful the intercept form is, use it to plot the graph for the equation $4y = 2x - 8$ (see Figure 7-19). Solving for y, you have $y = \frac{1}{2}x - 2$. The required line has a slope of $1/2$ and intersects the y axis at the point $(0, -2)$. Locate the point $(0, -2)$; with $m = \frac{1}{2} = \Delta y/\Delta x$, you know that the change in y will be

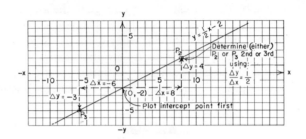

Figure 7-19

just half of any change in x. So if you move to the right (in the $+x$ direction) say 8 units, then you must move up 4 units (in the $+y$ direction); if $\Delta x = 8$, then $\Delta y = 4$. This process will locate a second point on the line. As a check, repeat this procedure but this time change x in the negative direction. For example, move to the left from the intercept point 6 units ($\Delta x = -6$) and then move downward 3 units ($\Delta y = -3$). This locates a third point on the line.

Figure 7-20 shows the relationship $E = 2W$ that exists between the elongation of a spring and the weight applied to the spring. Note that the slope has units. The slope $= 2$ in./lb, which is exactly the proportionality constant for this relationship.

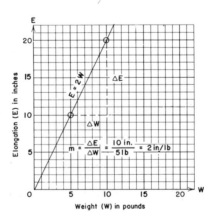

Figure 7-20   Elongation of a Spring as a Function of Weight

Slope of a Line Through Two Known Points

When a line passes through two points whose coordinates are known, it is possible to determine the slope of the line. According to the definition of slope ($\Delta y / \Delta x$), you only need to determine the change in y between the two known points and divide this by the change in x between these same two points. This procedure can be stated algebraically as

$$\text{slope } m = \frac{\Delta y}{\Delta x} = \frac{y_2 - y_1}{x_2 - x_1}$$

where (x_1, y_1) and (x_2, y_2) are the coordinates of any two points on the line.

example   Determine the slope of the line passing through the points $P_1(1, 2)$ and P_2 $(7, 8)$.

$$\text{slope} = \frac{\Delta y}{\Delta x} = \frac{y_2 - y_1}{x_2 - x_1}$$

$$\text{slope} = \frac{8 - 2}{7 - 1} = \frac{6}{6} = 1$$

EXERCISE 7.4

In Problems 1–10 duplicate and complete the following table.

	Given Equation	Slope-Intercept Form	Slope	y-Intercept point
0.	$3y + x = 12$	$y = -\frac{1}{3}x + 4$	$-\frac{1}{3}$	$(0, 4)$
1.	$2y - 4x = 10$			
2.	$x - y = 5$			
3.	$y - 3x = 9$			
4.	$x - 2y + 4 = 0$			
5.	$4x - 5y = 20$			
6.	$6 = 3x - 2y$			
7.	$y = 5x - 8$			
8.	$x = 3y + 6$			
9.	$2x + 2y = 2$			
10.	$3x - 3y = 3$			

In Problems 11–20, write an equation in the slope-intercept form, using y for the dependent variable and x for the independent variable, having the given slope (or proportionality constant) and y-intercept point.

11. $m = 3, (0, -3)$

12. $m = \frac{1}{2}, (0, 8)$

13. $k = m = \frac{3}{4}, (0, 2)$

14. $k = m = \frac{4}{5}, (0, 5)$

15. $k = \dfrac{-7}{8}, (0, -1)$

16. $k = \frac{2}{3}, (0, 6)$

17. $m = 2, (0, 0)$

18. $m = 5, (0, -8)$

19. $\dfrac{\Delta y}{\Delta x} = \dfrac{3}{4}, (0, 2)$

20. $\dfrac{\Delta y}{\Delta x} = 3, (0, -2)$

21. What would be the slope of a horizontal line?

22. What would be the slope of a vertical line?

In Problems 23–30, graph each of the given functions using the slope-intercept method. Determine three points as in Figure 7–19. Use a separate coordinate system for each problem.

23. $y = 4x - 5$

24. $y = 2x + 3$

25. $y = \dfrac{-2}{3}x - 2$

26. $y = \dfrac{-4}{5}x + 3$

27. $2y = 6x - 8$

28. $3y = -x + 6$

29. $2x + y = 12$

30. $3x - 2y = 4$

In Problems 31–36, determine the slope of the line passing through the two given points.

31. $(3, 4), (7, 16)$

32. $(0, 0), (5, 2)$

33. $(-3, 4), (5, -1)$

34. $(2, 0), (0, 4)$

35. $(0, -3), (6, 0)$

36. $(-1, -1), (3, -3)$

In Problems 37–40, plot the two given points, draw a straight line through them, determine the slope, estimate the value of the y intercept, and write the equation of the line in the slope-intercept form.

37. $P_1(-6, 4), P_2(12, -2)$

38. $P_1(-9, -7), P_2(9, 1)$

39. $P_1(-10, -1), P_2(5, 2)$

40. $P_1(-4, -4), P_2(2, 8)$

7.5 GRAPHIC SOLUTIONS OF LINEAR EQUATIONS

In Section 4.4, we briefly discussed the relationship between the algebraic solution and graphic solution of simultaneous equations. The graphic method of solving a system of simultaneous equations consists of graphing the equations using the same coordinate axes and then determining by inspection the coordinate values of the point (or points) of intersection. Consider the following examples.

example   Determine graphically the solution of the following system of equations:

$$\begin{cases} y = 2x - 3 \\ 2y = -x + 19 \end{cases}$$

Figure 7-21 shows the graph of these two equations and their point of

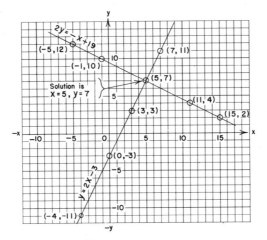

Figure 7-21   Graphic Solution of:

$$y = 2x - 3$$
$$2y = -x + 19$$

intersection; the values $x = 5$ and $y = 7$ constitute the solution. It is good practice to check the graphic solution by substituting the values for x and y in both equations.

If the graphs of two equations do not intersect at an integral point (a point whose coordinates are integers), then you must approximate the coordinate values. In this case, the algebraic check is approximate also.

example   Determine a graphic solution for the following system of equations:

$$\begin{cases} 3y = 4x - 19 \\ y = -x + 2 \end{cases}$$

Figure 7-22(a) shows the graphic solution. Figure 7-22(b) shows a graph of a small part of (a). You could repeat this (enlarging) process over and over until you obtained the desired accuracy. When checking the values of $x = 3.57$ and $y = -1.57$ in the equations, you obtain

and

$$3(-1.57) = 4(3.57) - 19$$
$$-4.71 \approx -4.72 \checkmark$$

$$-1.57 = -(3.57) + 2$$
$$-1.57 = -1.57 \checkmark$$

You cannot obtain an exact check using approximate values.

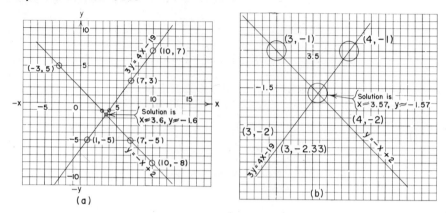

Figure 7-22   Graphic Solution of:

$$3y = 4x - 19$$
$$y = -x + 2$$

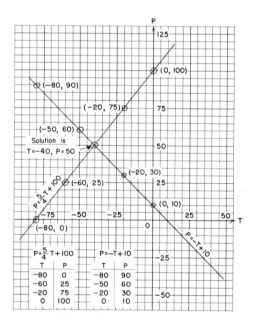

Figure 7-23   Graphic Solution of:

$$P = \tfrac{5}{4}T + 100$$
$$P = -T + 10$$

Even though you obtain approximate solutions by graphic methods, they are nevertheless valuable. They may be used as a check on the reasonableness of a calculated result, or they may indicate the existence or nonexistence of a practical solution. If in the last example you wanted to determine whether or not you could obtain a solution with the x and y quantities both positive, the graph shows that no such solution exists.

example Determine the graphic solution for the following system of equations:

$$P = \tfrac{5}{4}T + 100$$

$$P = -T + 10$$

Figure 7-23 shows the graphic solution for this system.

example Two ships are traveling the same course at 12:00 noon. Ship A is 100 miles ahead of ship B. A is traveling 20 knots and B is traveling 25 knots. By graphic means determine the time when ship B will overtake ship A. First, you need to write the relationships:

$$\text{distance for (ship } A) = \text{(velocity) (time)}$$
$$\text{distance for (ship } B) = \text{(velocity) (time)}$$
$$D_A = (20 \text{ knots}) (T) + 100 \text{ miles}$$
$$D_B = (25 \text{ knots}) (T), \quad (T = \text{time in hours})$$

Figure 7-24 shows the graphic solution for these two equations; 12:00 noon is $T = 0$ and ship A has a 100-mile lead.

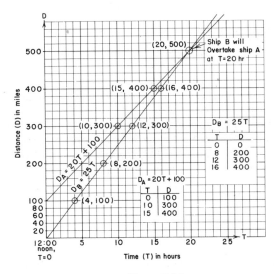

Figure 7-24

EXERCISE 7.5

Determine a graphic solution for each system of equations in Problems 1–8. Check by determining the algebraic solution and give reasons for any differences.

1. $\begin{cases} 3y = 2x \\ 3y + x = 9 \end{cases}$

2. $\begin{cases} y + x = 2 \\ y + 4 = 2x \end{cases}$

3. $\begin{cases} 2y - 10 = 2x \\ y = -x \end{cases}$

4. $\begin{cases} y = x - 6 \\ x = -y \end{cases}$

5. $\begin{cases} 5y = 2x + 5 \\ y = 2x - 7 \end{cases}$

6. $\begin{cases} R = \dfrac{25}{11}W + 60 \\ \\ R = \dfrac{100}{11}W - 90 \end{cases}$ (*Hint:* Use integral multiples of 11 for *W*.)

7. $\begin{cases} L = 30r - 400 \\ L = -100r + 900 \end{cases}$

8. $\begin{cases} 8y + 3x = 24 \\ 20y + 7.5x = 15 \end{cases}$

In Problems 9–11, formulate the necessary equations and solve graphically.

9. The sum of two numbers is 200 and their difference is 40. (*Hint:* Let $y =$ the larger number.)

10. Two numbers are so related that the larger number equals twice the smaller number, and three times the smaller number plus 15 equals four times the larger number. (*Hint:* Let $y =$ the larger number.)

11. What length of No. 30 gage aluminum wire having a resistance of 0.169 ohm/ft will have the same resistance as 240 ft of No. 29 gage wire whose resistance is 0.134 ohm/ft? (*Hint:* Plot the lines $R = 0.169L$ and $R = 0.134L$ where $R =$ resistance in ohms and $L =$ length in feet. The two lines do not intersect except at the origin. Nevertheless, you can use the graph to answer this question.)

12. The relationship between Fahrenheit and Centigrade temperature readings is given by the formula

$$F = \tfrac{9}{5}C + 32°$$

where $F =$ degrees Fahrenheit and $C =$ degrees Centigrade. At a certain temperature the two readings are the same. Plot the lines $F = \tfrac{9}{5}C + 32°$ and $F = C$ to determine this temperature.

7.6   GRAPHING QUADRATIC EQUATIONS

The procedure used to graph quadratic relationships is essentially the same as for linear relationships. Values for the dependent variable are determined by substituting values for the independent variable in the equation. Several points must be plotted to obtain a reasonable picture of the curve.

In contrast to linear equations producing straight-line graphs, quadratic equations of the form $y = ax^2 + bx + c$ turn out to be parabolas. Some quadratic relationships produce graphs which are circles, ellipses, hyberbolas, spirals, and numerous others of such variety that it is impossible to give each one a name. We are restricting our discussion to the graphs of the general quadratic equation because: (1) it is extremely important in technical work, and (2) a discussion of graphs of quadratics in general would take us far beyond the scope of this text.

Study Figure 7-25 which shows the graph of the equation $y = x^2 - 9$. Figure 7-26 shows the effect of adding a constant to the right member of the equation $y = x^2$. Note that the addition of a constant only changes the position of the parabola and not its shape.

EXERCISE 7.6A

Plot the graphs for the following equations.

1. $y = 2x^2$
2. $y = \frac{1}{2}x^2$
3. $y = x^2 - x - 6$
4. $y = x^2 + x - 12$
5. $y = x^2 - 25$
6. $y = x^2 - 36$
7. $y = -x^2$ (Graph will be concave down, i.e., open downward.)
8. $y = -x^2 + 16$
9. $2y = 4x^2 - 32$
10. $4y = 2x^2 - 8$
11. $H = S^2$
12. $R = -W^2$
13. $L = D^2 - 9$
14. $Z = T^2 + T - 20$

Roots of the Quadratic Function

When a quadratic equation is solved explicitly for the dependent variable, such as (1) $y = 3x^2 + 7x - 4$ or (2) $T = N^2 - 3N - 1$, the member containing the independent variable is called the quadratic member of the equation. You would say that $3x^2 + 7x - 4$ is the quadratic member in (1) and that $N^2 - 3N - 1$

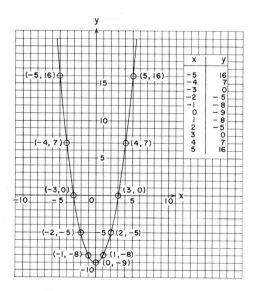

Figure 7-25   Graph of $y = x^2 - 9$

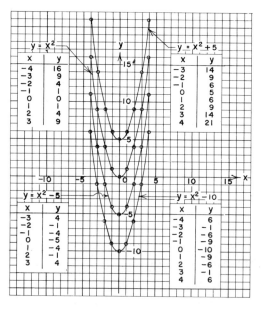

Figure 7-26   Graph Showing Effect of Adding a
Constant to Right Member of $y = x^2$

is the quadratic member in (2). The terms $3x^2$ and N^2 are responsible for the equations being quadratic; without these terms the two equations become linear. It is common practice to describe equations (1) and (2) by saying that y equals a quadratic function of x and T equals a quadratic function of N.

There are numerous applications of mathematical relationships expressed by the general quadratic function in all technical fields. It is important, therefore, to be well acquainted with some of the characteristics of this function.

1. The graph of the general quadratic equation $y = ax^2 + bx + c$ is a parabola.

2. Parabolas which open down have a high point and those which open up have a low point. The high or low point is called the vertex of the parabola.

3. A vertical* line drawn through the vertex divides the parabola into two equal symmetrical branches. Such a line is called the axis of symmetry.

4. The roots of a quadratic function are the values of the independent variable that give the function a zero value. They are also the abscissa values of the points where the graph of the function crosses the x axis.

These characteristics are illustrated in Figure 7-27.

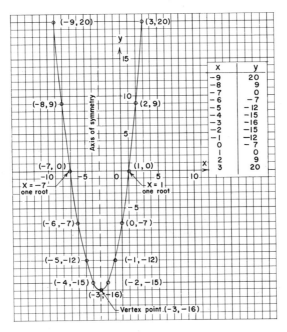

Figure 7-27   Graph of $y = x^2 + 6x - 7$ (*Note*: $x^2 + 6x - 7 =$
$(x+7)(x-1) = 0$, having roots of $x = -7$, $x = 1$)

* For parabolas opening up or down. For those opening right or left, the line must be horizontal.

EXERCISE 7.6B

Determine the roots of the following quadratic functions by graphing and check by either factoring or using the quadratic formula. Sometimes only approximate root values can be obtained by graphing. Care in plotting a sufficient number of points and in drawing the curves allows a fairly close approximation of the roots.

1. $y = x^2 + 3x - 10$
2. $y = -x^2 - x + 6$
3. $y = 2x^2 + x - 3$
4. $y = 2x^2 - x - 10$
5. $S = -2t^2 + 5t + 12$

6. $R = 3w^2 + 5w - 12$
7. $Z = R^2 - 10$
8. $V = -Z^2 + 8$
9. $W = V^2 - 4V - 2$
10. $T = -P^2 + P + 9$

Practical Significance of the Vertex

To understand the importance of the vertex of a parabola, consider the following application.

A projectile is shot vertically upwards with an initial velocity of 320 ft/sec. Its height above the ground as a function of the time t is given by $h = -16t^2 + 320t$. What is the maximum height obtained by the projectile, and what time is required to reach this maximum height? Figure 7-28 shows the graph of this function.

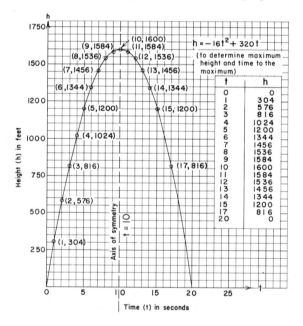

Figure 7-28   Graph of $h = -16t^2 + 320t$ (to determine maximum height and time to the maximum)

The graph shows that the maximum height of 1600 ft is reached 10 sec after the projectile is fired into the air. Note that the axis of symmetry, whose equation is $t = 10$, passes halfway between the two root values 0 and 20. In fact, $t = 10$ is the average of the two roots; $(0 + 20)/2 = 10$. This gives a clue how to determine the abscissa value of the vertex without graphing the function. All you need to do is determine the average of the two roots. The function $-16t^2 + 320t$ could be factored as $(-16t)(t - 20)$, giving the roots $t = 0$ and $t = 20$. Their average gives $t = 10$ for the equation of the axis of symmetry. This tells you that the vertex has an abscissa value of $t = 10$ sec; hence the maximum height occurs at this time. To solve for the maximum height, substitute $t = 10$ in the equation $h = -16t^2 + 320t$; this gives $h_{max} = 1600$ ft.

If you use the quadratic formula to determine the roots, how would you get the sum of the roots? Remember that the general quadratic equation

$$y = ax^2 + bx + c$$

has roots of

$$x_1 = \frac{-b + \sqrt{b^2 - 4ac}}{2a} \qquad x_2 = \frac{-b - \sqrt{b^2 - 4ac}}{2a}$$

If you add these two roots together, you get

$$x_1 + x_2 = \frac{-2b}{2a} = \frac{-b}{a}$$

(The radical parts add up to zero.) To get the average, you just divide by 2:

$$\frac{x_1 + x_2}{2} = \frac{-b}{2a}$$

It looks too simple, but it works; $-b/2a$ is the abscissa value of the vertex of the parabola. Just to convince yourself, go back to the last example. In $h = -16t^2 + 320t$, with $a = -16$ and $b = 320$:

$$\frac{-b}{2a} = \frac{-320}{2(-16)} = \frac{320}{32} = 10$$

Sure enough, $t = 10$ sec is the time to maximum height.

Let us try another. Figure 7-29 shows a sketch (not a plot*) of the function $h = -16t^2 + 96t + 12$. Using half-the-sum-of-the-roots method to determine

---

* A plot consists of locating several specific points, whereas a sketch only serves to indicate a rough approximation of the graph.

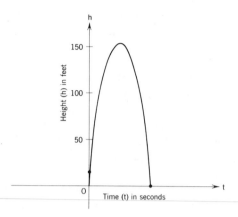

Figure 7-29   Sketch of $h = -16t^2 + 96t + 12$

the time to maximum:

$$t_{max} = \frac{-b}{2a} = \frac{-96}{2(-16)} = 3 \text{ sec}$$

$$\text{Maximum height} = -16(3)^2 + 96(3) + 12$$

$$h_{max} = 156 \text{ ft}$$

When a quadratic equation is solved explicitly for the dependent variable, its graph opens up when the sign on the squared term (independent variable squared) is positive and it opens down when this sign is negative. This information, along with the half-the-sum-of-the-roots to determine the vertex, indicates a simple but useful means of determining maximum or minimum values for many quadratic relationships.

example   A rectangular plate must have the maximum obtainable area with a fixed perimeter of 80 cm. Find the dimensions of this rectangle. If you designate the width of the rectangle by W cm and the length as $(40 - W)$ cm (Figure 7-30), then the area A is $A = (\text{length})(\text{width}) = (40 - W)(W)$ sq cm, or

$$A = (-W^2 + 40W) \text{ sq cm}$$

Using half the sum of the root, A_{max} occurs when

$$W = \frac{-b}{2a} = \frac{-40}{-2} = 20 \text{ cm}$$

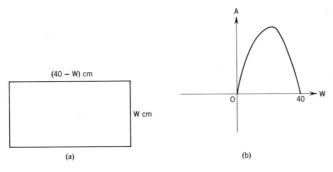

Figure 7-30 Sketch of $A = -W^2 + 40W$

The rectangle with maximum area (and $P = 80$ cm) is a square measuring 20 cm on each edge.

example A sheet metal company has a large surplus of metal strips, 16 in. wide, suitable for making troughs. It is desired to form a trough with a rectangular cross section of maximum area. What should be the dimensions of the

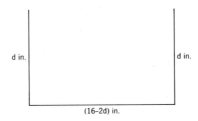

Figure 7-31

trough? If you designate the depth of the trough by d in. then the width of the trough will be $(16 - 2d)$ in. (See Figure 7-31.) Then:

$$\text{area} = (16 - 2d)(d) \text{ sq in.} = (-2d^2 + 16d) \text{ sq in.}$$

A_{max} will occur when

$$d = \frac{-b}{2a} = \frac{-16}{-4} = 4$$

Dimensions for maximum area are depth $= 4$ in. and width $= 8$ in.

EXERCISE 7.6C

In Problems 1–8, state whether the graph of the function will open up or down. Determine the coordinates of the vertex and plot this point. Determine the roots of the quadratic function by factoring, and plot the points where the parabola crosses the horizontal axis. Make a simple sketch through the three plotted points.

1. $y = x^2 - 25$ (Hint: $b = 0$)
2. $y = 36 - x^2$
3. $y = x^2 + x - 20$
4. $y = x^2 - x - 12$

5. $y = -6x^2 + 5x + 6$
6. $y = 2x^2 + 7x - 30$
7. $-y = -x^2 + 2x + 8$
8. $y + x^2 + 3x - 10 = 0$

In Problems 9–14, plot a graph of the given function and determine the coordinates of its maximum or minimum point (vertex point). State whether the point determined is a maximum or minimum point.

9. $h = -16t^2 + 160t + 16$
10. $h = -16t^2 + 64t + 8$
11. $4y = -x^2 + 10x - 1$

12. $6y = x^2 + 6x - 12$
13. $R = \frac{1}{2}S^2 + 2S - 2$
14. $T = 2P^2 - 4P + 1$

In Problems 15–20, develop a quadratic function from the given data; determine its maximum or minimum value as specified. Show a sketch of the function.

15. A 20-in. wide strip of metal is to be bent to form a trough (open top) with a rectangular cross section of maximum area. What should be the dimensions of the trough?
16. A farmer has $80\bar{0}$ rods of fencing to fence a rectangular pasture adjoining a river. If the river is to serve as one boundary, what dimensions will give the maximum area?
17. An object is thrown straight up from an initial height of 6.00 ft and has an initial velocity of 192 ft/sec. Its height h (in feet) is given by the function $h = -16t^2 + 192t + 6$ ($t =$ time in seconds). What time is required for the object to reach its maximum height? What is the maximum height? How long is the object in the air? (Hint: When the object hits the ground, $h = 0$.) Make a sketch.
18. Work Problem 17 using 224 ft/sec for the initial velocity and 12.0 ft for the initial height.
19. Find two integers whose sum is 60 and whose product is a maximum.
20. Determine two integers whose sum is 20 and whose sum of their squares is a minimum.

7.7 GRAPHIC INTERPRETATION OF DATA

It is frequently necessary to determine an exact mathematical expression which shows how one quantity varies with another while all other quantities are held constant. Graphic interpretation is one of the most useful methods of analyzing quantitative data.

We shall consider a few common relationships between physical quantities. Many of these important relationships fall into one of six categories:

Category	Example
Directly proportional*	$y = kx$
Inversely proportional	$y = \dfrac{k}{x}$
Directly proportional to the square	$y = kx^2$
Inversely proportional to the square	$y = \dfrac{k}{x^2}$
Directly proportional to the square root	$y = k\sqrt{x}$
Inversely proportional to the square root	$y = \dfrac{k}{\sqrt{x}}$

Experimental data contain errors which cause some scattering of points when plotted. By scattering, we mean that the points may not fall exactly on the curve which represents the mathematical relationship between the quantities. Therefore, to grasp the basic relationship between the variables, we shall only use "idealized" data (free from error). This procedure is justifiable; in actual practice we assume that if experimental data were free of all error, the plotted points would fall on the curve of some mathematical relationship.

Directly Proportional

This relationship is the simplest to discover, because the graph of two quantities that are directly proportional is a straight line. The slope of the line is the proportionality constant; the intercept (on the dependent variable axis) is the initial value of the dependent variable corresponding to an initial value of zero for the independent variable.

example Plot the following data and determine the mathematical statement of proportionality between the variable quantities. (Treat x as the independent variable.)

x	y
-5	-2
0	3
3	6
5	8

Figure 7-32 shows these points plotted and joined with a straight line.

---

* The term "directly proportional" implies a linear relationship between two quantities. Note that when the direct proportionality is nonlinear, it is so specified.

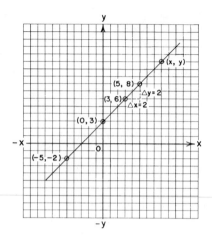

Figure 7-32   Direct Proportionality

You can write $y = mx + b$ for the equation of this line and proceed to determine the slope m (the proportionality constant) and the intercept value b. Using the relationship $m = \Delta y/\Delta x$, determine the corresponding changes in y and x between any two specific points. The graph shows these changes between the points $(3, 6)$ and $(5, 8)$. Therefore,

$$m = \frac{\Delta y}{\Delta x} = \frac{2}{2} = 1$$

From the intercept point $(0, 3)$, you know that $b = 3$. Therefore, write $y = 1x + 3$ or simply $y = x + 3$ for the mathematical statement of variation for the given data.

A second method of determining the equation in the above example follows:

1. Determine the slope, using the coordinates of the two known points.
2. Locate a point on the line and call its coordinates (x, y).
3. Determine the change in y (i.e., Δy) and the change in x (i.e., Δx) from any other point on the line to this point.
4. Equate $\Delta y/\Delta x$ in Step 3 to the slope in Step 1.
5. Clear fractions and solve for y.

Applying this second method for the data in the example, you have:

1. Slope $= 1$ (as before).
2. Point (x, y) located on line (Figure 7-32).
3. Using this point (x, y) and the point $(5, 8)$, $\Delta y = y - 8$ and $\Delta x = x - 5$.

4. $\dfrac{\Delta y}{\Delta x} = \dfrac{y - 8}{x - 5} = 1$

5. $(x - 5)$, $y - 8 = x - 5$
     $+8$, $y = x + 3$ (the required equation)

Inversely Proportional*

example Determine the mathematical statement of variation between N_1 (dependent variable) and N_2 for the following data.

N_2	N_1
1	48
2	24
3	16
4	12
6	8
8	6
12	4
16	3
24	2
48	1

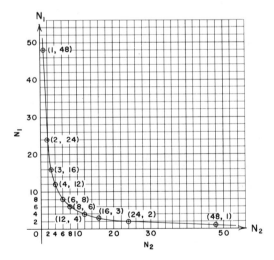

Figure 7-33 Inverse Proportionality

* The words "inversely proportional" imply that two quantities are both first degree variables; otherwise the statement will read "inversely proportional to the square—to the square root—or to the cube," etc.

Because the graph for the data (Figure 7-33) is a curve rather than a straight line, rule out the possibility of a direct linear variation. The next question is: "Is it an inverse proportionality?" The simplest check or clue is that the product of inversely related quantities must be a constant. Note that if you multiply N_2 by the corresponding value for N_1, you always obtain 48 for the product. From this you know that $N_1 N_2 = k$ or $N_1 = k/N_2$. In this case, $k = 48$ and $N_1 = 48/N_2$ is the required statement.

A second method of determining the inverse relationship between two quantities is to plot one quantity against the reciprocal of the other. If a straight

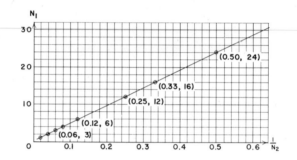

Figure 7-34   Plot of N_1 Against N_2^{-1}

line results, the two quantities vary inversely with each other. Figure 7-34 shows the graph for N_1 plotted against the reciprocal of N_2.

N_2	N_2^{-1}	N_1
1	1	48
2	0.50	24
3	0.33	16
4	0.25	12
6	0.17	8
8	0.12	6
12	0.08	4
16	0.06	3
24	0.04	2
48	0.02	1

When you use this second method to determine the inverse proportionality, you still need to determine the proportionality constant for $N_1 = k/N_2$. To determine k, substitute any two corresponding values for N_1 and N_2. Suppose you use $N_1 = 3$ when $N_2 = 16$. Then $3 = k/16$ and $k = (3)(16) = 48$. Therefore, $N_1 = 48/N_2$ as before. Or you can determine the slope $m = 48$ for the line in Figure 7-34.

Directly Proportional to the Square

example   Determine the mathematical relationship between the lateral surface area A and the edge E of a cube for the following data. (Treat E as the independent variable.)

E (in.)	A (in.^{2})
0	0
1	6
2	24
3	54
4	96
5	150

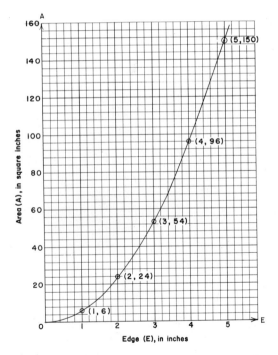

Figure 7-35   Plot of Lateral Surface area A of a Cube Against its Edge E

Figure 7-35 shows the area A plotted against the edge E. According to the graph, the relationship is neither a direct linear nor an inverse variation. The next question is: "Does A vary as the square of E?" You can answer this question in either of two ways, but first add E^2 to the data table.

E (in.)	E^2 (in.^{2})	A (in.^{2})
0	0	0
1	1	6
2	4	24
3	9	54
4	16	96
5	25	150

method I If the area A varies directly with the square of E, then $A = kE^2$ and $A/E^2 = k$; that is, the area A divided by the square of the edge must be constant. By dividing each A value by the corresponding entry for E^2, you always obtain $k = 6$. Therefore, $A = 6E^2$ is the required statement of variation.

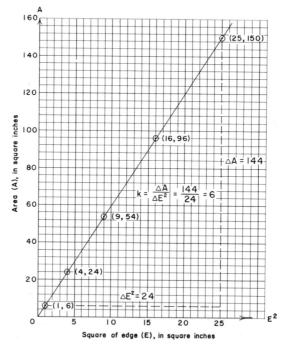

Figure 7-36 Plot of Lateral Surface Area A of a Cube Against the Square of its Edge

method II Plot *A* against E^2. See Figure 7-36. If a straight line results, $A = kE^2$ where k = slope of the line.

Inversely Proportional to the Square

example Determine the mathematical relationship between the resistance *R* and the diameter *d* of annealed copper wire. Use the following data. (Treat *d* as the independent variable.)

d (mm)	*R* (ohms/km)
0.912	26.4
1.15	16.6
1.29	13.2
1.45	10.5
1.63	8.30
1.83	6.60
2.05	5.24
2.31	4.13
2.59	3.28
2.91	2.60
3.26	2.07
4.12	1.30

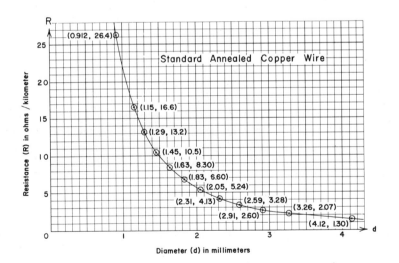

Figure 7-37 Plot of Resistance *R* Against Diameter *d* of Standard Annealed Copper Wire

Figure 7-37 shows the resistance R plotted against the diameter d. The graph shows that the relationship is nonlinear. Also, $(R)(d) \neq k$; therefore the relationship is not an inverse variation of d. However, R does decrease as d increases and therefore indicates the possibility of R being inversely proportional to the square of d, $(R = k/d^2)$. If this is the case, then plotting R against $1/d^2$ will produce a straight line whose slope is the proportionality constant k. Figure 7-38 shows R plotted against the reciprocal of d^2. When

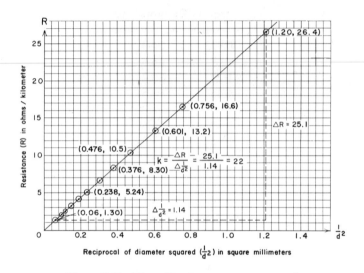

Figure 7-38   Plot of Resistance R Against $1/d^2$

the slope is to be determined from calculations based on measured data, less error is introduced if the coordinates of the first and last points *which fall on the line* are used rather than the coordinates of two points which are close together. Remember that in actual practice there will be some scattering of points due to error in the measured data. Thus, any curve or line drawn through the plotted points will be an approximation for the mathematical relationship between the variable quantities. Figure 7-38 shows the slope $k = 22$. Therefore, the statement of variation for the given data is $R = 22/d^2$ ohms.

Directly Proportional to the Square Root

example   Determine the mathematical relationship between the time t and the distance s for a freely falling body starting from rest. Use the following data. (Treat s as the independent variable.)

s (ft)	t (sec)
0	0
1	0.25
4	0.50
9	0.75
16	1.00
25	1.25
36	1.50
49	1.75
64	2.00

The data plotted in Figure 7-39 show that the relationship between *t* and *s* is nonlinear. Since *t* increases as *s* increases, rule out an inverse variation.

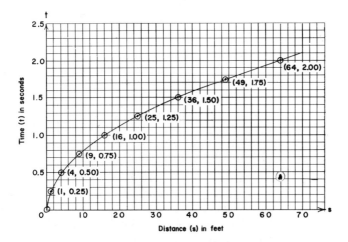

Figure 7-39   Plot of Time Against Distance for a Freely Falling Body

Recalling from Figure 7-35 that the curve showing the direct proportionality to the square of a variable opens up, also rule out this variation. The next question is: "Is *t* directly proportional to the square root of *s*, i.e., $t = k\sqrt{s}$?" If this is the case, then *t* plotted against the square root of *s* will produce a straight line and the slope of the line will be the proportionality constant *k*. Figure 7-40 shows the graph for *t* plotted against $\sqrt{s}$.

s (ft)	$\sqrt{s}$	t (sec)
0	0	0
1	1	0.25
4	2	0.50
9	3	0.75
16	4	1.00
25	5	1.25
36	6	1.50
49	7	1.75
64	8	2.00

The graph confirms a yes answer to the question; $t = \frac{1}{4}\sqrt{s}$ for the given data.

Inversely Proportional to the Square Root

example   Determine the mathematical relationship between the time t and the acceleration a for an object (initially at rest) which is to be uniformly accelerated a total distance of 1250 ft. Use the following data. (Treat a as the independent variable.)

a (ft/sec^2)	t (sec)
1	50.0
4	25.0
9	16.7
12	14.4
16	12.5
20	11.2
25	10.0
36	8.33
49	7.14
64	6.25

Plotting the graph for this data shows an inverse proportionality, Figure 7-41. Therefore, we must consider the three possibilities:

$$t \stackrel{?}{=} \frac{k}{a} \qquad t \stackrel{?}{=} \frac{k}{a^2} \qquad t \stackrel{?}{=} \frac{k}{\sqrt{a}}$$

If the first relationship exists, then plotting t against $1/a$ will yield a straight line whose slope $= k$. If the second relationship exists, then plotting t

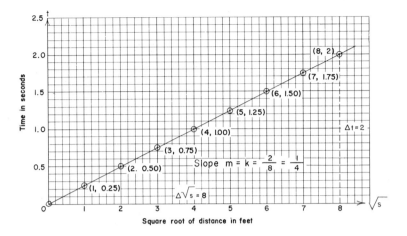

Figure 7-40   Plot of Time Against Square Root of Distance for a Freely
Falling Body

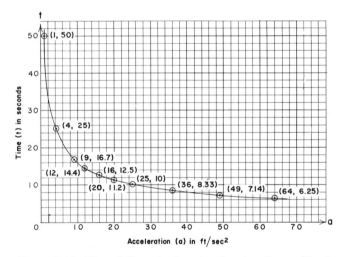

Figure 7-41   Plot of Time Against Acceleration Over a Fixed
Course Distance of 1250 ft

against $1/a^2$ will yield a straight line whose slope $= k$. If the third relationship exists, plotting t against $1/\sqrt{a}$ will yield a straight line whose slope equals the proportionality constant k. Only the function $t = 1/\sqrt{a}$ gives a straight line (Figure 7-42) which answers the question. With the slope $m = k = 50$, the required statement for the given data is $t = 50/\sqrt{a}$.

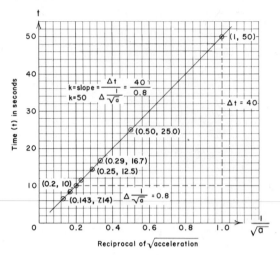

Figure 7-42   Plot of Time t Against $1/\sqrt{\text{Acceleration}}$

EXERCISE 7.7

   Follow the procedures used in the previous six examples to determine the mathematical relationship between the variable quantities listed in Problems 1–10. Treat the overscored variable as the independent variable. Consider all given data as exact.

1.

$\bar{t}$	s
0	2
3	11
4	14
6	20
8	26

2.

$\bar{t}$	h
0	1
2	2
4	3
6	4
10	6
14	8

3.

w	0	2	8	32	50	98
$\bar{r}$	0	1	2	4	5	7

4.

s	0	0.5	2	8	18	50
$\bar{t}$	0	1	2	4	6	10

5.

$\bar{v}$	h
1	12
1.5	8
2.4	5
2.5	4.8
4.8	2.5
5	2.4
6	2
8	1.5

6.

$\bar{x}$	y
0.5	1
1	0.5
2	0.25
5	0.1
10	0.05
25	0.02

7.

P	24	6	1.5	0.375	0.24	0.06
$\bar{i}$	0.5	1	2	4	5	10

8.

A	24	6	1.5	0.375	0.24
$\bar{r}$	1	2	4	8	10

9.

y	24	16	12	8	6	4
$\bar{x}$	4	9	16	36	64	144

10.

t	30	20	15	12	10	6
$\bar{s}$	4	9	16	25	36	100

7.8 LINEAR INEQUALITIES

Restrictions on variables are called *constraints*. Naturally occurring constraints accompany all quantities. For example, a certain material may expand proportionally to the increase in its temperature. When the material is heated hot enough, it may melt or burn or evaporate or explode and no longer expand according to the earlier observed variation at lower temperatures. Therefore, you must observe the natural constraints which accompany such quantities.

Consider the set of points in the coordinate plane such that $x > 4$ (x is greater than 4). This constraint limits the points (x, y) to those with abscissa values

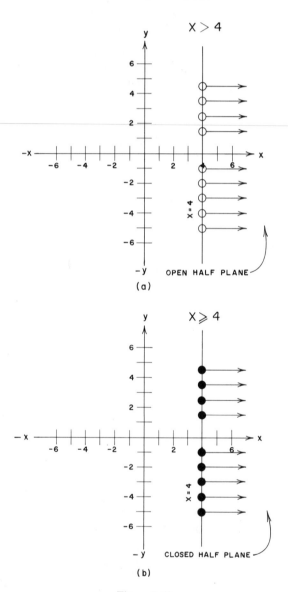

Figure 7-43

greater than 4. All of these points are located to the right of the line $x = 4$. See
Figure 7-43(a). Such a set of points (excluding the points on the line $x = 4$) is
called an *open half plane*. Figure 7-43(b) shows the *closed half plane* corresponding
to the statement $x \geqslant 4$ (x is greater than or equal to 4). Note that the points on the

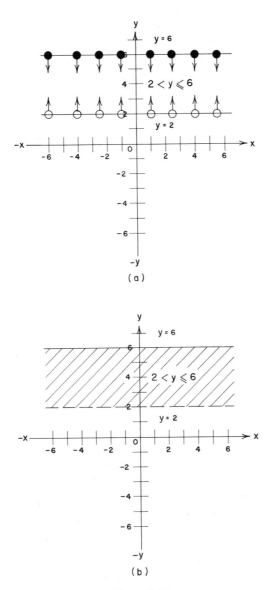

Figure 7-44

boundary line are included. The appropriate symbols for indicating half planes are ○→, which indicates that the initial points are not included, and ●→, which indicates that the initial points are included.

example   Determine the set of points in the coordinate plane whose coordinates satisfy the statement $2 < y \leqslant 6$ (y is greater than 2 and less than or equal to 6). Figure 7-44(a) shows the required region. Figure 7-44(b) shows an alternate method for showing the required region. The solid line $y = 6$ indicates that the points on the line are included in the region (i.e., closed on that boundary). The dotted line $y = 2$ indicates that the points on the line are not included in the region (i.e., open on that boundary). The shading takes the place of the arrows.

example   Determine the set of points in the coordinate plane such that

$$\begin{cases} 3 < x \leqslant 9 \\ 3 \leqslant y < 8 \end{cases}$$

The simplest way to determine this region is to graph the boundary lines $x = 3$, $x = 9$, $y = 3$, and $y = 8$. The lines $x = 3$ and $y = 8$ should be dotted if you use shading to show this region. Then shade in the area bounded by these four lines. See Figure 7-45.

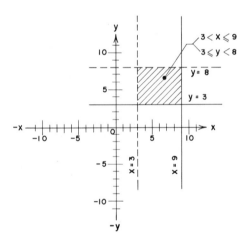

Figure 7-45

example   Show graphically the open half plane described by $y < x + 3$. First draw the line $y = x + 3$. Then indicate with appropriate arrows (or shading) the region below the line. See Figure 7-46.

In the last example, how did you know that the required region fell below the line? To answer this question, proceed as follows: pick any point P_1 (x_1, y_1) on the line $y = x + 3$. At this point the ordinate value y_1 equals the abscissa value x_1 plus 3. From this point move vertically downward to a second point P_2 (x_2, y_2). Here the ordinate y_2 is less than the ordinate y_1. Therefore, $y_2 < x_1 + 3$. Because P_1 can represent any point on the line $y = x + 3$ and P_2 can represent any point vertically below P_1, the region represented by $y < x + 3$ falls below the line $y = x + 3$.

example Show graphically (by shading) the region represented by the conditions

$$\begin{cases} y < -x + 10 \\ y > 2 \\ x \geqslant 3 \end{cases}$$

First draw the lines $y = -x + 10$ (dotted), $y = 2$ (dotted), and $x = 3$ (solid). Then shade the region below $y = -x + 10$, above $y = 2$, and to the right of $x = 3$. See Figure 7-47.

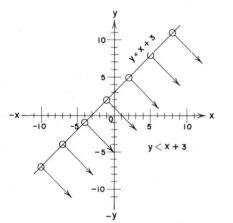

Figure 7-46

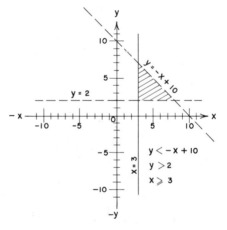

Figure 7-47

EXERCISE 7.8

In Problems 1–12, represent graphically the region represented by the given conditions.

1. $y < -2$ (arrows)
2. $x < 5$ (shading)

3. $\begin{cases} x > 0 \\ y > 0 \end{cases}$ (arrows)

4. $0 < x \leqslant 5$ (shading)

5. $-6 < y \leqslant -1$ (arrows)

6. $\begin{cases} -5 < y < 5 \\ \ \ 1 < x < 6 \end{cases}$ (shading)

7. $\begin{cases} -10 < y \leqslant -2 \\ -10 < x \leqslant 2 \end{cases}$ (arrows)

8. $y < x$ (arrows)

9. $y < -x$ (shading)

10. $\begin{cases} y < x + 5 \\ y \geqslant x - 5 \end{cases}$ (shading)

11. $\begin{cases} y < -x + 8 \\ y > -x - 8 \\ 0 \leqslant x < 5 \end{cases}$ (arrows)

12. $\begin{cases} 2y + 4x > 8 \\ 0 \leqslant x < 2 \\ y < 4 \end{cases}$ (shading)

13. Write the algebraic statement(s) corresponding to the region shown in Figure 7-48.
14. Write the algebraic statements corresponding to the region shown in Figure 7-49.

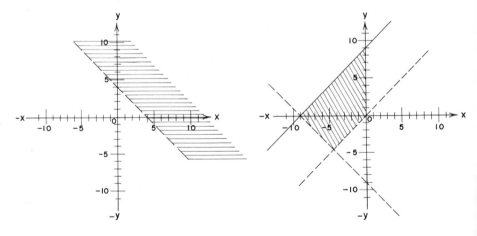

Figure 7-48　　　　　　　　**Figure 7-49**

⑧ Trigonometry

8.1 ARCS, ARC LENGTH, AND ANGULAR MEASURE

A large part of applied mathematics has developed from man's study of the triangle and the circle. Without considerable knowledge about the interrelationships of these two geometric figures, the technician would be lost.

Let us start with the unit of angular measurement used by the early Babylonians. They divided the circumference of a circle into 360 equal parts (Figure 8-1). Each of these unit parts, called a degree, is $\frac{1}{360}$ of the circumference. An arc

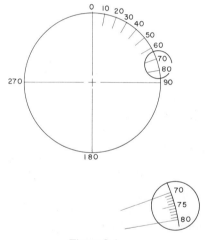

Figure 8-1

of 1 degree (written 1°) denotes an arc length which is $\frac{1}{360}$ of the circumference. The length of an arc of 1° will vary with the size of the circle. A 3° arc on a small circle will be of shorter length than a 3° arc on a larger circle. Yet, in both cases, the 3° arc denotes $\frac{3}{360}$ of their respective circumferences.

Although the degree is a rather small unit, it is not sufficiently small for many practical purposes. The degree is divided into 60 equal parts called minutes; 1° = 60′ (read 60 minutes). Therefore 1′ is 1/60 of 1/360, or $\frac{1}{21,600}$ of the circumference. The minute is further divided into 60 equal parts called seconds; 1′ = 60″

(read 60 seconds). Therefore, 1″ of arc is 1/60 of 1/60 of 1/360 or $\frac{1}{1,296,000}$ of the circumference.

The Measure of an Angle

The degree is used not only for measuring arcs but also to measure or describe the relative positions of objects or lines which have been or could be moved in a circular motion with respect to each other. When we talk about an angle of 45°

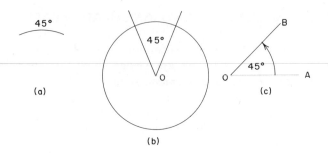

Figure 8-2

(Figure 8-2), we could be interested in any one (or all) of three closely related properties.

1. In Figure 8-2(a), the 45° means an arc of a circle which is 45°/360° or $\frac{1}{8}$ of the entire circumference. Thus, a specific part of a circumference may be expressed as an angle.

2. In Figure 8-2(b), the 45° angle refers to the relative position of the two half lines (or rays), having a common point at the center of a circle, such that the two rays (half lines) subtend (cut) an arc of 45° on the circle. The two half lines are said to form a central angle of 45°.

3. In Figure 8-2(c), the 45° angle refers to the amount of turning (about a point) required for a line (or object) to move from one position to another position. Ray OA would have to turn through (or generate) a 45° angle to reach the position of ray OB.

The concept of turning, or circular motion, expressed as an angle (or angular velocity) is very important when working with motion along circular paths such as cars going around curves, moving pulleys, belts, armatures, flywheels, grind-stones, lathes, and centrifuges. Consider a point such as the tip of a propeller blade moving in a circular path (Figure 8-3). When the tip has moved one-fourth of its complete circular path, it will have turned one-fourth of 360°, or 90°. When the tip at A gets to position C, the blade will have turned through an angle of 180°; when it is at D, it will have turned through an angle of 270°. When it again

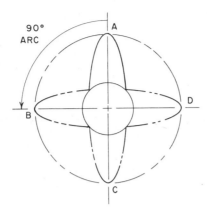

Figure 8-3

reaches point *A*, it will have turned through an arc of 360°. One complete turn or revolution (rev) is equivalent to 360° (1 rev = 360°).

The degree as a unit of angular measure enables us to set up the following important relationship between the length of an arc in linear units and the measure of the corresponding central angle in degrees:

$$S = \frac{2\pi r\theta}{360°} = \frac{\pi r\theta}{180°} \qquad \text{where} \begin{cases} S = \text{arc length} \\ r = \text{radius of circular path} \\ \theta = \text{angle in degrees} \end{cases}$$

This relationship allows the computation of arc length when the radius *r* and the angle *θ* are known.

example   The rotor in the distributor of an engine has an effective radius of 1.05 in. How far does the tip of the rotor move in traveling the 60° arc between two adjacent contact points? Using the formula $S = \pi r\theta/180°$ and substituting 1.05 in. for *r* and 60° for *θ*, you have

$$S = \frac{(\pi)(1.05 \text{ in.})(\cancel{60°})}{\cancel{180°}\atop 3} = \frac{(\pi)(1.05 \text{ in.})}{3} = 1.10 \text{ in.}$$

example   How long is an arc of 35.0° if the diameter of the circle is 10.0 ft?

$$\text{Arc length } S = \frac{\pi r\theta}{180°} = \frac{\pi(5.0 \text{ ft})(\overset{7}{\cancel{35°}})}{\underset{36}{\cancel{180°}}} = \frac{35\pi}{36} \text{ ft} = 3.05 \text{ ft}$$

EXERCISE 8.1A

1. Add and simplify:

    37° 52′ 86″
    29° 37′ 14″
    52° 83′ 41″
    ‾‾‾‾‾‾‾‾‾‾

2. Add and simplify:

    128° 31′ 43″
    72° 58′ 39″
    ‾‾‾‾‾‾‾‾‾‾

3. Determine the complement of:

    (a)   35° 12′ 42″
    (b)   19° 28′ 41″
    (c)   43° 5′ 9″

4. Determine the supplement of:

    (a)   58° 3′ 45″
    (b)   135°
    (c)   30°
    (d)   62° 36′ 55″

5. Using the fact that one revolution (rev) corresponds to 360° of arc, determine the corresponding revolutions for:
    (a)   3600°   (c)   7920°   (e)   45°   (g)   180°
    (b)   5400°   (d)   31,500°   (f)   90°   (h)   270°

6. Determine the degrees of arc corresponding to the given revolutions (1 rev = 360°):
    (a)   42.5 rev   (d)   2000 rev
    (b)   42 rev   (e)   0.01 rev
    (c)   0.75 rev

7. Construct a right triangle with sides of 3.0 in., 4.0 in., and 5.0 in., and measure the acute angles with a protractor.

8. Construct a quadrilateral having angles of 52°, 120°, 97°, and 91°. Draw in the two diagonals and measure the supplementary angles formed.

9. How long is an arc of 48° on a circle with a radius of 6.0 ft?

10. How many miles is it from a point on the earth to a second point due north of the first if the difference in their latitudes is 36.0°? The mean polar radius of the earth is 3950 miles.

11. Find the distance traveled by the tip of a distributor rotor of radius 1.32 in. as it moves between adjacent spark plug contact points for:
    (a)   a 4-cylinder, 4-cycle engine.
    (b)   an 8-cylinder, 4-cycle engine.

12. If the 30.0° arc between adjacent seats on a ferris wheel measures 10.5 ft, what is the diameter of the ferris wheel?

13. If the spokes of a 26-in. diameter wheel are to be uniformly spaced 15° apart, how many spokes are required for the wheel? What is the arc length (along the rim) between adjacent spokes?

14. Through how many degrees does a 6.00-in. diameter pulley turn when 40.0 ft of the driving belt have passed over the pulley? (Assume no slippage.)

15. A 12° arc of a certain circle measures 4.8 in. in length. Determine the radius of the circle.

16. A certain circular section of a modern freeway forms a 45.0° arc. If a car travels 2.70×10^3 ft in executing this turn, what is the radius of the circular section?

17. What is the magnitude of the smaller angle formed by the lines $y = 2x - 4$ and $y = -x + 3$? (Graph the lines and measure the angle with a protractor.)

18. What angles does the line passing through the points $(-4, 5)$ and $(7, -2)$ form with the x axis? The y axis?

The Radian

To facilitate calculations involving angular velocities and linear velocities along circular paths, we introduce the *radian* which is another unit of angular measure. Consider a circle of radius r (Figure 8-4) with central $\sphericalangle \theta$ which subtends an arc of length r. This particular angle defines the angular unit called a *radian*.

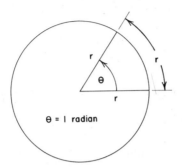

$\theta = 1$ radian

Figure 8-4

The radian is the angle measured by a circular arc whose length is equal to the radius of the circle. Compared to the degree, a radian is a rather large unit.

How many radians are there in one complete circle? Suppose you start dividing the circumference of a circle by its radius. With $C = 2\pi r$,

$$\frac{C}{r} = \frac{2\pi r}{r} = 2\pi$$

This tells you that there are 2π (approximately 6.28) radians in 360°. If

$$2\pi \text{ radians} = 360°$$

then

$$\pi \text{ radians} = 180°$$

$$\frac{\pi}{2} \text{ radians} = 90°$$

$$\frac{\pi}{180} \text{ radians} = 1°$$

and

$$\frac{180}{\pi} = 1 \text{ radian (approximately } 57.3°)$$

To convert degrees to radians, multiply the angle in degrees by π radians/180°.

example   Change 270° to radians:

$$\overset{3}{\cancel{270°}} \cdot \frac{\pi \text{ radians}}{\underset{2}{\cancel{180°}}} = \frac{3\pi}{2} \text{ radians}$$

example   Change 450° to radians:

$$\overset{5}{\cancel{450°}} \cdot \frac{\pi \text{ radians}}{\underset{2}{\cancel{180°}}} = \frac{5\pi}{2} \text{ radians}$$

To convert radians to degrees, multiply the angle in radians by 180°/π *radians.*

example   Change 8π radians to degrees:

$$8\cancel{\pi \text{ radians}} \cdot \frac{180°}{\cancel{\pi \text{ radians}}} = 1440°$$

example   Change $\pi/4$ radians to degrees:

$$\frac{\cancel{\pi \text{ radians}}}{\underset{1}{\cancel{4}}} \cdot \frac{\overset{45°}{\cancel{180°}}}{\cancel{\pi \text{ radians}}} = 45°$$

Circular Motion

Consider a point on the circumference of a wheel with a radius of 4 ft. How far does the point move when the wheel turns through an angle of 1 radian? The point would turn through an arc of 1 radian. Therefore, it must move a distance equal to the radius—4 ft. How far will the point move in turning through 2 radians? Because the point will move a distance of 4 ft for each radian of arc, it must move a distance of 8 ft to generate an arc of 2 radians. This example illustrates an important and useful algebraic statement relating distance along a circular path (arc length) and the central angle measured by the arc:

$$S = r\theta \qquad \text{where} \begin{cases} S = \text{arc length} \\ r = \text{radius of circular path} \\ \theta = \text{number of radians in the central angle} \end{cases}$$

This equation says that the length of any circular arc of any circle is equal to the product of the radius of the circle and the central angle (in radians).

example   How far does the tip of an exhaust fan blade move if the blade has a radius of 8.0 ft and turns through 45 radians? Using $S = r\theta$, with $r = 8.0$ ft and $\theta = 45$:

$$S = (8.0 \text{ ft})(45)$$

$$S = 360 \text{ ft}$$

example Through what angle must pulley A with an 8.0-ft radius be turned to move point P on the belt a distance of 960 ft? See Figure 8-5. Using $S = r\theta$,

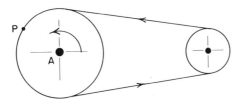

Figure 8-5

with $S = 960 \text{ ft}$ and $r = 8.0 \text{ ft}$:

$$960 \text{ ft} = (8.0 \text{ ft})(\theta) \tag{1}$$

$$\div 8.0 \text{ ft}, \qquad 120 = \theta \tag{2}$$

$$\therefore \theta = 120 \text{ radians} \tag{3}$$

Angular and Linear Velocities

If you start with the statement $S = r\theta$ and divide each member by time t, you obtain

$$\frac{S}{t} = \frac{r\theta}{t} = r\frac{\theta}{t}$$

where

$$\frac{S}{t} \text{ is } \frac{\text{distance}}{\text{unit of time}} = \text{velocity } V$$

and

$$\frac{\theta}{t} \text{ is angular velocity}$$

θ/t is usually represented by the Greek letter ω (omega). Replacing S/t with V and θ/t with ω, you have $V = r\omega$. According to this equation, the velocity of a point along a circular path is equal to the product of the radius of the path and the angular velocity ω (in radians/unit of time).

For a better concept of an angular velocity in radians/unit of time, consider a point on the circumference of a flywheel. If the point moves a distance equal to the radius of the wheel in 1 sec, then the point has generated (or moved through) an angle of 1 radian in 1 sec. The angular velocity ω of the point or of the wheel would be 1 radian/sec. If the point moved 5 radians in 1 sec, then $\omega = 5$ radians/sec. Angular velocity ω is the rate at which an angle is being generated.

example   Find the surface velocity of a steel shaft 20.0 in. in diameter if the shaft has an angular velocity ω of 250 radians/sec. Using $V = r\omega$, with $r = 10$ in. and $\omega = 250$:

$$V = (10)(250) \text{ in./sec} = 2500 \text{ in./sec}$$

example   A boy rides a bicycle 15 mph (22 ft/sec). If the cycle has wheels of 28-in. diameter, what is the angular velocity of the wheels? Using $V = r\omega$ and solving for ω:

$$\omega = \frac{V}{r} \qquad \text{with } V = 22 \text{ ft/sec and } r = \frac{14 \text{ ft}}{12} = \frac{7}{6} \text{ ft}$$

$$\omega = \frac{22 \text{ ft/sec}}{\frac{7}{6} \text{ ft}} = 18\tfrac{6}{7} \text{ radians/sec}$$

example   The machine cutting speed for a certain alloy is 1.50 ft/sec. If a part made from this alloy is being machined in a lathe and the part has a diameter of 10.0 in., find the maximum rpm allowed for turning this part. Using $V = r\omega$, with $V = 1.50$ ft/sec and $r = 5/12$ ft:

$$\omega = \frac{V}{r} = \frac{1.50 \text{ ft/sec}}{5/12 \text{ ft}} = 3.60 \text{ radians/sec} = 216 \text{ radians/min}$$

$$(1 \text{ rev} = 2\pi \text{ radians})$$

$$\text{rpm} = \frac{\omega}{2\pi} = (216 \text{ radians/min})\frac{(1 \text{ rev})}{2\pi \text{ radians}} = 34.4 \text{ rpm maximum}$$

Area of Circular Sectors

We frequently have to calculate the area of circular sectors such as the area of the walls of a cone formed from a circular sector (Figure 8-6). The formula $A = \frac{1}{2}r^2\theta$ is derived from the area of a circle, $A = \pi r^2$. This formula gives the area of a circular sector of radius r and central angle θ in radians.

example   Find the outside surface area of a cone formed from a circle of 4.00-ft diameter by cutting a circular sector with a central angle of 1.00 radian from the circle and joining the cut edges (of the remaining sector), as shown in Figure 8-6. The entire circle has a central angle of 2π radians; the remaining circular sector has a central angle of $(2\pi - 1)$ radians. Using the

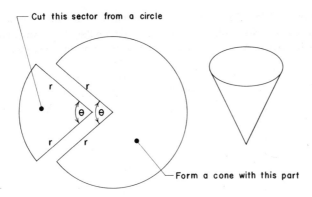

Figure 8-6

formula $A = \frac{1}{2}r^2\theta$, with $r = 2$ ft and $\theta = (2\pi - 1)$:

$$A = \frac{1}{2}(2 \text{ ft})^2(2\pi - 1) = 2(2\pi - 1) \text{ sq ft} = (4\pi - 2) \text{ sq ft} \approx 10.6 \text{ sq ft}$$

EXERCISE 8.1B

1. Express angles (a)–(d) in degrees and angles (e)–(h) in radians:

    (a) 2π (c) $\dfrac{3\pi}{2}$ (e) $45°$ (g) $90°$

    (b) $\dfrac{\pi}{4}$ (d) $\dfrac{5\pi}{6}$ (f) $60°$ (h) $180°$

2. Express angles (a)–(d) in degrees and angles (e)–(h) in radians:

    (a) 10π (c) $\dfrac{\pi}{3}$ (e) $360°$ (g) $1080°$

    (b) $\dfrac{\pi}{2}$ (d) π (f) $720°$ (h) $540°$

    Using $S = r\theta$, in Problems 3–6, calculate the length of the arc which subtends the given central angle θ.

3. $r = 8.0$ ft, $\theta = \dfrac{\pi}{4}$ radians   5. $r = 800.0$ miles, $\theta = \dfrac{\pi}{50}$ radians

4. $r = 10.0$ in., $\theta = \dfrac{\pi}{2}$ radians   6. $r = 168$ ft, $\theta = 5.00$ radians

    In Problems 7–10, the radius r and arc length S are given. Find the central angle measured by the given arc.

7. $r = 12.0$ ft, $S = 4.00$ ft   8. $r = 6.0$ in., $S = 27$ in.

9. $r = 3.075$ in., $S = 15.375$ in.   10. $r = 4.00$ ft, $S = 30.0$ in.
11.   Find the linear velocity V of point P (Figure 8-7) if pulley B has an angular velocity of $2\tilde{0}$ radians/sec. (*Hint :* All points on the belt have the same linear velocity.)

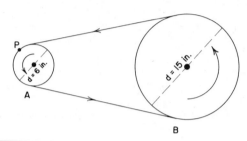

Figure 8-7

12.   The front wheel of a bicycle has a diameter of 16 in. and the back wheel has a diameter of 26 in. Through what angles does the front wheel turn if the rear wheel turns through 24 radians?
13.   If the velocity of sound in air is 950 ft/sec, find the angular velocity in radians/second at which the tip of a propeller blade ($r = 8.00$ ft) enters the supersonic range.
14.   What is the linear velocity in feet/second of the tip of a fan blade ($r = 6.00$ ft) when the blade is turning 60.0 rpm?
15.   If 88.0 ft of belt pass around a wheel 9.00 in. in diameter, through what angle does the wheel turn?
16.   A pipe cutting and threading machine turns the pipe 1 rev/sec. What is the surface speed of a pipe whose outside diameter is $1\frac{7}{8}$ in.?
17.   Find $\sphericalangle ABC$ (in radians) in Figure 8-8 if the circle has a radius of 20.0 in. and the arc $S = 10.0$ in. (*Hint :* See Appendix B, Facts 65 and 66.)
18.   If $\sphericalangle ABC = \pi/4$ radians in Figure 8-8, what is the length of arc S (radius $= 20.0$ in.)?

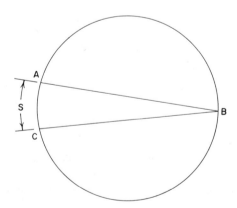

Figure 8-8

Challenge Problems

19. A certain carborundum grindstone has a maximum surface speed of 60π ft/sec. A V-belt pulley, 3.00-in. in diameter, on the shaft of an electric motor drives a 2.00-in. V-belt pulley on the shaft of the grindstone. What should be the maximum rpm of the electric motor? See Figure 8-9.

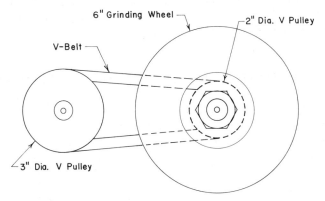

Figure 8-9

20. Three pulleys, 1, 2, and 3, are driven by a belt as shown in Figure 8-10. Calculate the linear speed of points *A* and *B* and the angular velocity of pulleys 2 and 3. (Assume 2 digit accuracy.)
21. Three meshing gears in a gear train have 22, 26, and 33 teeth, respectively. If the first gear has an angular velocity of 12 radians/sec, what is the angular velocity of the third gear?

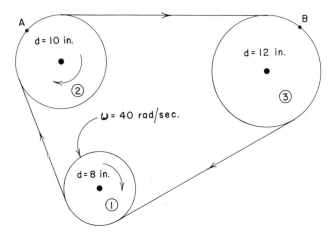

Figure 8-10

22. Four gears in a gear train have 16, 24, 32, and 36 teeth, respectively. If the first gear turns 1800 rpm, what is the rpm for the last gear?
23. Two meshing gears are to have (as close as possible) an rpm of 1250 and 1720, respectively. How many teeth should each gear have if the total number of teeth is to be a maximum but less than 100?
24. Calculate the area of a circular sector of 6.4-ft radius and with a central angle of 2.5 radians.
25. The area of the sector of a circle is 4.40×10^2 sq ft. If the arc of the sector is 44.0 ft, what is the radius of the circle?

8.2   TRIGONOMETRIC RATIOS OF THE RIGHT TRIANGLE

We already know that a right triangle contains one right angle and two acute angles. We shall need the following terms in our study of the right triangle (See Figure 8-11.)

Hypotenuse	Side of triangle opposite the 90° angle
Opposite side	Side of triangle opposite a specified angle
Adjacent side	Side of an angle other than the hypotenuse

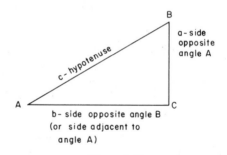

Figure 8-11

Using any two of the three sides a, b, and c of the triangle in Figure 8-11, you can form the ratios a/c, a/b, b/c, b/a, c/a, and c/b. If you chose to denote the angles as M, N, and O and the sides opposite these angles as m, n, and o, these same six ratios would be m/o, m/n, n/o, n/m, o/m, and o/n, respectively. The ratio a/c in Figure 8-11 as it relates to ∡ A is described as the ratio formed by the length of the side opposite ∡ A over the length of the hypotenuse.

Because any particular ratio depends upon the magnitude of an acute angle, these ratios are called trigonometric functions. These ratios depend only on the magnitude of the angle and not on the size of the triangle. It makes little difference whether you consider the ratios as functions of angles or simply as names of specific ratios, as long as you remember that these trigonometric functions are ratios.

When learning the following ratios, refrain from using the same letter notation on every triangle. The following table lists the names of the six functions for an acute angle of a right triangle.

Name of Trigonometric Function or Ratio	Defined by
sine A (sin A) =	$\dfrac{\text{side opposite angle } A}{\text{hypotenuse side}}$
cosine A (cos A) =	$\dfrac{\text{side adjacent angle } A}{\text{hypotenuse side}}$
tangent A (tan A) =	$\dfrac{\text{side opposite angle } A}{\text{side adjacent angle } A}$
cotangent A (cot A) =	$\dfrac{\text{side adjacent angle } A}{\text{side opposite angle } A}$
secant A (sec A) =	$\dfrac{\text{hypotenuse side}}{\text{side adjacent angle } A}$
cosecant A (csc A) =	$\dfrac{\text{hypotenuse side}}{\text{side opposite angle } A}$

Note that certain pairs of the functions are reciprocals of each other:

1. Sine and cosecant are reciprocal ratios.
2. Cosine and secant are reciprocal ratios.
3. Tangent and cotangent are reciprocal ratios.

If you know sine A, you can determine cosecant A, because csc $A = 1/\sin A$. Likewise, sec $A = 1/\cos A$ and cot $A = 1/\tan A$.

With these definitions let us establish specific ratios using lengths of the sides of the right triangle in Figure 8-12:

$$\sin A = \frac{12 \text{ in.}}{20 \text{ in.}} = \frac{3}{5} \qquad \csc A = \frac{20 \text{ in.}}{12 \text{ in.}} = \frac{5}{3}$$

$$\cos A = \frac{16 \text{ in.}}{20 \text{ in.}} = \frac{4}{5} \qquad \sec A = \frac{20 \text{ in.}}{16 \text{ in.}} = \frac{5}{4}$$

$$\tan A = \frac{12 \text{ in.}}{16 \text{ in.}} = \frac{3}{4} \qquad \cot A = \frac{16 \text{ in.}}{12 \text{ in.}} = \frac{4}{3}$$

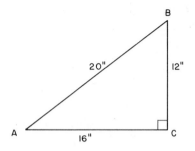

Figure 8-12

Setting up the ratios for ∢ B in Figure 8-12, you obtain:

$$\sin B = \frac{16 \text{ in.}}{20 \text{ in.}} = \frac{4}{5} \qquad \csc B = \frac{20 \text{ in.}}{16 \text{ in.}} = \frac{5}{4}$$

$$\cos B = \frac{12 \text{ in.}}{20 \text{ in.}} = \frac{3}{5} \qquad \sec B = \frac{20 \text{ in.}}{12 \text{ in.}} = \frac{5}{3}$$

$$\tan B = \frac{16 \text{ in.}}{12 \text{ in.}} = \frac{4}{3} \qquad \cot B = \frac{12 \text{ in.}}{16 \text{ in.}} = \frac{3}{4}$$

It is important to observe that the side opposite one of the acute angles is the adjacent side for the other acute angle. The two ratios sin A and cos B (for the complementary ∢'s A and B) both equal $\frac{12}{20}$. Also, tan $A = \cot B = \frac{12}{16}$ and sec $A = \csc B = \frac{20}{16}$. In other words the *cofunctions of complementary angles are equal*; *co*sine, *co*tangent and *co*secant are the cofunctions of sine, tangent, and secant, respectively. Therefore, sin $A = \cos B$, tan $A = \cot B$, and sec $A = \csc B$ when A and B are complementary angles.

example How does the tangent of a 30° angle compare with the cotangent of a 60° angle? Because cofunctions of complementary angles are equal,
$$\tan 30° = \cot 60°$$

example If the sine of 30° $= \frac{1}{2}$, what is the value of the cosine 60°? *Answer:* $\frac{1}{2}$.

example If a trigonometric table shows secant 43° $= 1.3673$, what is the value of cosecant 47°? *Answer:* 1.3673.

example If the sine 30° $= \frac{1}{2}$, what is the value of the cosecant 30°?
$$\csc 30° = 1/\sin 30° = \frac{1}{1/2} = 2$$

A close look at *similar right triangles* will give you a good background for understanding and using trigonometric ratios. The beauty of similar right triangles is that not only are corresponding angles equal, but corresponding sides are proportional. Thus, the trigonometric ratios formed by corresponding sides are equal.

Consider the similar right triangles in Figure 8-13. Starting with the proportion 4/8 = 5/10, you can state that sin A = sin A', because 4/8 = 1/2 = sin A and 5/10 = 1/2 = sin A'. If only one side of $\triangle ABC$ is known, the other two sides can

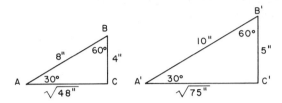

Figure 8-13

be found by using corresponding ratios obtained from $\triangle A'B'C'$. For example, if sides AC and BC are unknown, you could equate the sine ratios for angles A and A' obtaining BC/8 in. = 5 in./10 in. = $\frac{1}{2}$. Solving for BC gives BC = 4 in.

example Refer to the similar right triangles in Figure 8-14. A telegraph pole casts a shadow 24 ft long at the same time that a 4.0-ft post casts a shadow 2.4 ft

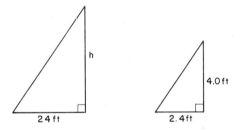

Figure 8-14

long. How tall is the pole? Using the fact that corresponding trigonometric ratios of similar right triangles are equal, write:

$$\frac{h}{24 \text{ ft}} = \frac{4.0 \text{ ft}}{2.4 \text{ ft}}$$

Solving for h gives

$$h = 4\tilde{0} \text{ ft}$$

Or you might observe that the pole shadow is 10 times as long as the post shadow and calculate the height of the pole equal to 10 times the height of the post.

EXERCISE 8.2

1. Set up the six trigonometric ratios for ∢ R in Figure 8-15.
2. Using the triangle in Figure 8-15, set up the six trigonometric ratios for ∢ P.
3. Use the Pythagorean theorem to find the length of the rafter PH in Figure 8-16. Set up the six trigonometric ratios for ∢ H. Then use the slide rule to express these ratios as decimals. Assume three-figure accuracy for given data.

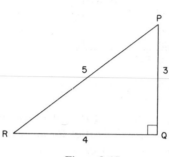

Figure 8-15

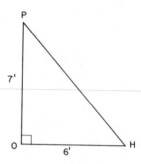

Figure 8-16

4. Use the Pythagorean theorem to find the length of the missing side of the template in Figure 8-17. Then set up the six trigonometric ratios for ∢ Q. Express these ratios as decimals by using the slide rule. (Assume 3 digit accuracy.)

Figure 8-17

5. Without using tables, tell what ratio should be listed for cos 72° if a table lists sin 18° = 0.30902. What ratio should be listed for cot 49° if a table lists tan 41° = 0.86929? If sec 62° = 2.1301, what should be the listing for csc 28°?
6. The following values are taken from tables:

$$\cos 27° = 0.89101 \qquad \cot 63° = 0.50953 \qquad \csc 12° = 4.8097$$

Without using tables, tell what values should be listed for sin 63°, tan 27°, and sec 78°.

7. Explain why the values of the sine and cosine functions of an acute angle of a right triangle will always be less than 1.
8. Explain why the values of the secant and cosecant functions of an acute angle will always be greater than 1.
9. Given $\tan 31° = 0.60086$, $\tan 35° = 0.70021$, and $\tan 39° = 0.80978$, determine the approximate magnitude of $\measuredangle B$ in Figure 8-18.

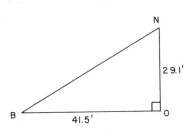

Figure 8-18

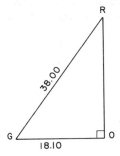

Figure 8-19

10. Given $\cos 59° = 0.51504$, $\tan 41° = 0.86929$, and $\sec 62° = 2.1301$, determine the approximate magnitude of $\measuredangle G$ in Figure 8-19.
11. Using the figure and data in Problem 10, determine the approximate magnitude of $\measuredangle R$ if GR was unknown and $RO = 15.71$.
12. In the triangle shown in Problem 10, what should the trigonometric table list for tan G?

8.3   CONSTRUCTING SPECIFIC ANGLES

An angle is said to be in *standard position* (in the rectangular coordinate system) when its vertex is at the origin and its initial side lies on the positive x axis. Figure 8-20 shows angles in standard position. Because an angle of any magnitude may

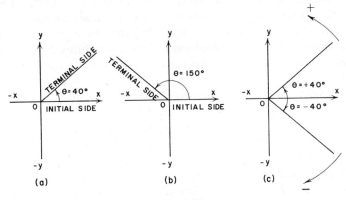

Figure 8-20

be generated, the terminal side may fall in any of the four quadrants. Figure 8-20(b) shows an angle in standard position terminating in the second quadrant. Angles generated in the counterclockwise direction are called positive angles and those generated in the clockwise direction are called negative angles. See Figure 8-20(c).

Angle θ in Figure 8-21 is called the *angle of elevation* when P, the point being observed, lies above the x axis, but it is called the *angle of depression* when point P lies below the x axis.

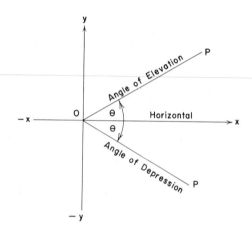

Figure 8-21

The orientation of an angle in a coordinate plane facilitates the construction of a desired angle. For example, suppose a man wished to construct an angle of 33° 20′. By reference to tables, he would find tan 33° 20′ = 0.6577. The ratio 0.6577 could be written as 6.577/10. By constructing a right triangle so that the leg opposite one acute angle was 6.577 units and the side adjacent to the angle was 10 units, he would have the desired angle. To construct this angle in standard position, he need only locate a point with coordinates of $x = 10$ and $y = 6.577$ units (Figure 8-22).

Specific angles can also be constructed by using a carpenter's framing square. Figure 8-23 shows how the square can be used to construct an angle of 33° 20′. Referring to the ratio 6.577/10 just mentioned, measure out 10 in. on the body of the square and $6\frac{9}{16}$ in. on the tongue of the square. (Table of decimal equivalents shows $0.577 \approx \frac{9}{16}$.) The line joining these two points forms the required angle with the body of the square. Usually the carpenter knows the y and x values. He refers to these values as *rise* and *run*, respectively. The ratio of rise/run is the tangent of the *angle of elevation*.

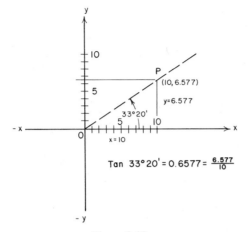

Figure 8-22

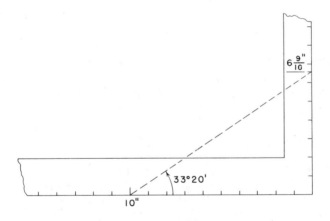

Figure 8-23

Another common and practical method of constructing a required angle is the *sine method*. Suppose a technician needed to construct an angle of 24° 50′. He finds in the tables that sin 24° 50′ = 0.4200, which can be written as 4.200/10. On a coordinate system he draws a line representing $y = 4.200$ (Figure 8-24). Then, using the origin as center, he swings an arc of radius = 10 units which intersects the line $y = 4.200$ in the first quadrant. The $\angle xOP$ is the required angle.

A machinist would construct this same angle with much greater accuracy, but his method would parallel the procedure just described. The machinist would

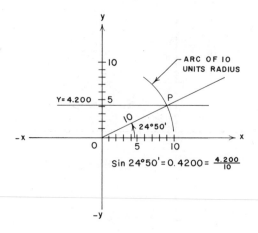

Figure 8-24

use an instrument called a sine bar which gets its name from the sine ratio. The two standard lengths for sine bars are 5 in. and 10 in. (Figure 8-25). He would place the sine bar on a very accurate surface plate or "master flat" and support the upper end with precision gage blocks with combined length of 4.200 in. This procedure actually constructs a right triangle whose hypotenuse is 10.00 in. and with the side opposite the desired angle equal to 4.200 in. The entire physical triangle can be placed or bolted into position for purposes of gaging or milling the desired angle on the work involved.

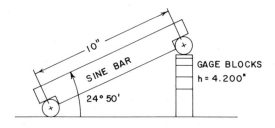

Figure 8-25

Construction technicians frequently use variations of the tangent and sine methods to construct a desired angle. Measurements may be made with 100-ft tapes. For example, a 90° angle can be constructed quite easily by establishing two points (nails driven in the top of two stakes) 80 ft apart on a line forming one side of the desired angle. Using two tapes simultaneously, hooking one on one nail and the other on the second nail while keeping the tapes taut (Figure 8-26), align the 60-ft mark on one tape with the 100-ft mark on the other. This process constructs a

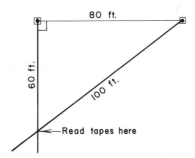

Figure 8-26

3-4-5 right triangle. After the 90° angle is established, the tangent method can be used to construct any desired acute angle. The *x* and *y* coordinate distances may be measured with the tapes along the perpendicular sides of the 90° angle.

EXERCISE 8.3

1. Using (5 squares/in.) graph paper and knowing that the tan 45° = 1.000, construct an angle of 45° in standard position. Check your work with a protractor. What is the equation of the line forming the terminal side?
2. Construct a 60° angle in standard position, using tan 60° = 1.732. Check with a protractor.
3. Construct a 38° 40′ angle in standard position, using tan 38° 40′ = 0.8002. Check with a protractor.
4. Construct an angle of − 55°, using tan 55° = 1.428.
5. If you have access to a carpenter's framing square, construct the angles given in Problems 1–4. Check with a protractor.
6. Given sin 45° = 0.70711, sin 60° = 0.86603, sin 38° 40′ = 0.62479, and sin 55° = 0.81915, use the sine method and graph paper to construct the angles in Problems 1–4.
7. If you have access to a sine bar and gage blocks, construct angles with the following magnitudes:
   (a) 22° 20′ given sin 22° 20′ = 0.37999.
   (b) 16° 35′ given sin 16° 35′ = 0.28541.
   (c) 28° 52′ given sin 28° 52′ = 0.48277.
   (d) 36° 41′ given sin 36° 41′ = 0.59739.

8.4 FUNCTIONS OF SPECIAL ANGLES

Certain angles are used so frequently that it pays to become familiar with them.

The 30°–60°–90° triangle gets its name from the magnitudes of the angles involved. This triangle should not be confused with the 3–4–5 triangle. Figure

8-27 shows a 30°–60°–90° triangle with sides of 1, 2 and $\sqrt{3}$ units. Setting up the trigonometric functions for the 30° and 60° angles, you get:

$$\sin 30° = \frac{1}{2} \qquad\qquad \sin 60° = \frac{\sqrt{3}}{2}$$

$$\cos 30° = \frac{\sqrt{3}}{2} \qquad\qquad \cos 60° = \frac{1}{2}$$

$$\tan 30° = \frac{1}{\sqrt{3}} = \frac{\sqrt{3}}{3} \qquad\qquad \tan 60° = \frac{\sqrt{3}}{1} = \sqrt{3}$$

$$\cot 30° = \frac{\sqrt{3}}{1} = \sqrt{3} \qquad\qquad \cot 60° = \frac{1}{\sqrt{3}} = \frac{\sqrt{3}}{3}$$

$$\sec 30° = \frac{2}{\sqrt{3}} = \frac{2\sqrt{3}}{3} \qquad\qquad \sec 60° = \frac{2}{1} = 2$$

$$\csc 30° = \frac{2}{1} = 2 \qquad\qquad \csc 60° = \frac{2}{\sqrt{3}} = \frac{2\sqrt{3}}{3}$$

Although you will encounter all of these specific ratios at some future time, you can get by quite well if you remember only the basic figure. You can then mentally picture the triangle shown in Figure 8-27 and set up in your mind the sine, tangent, and cotangent ratios.

The 45°–45°–90° triangle is shown in Figure 8-28 with sides of 1, 1, and $\sqrt{2}$ units. Setting up the trigonometric ratios for the 45° angle, you obtain:

$$\sin 45° = \cos 45° = \frac{1}{\sqrt{2}} = \frac{\sqrt{2}}{2} \qquad \text{(approximately 0.707)}$$

$$\tan 45° = \cot 45° = \frac{1}{1} = 1$$

$$\sec 45° = \csc 45° = \frac{\sqrt{2}}{1} = \sqrt{2} \qquad \text{(approximately 1.414)}$$

Constructing these two basic triangles on paper and writing the six ratios several times during the next few days will help fix these special ratios in your mind.

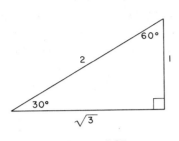

Figure 8-27

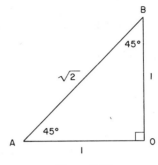

Figure 8-28

example A certain cam has the shape of a rhombus as shown in Figure 8-29(a). Draw in the diagonals and determine their lengths. Figure 8-29(b) shows the four right triangles formed by drawing in the diagonals. Recalling that the diagonals of a rhombus are perpendicular bisectors of each other, you

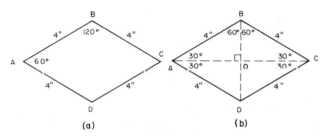

Figure 8-29

need only to find OB and double it to obtain DB. Then find AO and double it to get AC. In $\triangle AOB$,

$$\sin 30° = \frac{1}{2} = \frac{BO}{4 \text{ in.}}$$

Solving for BO, you find that $BO = 2$ in. and diagonal $BD = 4$ in. In the same triangle,

$$\tan 30° = \frac{BO}{AO} = \frac{1}{\sqrt{3}} = \frac{2 \text{ in.}}{AO}$$

Solving for AO, you find that $AO = 2\sqrt{3}$ in. and diagonal $AC = 4\sqrt{3}$ in. Thus, *diagonal BD = 4 in. and diagonal AC = $4\sqrt{3}$ in.*

example   A shipping box for a missile part has a square cross section (Figure 8-30(a)). The inside edge of the square is 3.74 ft. What is the length of the diagonal support? As shown in Figure 8-30(b), the diagonal forms two 45°–45°–90° triangles with legs of length 3.74 ft. Remember that the

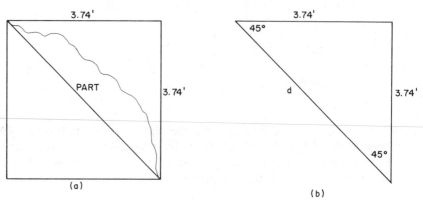

(a)   (b)

Figure 8-30

$\sin 45° = 1/\sqrt{2}$. In this figure, $\sin 45° = 3.74\,\text{ft}/d$. Thus, you can write

$$\frac{1}{\sqrt{2}} = \frac{3.74\,\text{ft}}{d}$$

Solving for d, you have

$$d = 3.74\sqrt{2}\,\text{ft} \approx 5.29\,\text{ft}$$

If the box measured 4.88 ft, its diagonal would equal $4.88\sqrt{2}$ ft. Again you see that the diagonal of a square equals the side times $\sqrt{2}$. Applying this relationship to any 45°–45°–90° triangle, you find that the hypotenuse always equals $\sqrt{2}$ times the length of a leg. From this relationship it is easy to find the edge of a square when the length of the diagonal is known by dividing the diagonal by $\sqrt{2}$. That is,

$$\text{side} = \frac{\text{diagonal}}{\sqrt{2}}$$

EXERCISE 8.4

Watch significant figures

1. Without using the text, sketch a basic 30°–60°–90° triangle and write the six trigonometric ratios for the 30° and 60° angles.

2. Sketch a basic 45°–45°–90° triangle and write the six trigonometric ratios for the 45° angle.

   Using the specific trigonometric ratios given in this section, solve the following problems. (Leave your answers exact.)

3. Referring to Figure 8-31 and assuming angle to be exact:
   (a) Find *AO* if *OB* = 10.0 ft.
   (b) Find *AB* if *OA* = 5.00 mm.
   (c) Find *OB* if *AB* = 15 cm.

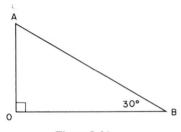

Figure 8-31

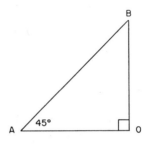

Figure 8-32

4. Referring to Figure 8-32 and assuming angle to be exact:
   (a) Find *BO* if *AO* = 14.415 ft.
   (b) Find *AO* if *AB* = 12.00 ft.
   (c) Find *AB* if *AO* = 6.25 ft.
5. Find the length of the water pipe *AOR* in Figure 8-33.

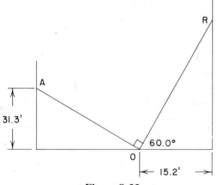

Figure 8-33

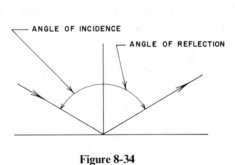

Figure 8-34

6. The most efficient operating angle of elevation for a conveyor used to elevate crushed ore in a smelter is 3̃0°. If the ore is to be elevated 215 ft, what length conveyor is needed?
7. Light is reflected from a plane surface so that the angle of reflection is equal to the angle of incidence (Figure 8-34). How high up on the side of the canal lock should a photoelectric

cell be located to be activated by a light source at point A which is reflected from the water surface at S when the water reaches a depth of 15 ft? See Figure 8-35.

8. If point S in Problem 7 was moved horizontally to a point 36.0 ft from point A, what would be the angle of incidence? What should be the distance h in this case? (Water depth remains 15 ft.)

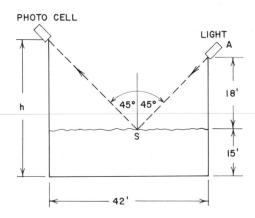

PHOTO CELL

Figure 8-35

9. The ore in Problem 6 is stored in a conical pile as shown in Figure 8-36.
   (a) What is the altitude of the pile when the diameter of the base is 80.0 ft and the vertex angle reaches 90.0°?
   (b) How many cubic yards of ore would the pile contain?
10. A surveyor at point P is trying to determine his position between two buildings of known heights (Figure 8-37). Determine his distance from building A. What is the distance between the two buildings? (Assume three-figure accuracy.)

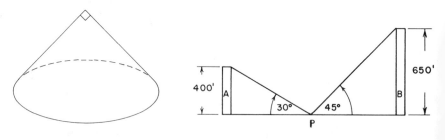

Figure 8-36   **Figure 8-37**

11. How far would the surveyor (Problem 10) have to move toward building A to make the angle of elevation for building A equal to 60.0°?
12. A surveyor gave this problem to his young son: When I was standing 30$\tilde{0}$ ft from the base of a building, the angle of elevation of the top of the building was 3$\tilde{0}$°. From the

same point the angle of elevation of the flagpole on top of the building was 45° (Figure 8-38). Find the following distances:

(a) *FO*.   (c) *TO*.   (e) *TF*.
(b) *PF*.   (d) *PT*.

13. Using Figure 8-38 but changing *PO* to 465 ft, find distances:
(a) *FO*.   (c) *TO*.   (e) *FT*.
(b) *PF*.   (d) *PT*.

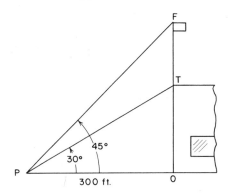

Figure 8-38

8.5 TABLES OF TRIGONOMETRIC RATIOS AND THEIR USES

Mathematicians have worked out tables which list the decimal values of the trigonometric ratios for various angles. See Table 16 in Appendix D. Let us look at a few sample problems.

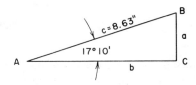

Figure 8-39

example   Find the distance between points *B* and *C* in Figure 8-39. You can use either the sine ratio or the cosecant ratio because they both are established using sides *a* and *c*. If you use the sine ratio,

$$\sin 17° \, 10' = \frac{a}{c} = \frac{a}{8.63 \text{ in.}}$$

In Table 16 under the column headed sin and on the line corresponding to 17° 10′, sin 17° 10′ = 0.2952. Therefore, write the proportion $a/8.63$ in. = 0.2952. Solving for a, you find $a = 2.55$ in.

example   Find the length of one edge of a jig, with the shape of a regular pentagon, cut from a circular plate of radius 4.212 ft (Figure 8-40). Because the figure is regular, you know that $\overset{\frown}{AC} = 360°/5 = 72°$ (exactly). Bisecting the 72° central angle in $\triangle AOC$ gives two right triangles, ABO and CBO. In $\triangle ABO$, $\angle AOB = 36°$:

$$\sin 36° = \frac{AB}{AO} = \frac{AB}{4.212 \text{ ft}}$$

In Table 16, under column headed sin and on the line corresponding to 36°, sin 36° = 0.5878. Therefore,

$$\frac{AB}{4.212 \text{ ft}} = 0.5878$$

Solving for AB, you have $AB = 2.476$ ft. The required edge = $(2)(2.476) = 4.952$ ft.

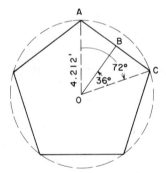

Figure 8-40

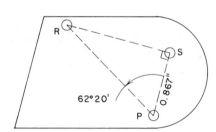

Figure 8-41

example   Find the distance between the centers of the holes R and S to be drilled in the template shown in Figure 8-41. You have

$$\tan 62° 20′ = \frac{RS}{SP} = \frac{RS}{0.867 \text{ in.}}$$

When you attempt to find tan 62° 20′ in Table 16, you discover that the table only goes to 45°. However, the table makes use of the fact that cofunctions of complementary angles are equal. If you start on the last page of the table and read from the bottom up, you see that the table actually contains a listing of ratios for angles up to 89° 60′ (90°). When reading the table in the range 45°–90°, use the headings (called *footings*) at the bottom of the columns and the angle listed in the extreme right-hand column. For this example, read up the tan column (tan at the bottom) to the line with the 60° 20′ entry: tan 62° 20′ = 1.907. Note that 1.907 is also the value listed for cot 27° 40′. (How is 27° 40′ related to 62° 20′?) You can now write the proportion

$$\frac{RS}{0.867 \text{ in.}} = 1.907$$

Solving for *RS* gives *RS* = 1.65 in.

Linear Interpolation and Significant Figures

A man once described interpolation as a necessary headache. This description is only partly true. It is necessary, but not necessarily a headache. Actually, you already know how to interpolate and, in fact, do it frequently.

example   If a certain grade and cut of meat sells at $0.60 for 1 lb and $1.20 for 2 lb, how much would you expect to pay for $1\frac{1}{2}$ lb? Answer: $0.90.

example   If a building board 2 in. thick costs $0.48/sq ft and another board of the same material 4 in. thick costs $0.96/sq ft, what would you expect a board 3 in. thick to cost? (Assume all other factors to be equal.) You would probably think something like this: 3 in. is halfway between 2 in. and 4 in., so the cost would be halfway between $0.48/sq ft and $0.96/sq ft, which comes to $0.72/sq ft.

example   A sales chart beside a barrel of paint thinner reads as follows:

Solvent XXX

1 gal	$0.50
2 gal	$0.95
3 gal	$1.35
4 gal	$1.70
5 gal	$2.00

A pump on the barrel is calibrated to dispense the thinner in gallons and tenths of a gallon. How much should a person pay for 3.3 gal? He would certainly pay $1.35 for the 3 gal. The question to answer is what does it cost for 0.3 gal figured at the rate/gallon between the 3 and 4 gal entries. Because he is buying 0.3 of the fourth gallon, this rate is the difference between the $1.35 and $1.70 figures which is $0.35. Now 0.3 of $0.35 is $0.11 (to the nearest cent). So he must add $0.11 to the $1.35 figure which comes to $1.46 for the 3.3 gal.

The procedure used in the above examples to determine a number that exists between two given numbers is called *interpolation*. Before looking at more examples, let us determine the situations in which interpolation becomes necessary. The following table will save a great deal of unnecessary interpolation.

Significant Figures in Data	Express Angle to Nearest
2	Degree
3	0.1 Degree*
4	Minute
5	0.1 Minute
6	Second

Note: If five-figure accuracy is needed, use at least five-place tables. For six-figure accuracy, use at least six-place tables.

example   Determine tan 30° 33′, using Table 16. In Table 16:

$$\left.\begin{array}{l} \tan 30° \ 40' = 0.5930 \\[6pt] \tan 30° \ 33' = \\[6pt] \tan 30° \ 30' = 0.5890 \end{array}\right\} \text{difference} = 0.0040$$

Because 33′ falls 0.3 of the way between 30′ and 40′, determine 0.3 of 0.0040 which = 0.0012 and add it to the entry for 30° 30′:

$$0.5890 + 0.0012 = 0.5902$$

for the calculated value of tan 30° 33′.

example   Determine tan 57° 16′. From Table 16:

$$\left.\begin{array}{l} \tan 57° \ 10' = 1.550 \\[18pt] \tan 57° \ 16' = \\[12pt] \tan 57° \ 20' = 1.560 \end{array}\right\} \text{difference} = 0.010$$

* For practical purposes in using Table 16, it is only necessary to convert tenths of a degree to minutes and use the entry to the nearest 10′. For example, 22.4° = 22° 24′ so use entry for 22° 20′.

Because 6' is $\frac{6}{10}$ of the 10' interval, determine $\frac{6}{10}$ of (0.010) which $= 0.006$ and add it to the 1.550 entry, obtaining tan 57° 16' $= 1.556$.

example Determine $\sphericalangle\,\theta$ if sin $\theta = 0.5404$. In Table 16, find the sin entries on either side of 0.5404:

$$\begin{array}{l} \text{sin } 32°\ 40' = 0.5398 \\[4pt] \qquad\qquad\quad -0.0006 \\[4pt] \text{sin } \theta \qquad = 0.5404 \qquad\qquad -0.0024 \\[4pt] \text{sin } 32°\ 50' = 0.5422 \end{array}$$

The entry for $\sphericalangle\,\theta$ lies 0.0006/0.0024 or 1/4 of the way between 40' and 50'; 1/4 of 10' $\approx 3'$. Therefore, add 3' to 32° 40', obtaining $\theta = 32°\ 43'$.

example Determine cot 35° 16'. From Table 16:

$$\begin{array}{l} \text{cot } 35°\ 10' \qquad = 1.419 \\[4pt] \qquad\qquad -6' \\[4pt] \text{cot } 35°\ 16' \qquad\qquad\qquad -0.008 \\[4pt] \text{cot } 35°\ 20' \quad = 1.411 \end{array}$$

First note that the entries in the cot column decrease as the angle increases. Because 16' lies $\frac{6}{10}$ of the way between 10' and 20', decrease 1.419 by $\frac{6}{10}$ of 0.008. 1.419–0.005* $= 1.414$. Therefore, cot 35° 16' $= 1.414$.

example Determine $\sphericalangle\,\theta$ if sec $\theta = 1.762$. In Table 16:

$$\begin{array}{l} \text{sec } 55°\ 20' = 1.758 \\[4pt] \qquad\qquad\quad -0.004 \\[4pt] \text{sec } \theta \qquad = 1.762 \qquad\qquad -0.008 \\[4pt] \text{sec } 55°\ 30' = 1.766 \end{array}$$

You see that 1.762 lies 1/2 of the way between the entries for sec 55° 20' and sec 55° 30'. Therefore, θ has a magnitude halfway between 55° 20' and 55° 30':

$$\theta = 55°\ 25'$$

* $\frac{6}{10} \times 0.008 = 0.0048 \approx 0.005$. You cannot interpolate to increase the number of significant figures.

EXERCISE 8.5

Use Table 16, Appendix D, and interpolate when necessary to work the following problems.

1. $\sin 24° 30' =$
2. $\tan 65° 50' =$
3. $\cos 38° 50' =$
4. $\sec 55° 30' =$
5. $\cos 82° 10' =$
6. $\sin 33° 33' =$
7. $\cos 42° 12' =$
8. $\tan 53° 27' =$
9. Determine θ if $\sin \theta = 0.4937$.
10. Determine θ if $\tan \theta = 1.415$.
11. Determine θ if $\cot \theta = 2.341$.
12. Determine θ if $\cot \theta = 0.4781$.
13. Find the length of one edge of a regular octagon inscribed in a circle whose diameter is 12.00 ft.
14. Find the length of one edge of a regular octagon circumscribed about the circle in Problem 13.
15. The end plate of a boiler is to be riveted as shown in Figure 8-42. Find the distance between centers of adjacent rivets. Think of the rivets as being located at the vertices of a regular 15-sided polygon.

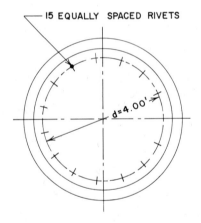

15 EQUALLY SPACED RIVETS

$d=4.00'$

Figure 8-42

16. How long is the supporting wire AB in Figure 8-43? How long is AC?
17. In Figure 8-43, find:
    (a) AB if $BC = 15.5$ ft.
    (b) BC if $AB = 22.2$ ft.
    (c) AC if $BC = 10.8$ ft.
18. Find the length of gas pipe needed to run from point N to point W in Figure 8-44.

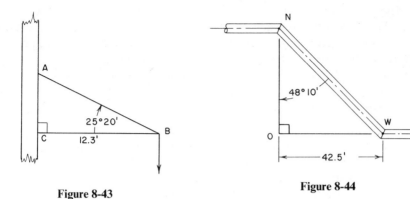

Figure 8-43

Figure 8-44

19. Compute the length of NW in Figure 8-44 if $OW = 5280$ ft.
20. (a) Find the area of the triangular opening to be cut from a metal plate as shown in Figure 8-45. (*Hint:* Drop a perpendicular from P; solve for altitude.)
    (b) Estimate the weight of the triangular piece and then calculate its weight if the original plate weighs 99.4 lb.

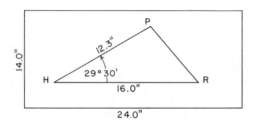

Figure 8-45

8.6 TRIGONOMETRIC SCALES ON THE SLIDE RULE

To understand the structure and use of the S (sine), T (tangent), and ST (sine-tangent) scales on the slide rule, turn to Table 16, Appendix D, and observe the entries in the sine and tangent columns. Starting with 0°, the entries in these two columns are the same down to 2° 20′; rounded to thousandths, they remain the same down to 6°. Because slide rule accuracy usually denotes three-figure accuracy, the sine and tangent scales are combined into one scale—the ST scale—for angles less than 5.74°. For angles greater than 5.74°, the sine and tangent scales are separate.

The *ST*, *S*, and *T* scales operate in conjunction with the *C*, *D*, *CI*, and *DI* scales. The angle entry is made by placing the hairline (*H*) over the angle on the appropriate scale; the value of the ratio is read below the hairline on the appropriate *C*, *D*, *CI*, or *DI* scale. Most makes of slide rules show the range in sine and tangent values for the *ST*, *S*, and *T* scales at the ends of the respective scales.

Changing Minutes to a Decimal Part of a Degree

When working with angles in degrees and minutes, the minutes must be changed to a decimal part of a degree (for setting on most makes of slide rules). First change minutes to hundredths of a degree. Remember that $0.1° = 6'$; it is only necessary to divide mentally the given number of minutes by 60. To do this, first divide by 10 (shifting the decimal point left one place) and then divide by 6. For example, to change 24° 44′, think $\frac{4.4}{6} \approx 0.73$:

$$24° \, 44' \approx 24.73°$$

The Sine-Tangent (*ST*) Scale

The *ST* scale runs from 34.38′ ($\approx 0.57°$) to 5.74°. Sine and tangent ratios for angles on this scale range in value from 0.01 to 0.1. This fact facilitates keeping track of the decimal point when calculating with this scale.

The Sine (*S*) Scale

Observe that on the *S* scale the double entries at each integral mark are complementary angles. The *S* scale runs from 5.74° to 90°. Sine and cosine values for angles in this interval fall between 0.1 and 1.0. The fact that the sine of an angle equals the cosine of its complement allows the determination of sine and cosine values from the same scale. When working with the sine function, read the *S* scale from left to right; when working with the cosine function read the *S* scale from right to left. The *S* scale is graduated in degrees and tenths of degrees for angles less than 20°. For angles between 20° and 30°, each small subdivision is 0.2°. Between 30° and 60°, each small division is 0.5°. Between 60° and 80°, each subdivision is 1°. The last mark before 90° is 85°. Do not make the mistake of thinking that the last bit of the *S* scale from 85° to 90° is worthless. Even though it seems that there is considerable guesswork in making a setting in this interval, readings from the *D* scale are usually within 0.001 of the value obtained from the table.

The *T* Scale

The *T* scale (on most slide rules) is double numbered, covering the range from 5.71° to 45°, left to right, and from 45° to 84.29°, right to left. In the 5.71° to 45° interval, the tangent ratio falls between 0.1 and 1.0. At 45° the tangent value becomes equal to 1. In the 45° to 84.29° interval, the values for the tangent ratio fall between 1 and 10. As in the case of the S scale the dual entries at the integral points are complementary angles, thus allowing the same scale to be used to find cotangent values.

The list below shows settings for determining trigonometric values for several angles.

Angle	Slide Rule Setting
$\sin 42' = \sin 0.7°$	H over 0.7° on ST; read 0.0122 on D
$\tan 54' = \tan 0.9°$	H over 0.9° on ST; read 0.01571 on D
$\sin 6° \ 45' = \sin 6.75°$	H over 6.75° on S; read 0.1175 on D
$\tan 12° \ 24' = \tan 12.4°$	H over 12.4° on T; read 0.220 on D
$\cos 39° \ 42' = \cos 39.7°$	H over 39.7° on S (reading right to left); read 0.769 on D
$\cos 87° \ 8' = \sin 2° \ 52' = \sin 2.87°$	H over 2.87° on ST; read 0.0500 on D
$\cot 82° \ 12' = \cot 82.2°$	H over 82.2° on T; read 0.1370 on D
$\tan 75° \ 32' = \tan 75.5°$	H over 75.5° on T; read 3.87 on DI

Scale Information

At this point a few observations about the construction of slide rule scales would be helpful.

1. The C and D scales form the basic pair of scales. All other scales relate to this pair.

2. When the numbers and graduations are color coded, black numbers always increase from left to right and colored (usually red) numbers increase from right to left.

3. The sine, tangent, and secant functions (in other words the direct trigonometric functions) are read on like colors, black to black or red to red, but never on opposite colors. The cofunctions (cosine, cotangent, and cosecant) are always read on opposite colors.

4. *Locating the decimal point*, when the angle setting is on the ST scale and you are reading:

(a) The C (or D scale), the value of the function is between 0.01 and 0.1.
(b) The CI (or DI) scale, the value of the function is between 10 and 100.

When the angle setting is on the S or T scale and you are reading:

(a) The C (or D) scale, the value of the function is between 0.1 and 1.0 (except $\tan 45° = 1$ and $\sin 90° = 1$).
(b) The CI (or DI) scale, the value of the function is between 1 and 10.

FUNCTION	HAIRLINE OVER ANGLE θ ON ──▶	SCALE	READ SCALE	VALUE OF FUNCTION	SPECIAL ANGLES
Sine or Tangent	0.57° ⩽ θ ⩽ 5.74°	ST	C or D	0.01 to 0.1	
Sine	5.74° ⩽ θ ⩽ 90°	S	C or D	0.1 to 1	Sin 90° = 1
Cosine	0° ⩽ θ ⩽ 84.26°	S (Reading right to left)	C or D	0.1 to 1 ◄──	Cos 0° = 1
Cosine	84.26° ⩽ θ ⩽ 89.43° Use Sin(90 − θ)°	ST	C or D	0.01 to 0.1 ◄──	
Tangent	5.71° ⩽ θ ⩽ 45°	T	C or D	0.1 to 1	Tan 45° = 1
Tangent	45° ⩽ θ ⩽ 84.29°	T (Reading right to left)	CI or DI	1 to 10	
Cotangent	0.57° ⩽ θ ⩽ 5.74°	ST	CI or DI	10 to 100	Cot 45° = 1
	5.71° ⩽ θ ⩽ 45°	T	CI or DI	1 to 10	
	45° ⩽ θ ⩽ 84.29°	T (rt. to left)	C or D	0.1 to 1	
Secant	0° ⩽ θ ⩽ 84.26°	S (Reading right to left)	CI or DI	1 to 10	Sec 0° = 1
∗{	84.26° ⩽ θ ⩽ 89.43°	ST	CI or DI	10 to 100	
Cosecant	0.57° ⩽ θ ⩽ 5.74°	ST	CI or DI	10 to 100	Csc 90° = 1
	5.74° ⩽ θ ⩽ 90°	S	CI or DI	1 to 10	

∗ Use Sec θ = $\frac{1}{\text{Cos }\theta}$ = $\frac{1}{\text{Sin }(90-\theta)°}$, Set up Sin(90 − θ)° on ST scale.

Figure 8-46   Data for Sine and Tangent Scales

EXERCISE 8.6

Use the slide rule to determine the values in Problems 1–20. Check, using Table 16, Appendix D.

1. sin 0.620°
2. sin 6.20°
3. sin 62.0°
4. sin 87.5°
5. sin 90°
6. tan 0.740°
7. tan 7.40°

8. tan 28.6°
9. tan 45°
10. tan 54°
11. tan 84.25°
12. cos 2.5°
13. cos 25.5°
14. cos 48° 48′

15. cos 62° 12′
16. cos 88.4°
17. csc 0.620°
18. csc 6.20°
19. csc 62.0°
20. csc 87.5°

Use the slide rule to determine ∡ θ in Problems 21–30.

21. sin θ = 0.129
22. cos θ = 0.208
23. tan θ = 0.352
24. tan θ = 2.280

25. sin θ = 0.0515
26. cos θ = 0.895
27. tan θ = 0.0166
28. cot θ = 0.216

29. cot θ = 2.760
30. sin θ = 0.0715

8.7 SLIDE RULE OPERATIONS WITH TRIGONOMETRIC RATIOS

Consider the multiplication of 175 by 0.416 using the slide rule. With the left index of C over 175 on D and hairline (H) over 416 on C, observe that 22.6° is found under H on the T scale and 24.6° is found under H on the S scale. In effect, you are really working the following three problems.

$$(175)(0.416) = 72.8$$

$$(175)(\tan 22.6°) = (175)(0.416) = 72.8$$

$$(175)(\sin 24.6°) = (175)(0.416) = 72.8$$

To multiply by $\sin \theta$, set the left index of C (or S) over the multiplicand on D and place H over θ on the S scale. This setting places the value of $\sin \theta$ under H on C (which is the appropriate position for a multiplier when using the C and D scales). The product is read under H on D.

example To multiply $(256)(\sin 22.0°)$, set left index of S over 256 on D and set H over 22.0° on S. The product 95.9 is found under H on D. (*Note:* $\sin 22.0° = 0.375$ is found under H on C.)

example To multiply 75.3 by $\cos 38.2°$, set right index of S over 75.3 on D. Move H to 38.2° on S (reading right to left for cosine); read "592" under H on D. With the value of $\cos 38.2°$ between 0.1 and 1, estimate the product to fall between 7.53 and 75.3 and locate the decimal point, obtaining 59.2 for the product.

example To multiply 13.5 by $\tan 31.4°$, set left index of T over 13.5 on D; move H to 31.4° on T. Read "824" under H on D. With the value of $\tan 31.4°$ between 0.1 and 1, estimate the product to fall between 1.35 and 13.5. Then locate the decimal point, obtaining 8.24 for the product.

example To multiply $(59.5)(\tan 72.6°)$*, work as $59.5/\cot 72.6°$. Set H over 59.5 on D. Move slide to place 72.6° on T (reading right to left) under H. Read "1899" under left index of S on D. $\tan \theta$ ($45° < \theta < 84.29°$) is between 1 and 10; therefore the product lies between 59.5 and 595. This gives $19\tilde{0}$ for the product.

example To multiply 38.6 by $\cos 41.2°$, set right index of S over 38.6 on D and place H over 41.2° on S (reading right to left). Read "290" under H on D. Because $\cos 41.2°$ is between 0.1 and 1, you know that the product must be between 3.86 and 38.6. Therefore, you obtain 29.0 for the product.

---

* A direct method for this calculation is to set the left index of T (use H for alignment) over 59.5 on DI. Move H to 72.6° on T; read "1899" under H on DI. Locate the decimal point as before.

example Multiplying by cos θ (84.29° < θ < 89.4°). Multiply 154 by cos 88.6°.

1. Set up the problem in factored form:

$$(154)(\cos 88.6°) = \text{product}$$

2. Replace cos 88.6° with sin (90 − 88.6)° = sin 1.4°, because 88.6° is not found on the S scale (right to left). You then have

$$(154)(\sin 1.4°) = \text{product}$$

3. Set left index of S over 154 on D and move H to 1.4° on the ST scale. Read "376" under H on D.
4. Because sin 1.4° is between 0.01 and 0.1, you know that the product must be between 1.54 and 15.4. You thus obtain 3.76 for the product.

Consider the slide rule solution of the following division problem: 54.0 ÷ 0.360. The divisor 0.360 on the C scale is located directly above the dividend 54.0 on the D scale. The quotient 150 is found under the left index of C on D. Note that under H on S, you find 21.1° (in black) and 68.9° (in red); under H on T, you find 19.8° (in black) and 70.2° (in red). In effect, you are working the following problems:

$$\frac{54.0}{0.360} = 15\tilde{0}$$

$$\frac{54.0}{\sin 21.1°} = \frac{54.0}{\cos 68.9°} = 15\tilde{0}$$

$$\frac{54.0}{\tan 19.8°} = \frac{54.0}{\cot 70.2°} = 15\tilde{0}$$

example To divide 212 by sin 35.0°, place H over 212 on D. Move the slide to place 35.0° under H on S. Read the quotient digits "370" under the right index of S on D. Knowing that sin 35.0° ≈ ½, you can approximate the answer as about 400. Locating the decimal point gives 37$\tilde{0}$ for the calculated quotient.

example Divide 1880 by tan 42.0°.

1. Set H over 1880 on D.
2. Move slide to place 42° on T under H.
3. Read digits of the quotient "209" under the right index of S on D.
4. With tan 42° ≈ 1, approximate your answer as about 1900. Then, locating the decimal point, obtain 2090 for the quotient.

example   Divide $15\tilde{0}$ by sin 2.5°.

1. Set H over 150 on D.
2. Move slide to place 2.5 under H on *ST*.
3. Read the quotient digits "344" under the right index of S on D.
4. With sin 2.5° falling between 0.01 and 0.1, the quotient will fall between 1500 and 15,000. Locating the decimal point gives you 3440 for the quotient.

In summary, to divide a number N by sin θ or tan θ ($\theta \leqslant 45°$), set H over N on the D scale and move the slide to place $\measuredangle \theta$ under H on S or T. Read the quotient under the left (or right) index of S on D.

Calculations with Tangent, Secant, Cotangent, and Cosecant Values

We are not going to emphasize any calculations with secant, cotangent, or cosecant functions. For practical purposes, make the following conversions:

1. For $\theta < 45°$, replace cot θ with 1/tan θ and proceed to operate with tan θ as a factor.
2. For $\theta > 45°$, replace tan θ with 1/cot and proceed to operate with cot θ as a factor.
3. Replace sec θ with 1/cos θ and proceed to operate with cos θ as a factor.
4. Replace csc θ with 1/sin θ and proceed to operate with sin θ as a factor.

For example:

$$\frac{26.4}{\csc 35°} = \frac{26.4}{\dfrac{1}{\sin 35°}} \cdot = (26.4)(\sin 35°)$$

$$(48.2)(\csc 42.2°) = (48.2)\frac{1}{\sin 42.2°} = \frac{48.2}{\sin 42.2°}$$

$$(128)(\sec 52°) = (128)\frac{1}{\cos 52°} = \frac{128}{\cos 52°}$$

$$\frac{46.8}{\sec 36°} = \frac{46.8}{\dfrac{1}{\cos 36°}} \cdot = (46.8)(\cos 36°)$$

$$(54.1)(\cot 32.2°) = (54.1)\frac{1}{\tan 32.2°} = \frac{54.1}{\tan 32.2°}$$

$$\frac{9.23}{\cot 14.6°} = \frac{9.23}{\dfrac{1}{\tan 14.6°}} \cdot = (9.23)(\tan 14.6°)$$

Dividing by tan θ (45° < θ < 84.29°)

example   Divide 65.5 by tan 73.4°.

method I   Work as (65.5) (cot 73.4°).

  1. With cot 73.4° between 0.1 and 1, the answer will fall between 6.55 and 65.5.
  2. Set right index of *T* over 65.5 on *D* and move *H* to 73.4° on *T*.
  3. Read "195" under *H* on *D* and locate the decimal point, obtaining 19.5.

method II

  1. Estimate the quotient to be between 6.55 and 65.5.
  2. Set *H* over 65.5 on the *DI* scale.
  3. Move *slide* to place 73.4° on *T* under *H*.
  4. Read quotient digits "195" under right index of *S* on *DI*. (Use *H* for alignment.)
  5. Locate decimal point, obtaining 19.5.

method III   Another method worth knowing comes from the trigonometric identity $\tan \theta = \sin \theta / \cos \theta$. If you replace tan 73.4° with

$$\frac{\sin 73.4°}{\cos 73.4°}$$

the problem becomes

$$\frac{(65.5)}{\dfrac{\sin 73.4°}{\cos 73.4°}} \cdot = \frac{(65.5)(\cos 73.4°)}{\sin 73.4°}$$

  1. Set *H* over 65.5 on *D*.
  2. Move *slide* to place 73.4° on *S* under *H* (reading left to right).
  3. Move *H* to 73.4° on *S* (reading right to left).
  4. Read "195" under *H* on *D*.
  5. Estimate quotient and locate decimal point, obtaining 19.5 as before.

EXERCISE 8.7

Perform the following indicated operations with the slide rule. Then determine values for the trigonometric factors from Table 16 and recalculate as a check.

  1. (120)(sin 20°)
  2. (17.5)(sin 32.5°)
  3. (25.0)(sin 41.2°)
  4. (3.40)(sin 5.1°)
  5. (1.40)(tan 15° 43′)
  6. (2.50)(tan 21° 7′)

7. $(31.4)(\tan 36.4°)$

8. $(408)(\tan 42.6°)$

9. $(12.6)(\sin 2.1°)$

10. $(1.44)(\sin 1.5°)$

11. $(4.20)(\tan 1.6°)$

12. $(5.10)(\tan 0.75°)$

13. $\dfrac{17.5}{\sin 65.0°}$

14. $\dfrac{21.4}{\sin 16° 20'}$

15. $348 \div (\tan 16° 12')$

16. $0.520 \div \tan 2.1°$

17. $8.20 \div \sin 0.88°$

18. $0.0025 \div (\tan 0.61°)$

19. $(188)(\csc 4.0°)$

20. $(266)(\cot 14.6°)$

21. $(68.2)(\cos 42.0°)$

22. $\dfrac{(456)}{\cos 65° 5'}$

23. $(29.2)(\tan 72.4°)$ (Work two ways.)

24. $\dfrac{4650}{\tan 35.6°}$ (Work two ways.)

25. $\dfrac{(31.5)(\sin 82.4°)}{\cos 46.6°}$

26. $(8.28)(\cot 44.6°) = \dfrac{(8.28)(\cos 44.6°)}{\sin 44.6°} = \dfrac{8.28}{\tan 44.6°}$

                                      Work this   Work this

27. $(56.6)(\csc 35.5°)$ (Use sine factor.)

28. $\dfrac{1295}{\sec 41.1°}$ (Use cosine factor.)

29. $\dfrac{1050}{\cot 72.4°}$ (Use tangent factor.)

30. $(1640)(\sec 44.8°)$ (Use cosine factor.)

8.8 SOLUTIONS OF RIGHT TRIANGLES

Before considering any specific right triangle solutions, review the following information.

1. A right triangle has: (a) three sides—one hypotenuse and two legs; (b) one right angle and two complementary (acute) angles; (c) the longer leg opposite the larger acute angle; (d) the shorter leg opposite the smaller acute angle; and (e) the hypotenuse opposite the 90° angle.

2. To solve a right triangle for an unknown side or angle, you must know: (a) any two sides, or (b) any side and one acute angle.

3. Whenever condition 2(a) or 2(b) is known, any unknown part of the right triangle can be determined by using only the sine and/or tangent ratios.

Basic Procedures

When two sides are known, set up a ratio with the shorter known side over the longer known side.

case I When both known sides are legs, use the tangent ratio for the angle opposite the shorter known side.

case II When the longer known side is the hypotenuse, use the sine ratio for the angle opposite the shorter known side.

When one side and one acute angle (θ) are known, set up one of the following proportions.

case III When the known side is the hypotenuse, use

$$\sin \theta = \frac{\text{side opposite } \theta}{\text{hypotenuse}}$$

case IV When the known side is a leg, use

$$\tan \theta = \frac{\text{side opposite } \theta}{\text{known leg}} \text{ or } \frac{\text{known leg}}{\text{leg adjacent } \theta}$$

The next four examples illustrate the four cases above, respectively. The slide rule procedure is given in each example. (Note the similarity of the slide rule procedure in all four examples.)

example *Two legs known.* A metal plate is to have a corner cut off as shown in Figure 8-47. Find angle θ. Following the procedure in Case I, set up the ratio

$$\tan \theta = \frac{12.4 \text{ in.}}{26.2 \text{ in.}}$$

Express $\tan \theta = 12.4/26.2$ as the decimal 0.4733. Then, using Table 16, Appendix D, for $\tan \theta = 0.4733$, $\theta = 25° 20'* = 25.3°$.

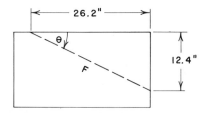

Figure 8-47

Slide rule solution. Set H over the shorter leg (12.4) on D and the right index of S over the longer leg (26.2) on D. Both acute angles of the triangle can

---

* Angle θ may be determined to nearest 10' for three-figure accuracy. Interpolation gives $\theta = 25° 19'$ which also equals 25.3° to nearest 0.1°.

be read under *H* on *T*. When the shorter of the two legs is opposite the desired angle (as in this case), read the smaller angle. But when the longer leg is opposite the desired angle, read the larger angle. For θ, read the smaller angle 25.3°.

example   *Hypotenuse and one leg known.* An electrical conduit, 152 ft long, is to be placed across the corner of a concrete form (Figure 8-48). Find angles *A* and *B*. Following the procedure outlined in Case II above, first set up the ratio

$$\frac{\text{leg opposite } B}{\text{hypotenuse}} = \frac{48.5 \text{ft}}{152 \text{ft}} = \sin B$$

Express $\sin B = 48.5/152$ as the decimal 0.3191. Then, using Table 16, determine $\sphericalangle B$ for $\sin B = 0.3191$. By some mental interpolation,

$$\sphericalangle B \approx 18° \ 37' = 18.6°$$

$$\sphericalangle A = 90° - 18.6° = 71.4°$$

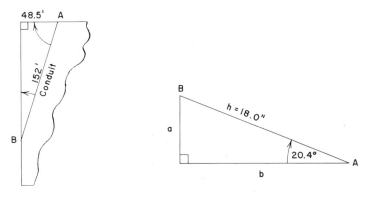

Figure 8-48   Figure 8-49

Slide rule solution. Set *H* over the shorter side (48.5) on *D* and the left index of *S* over the longer side (152) on *D*. Both acute angles can be read under *H* on *S*. Angle *B* (the angle opposite the 48.5 ft leg) is read on *S*, reading left to right. Angle *A* is read on *S*, reading right to left:

$$\sphericalangle B = 18.6°$$

$$\sphericalangle A = 71.4°$$

example   *Hypotenuse and one acute angle known.* Find side *a* in Figure 8-49. Following the procedure in Case III, set up the sine ratio involving the

20.4° angle:

$$\sin 20.4° = \frac{a}{18.0 \text{ in.}} \tag{1}$$

Determine the value of sin 20.4° (Table 16) = 0.3486. Then rewrite the proportion as

$$0.3486 = \frac{a}{18.0 \text{ in.}} \tag{2}$$

Solve this equation for side a:

$$a = (0.3486)(18.0) \text{ in.} \tag{3}$$

$$a = 6.27 \text{ in.} \tag{4}$$

Slide rule solution. Set the left index of S over the longer side (18.0) on D. If you did know side a, you would place H over a on D, and 20.4° would be found under H on S. Because you do not know a but you do know 20.4°, place H on 20.4° on S and read $a = 6.27$ in. under H on D.

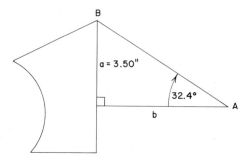

Figure 8-50

example   *One leg and an acute angle known.* Determine side b of the template shown in Figure 8-50. Following the procedure given in Case IV, set up the proportion

$$\tan 32.4° = \frac{3.50 \text{ in.}}{b} \tag{1}$$

Solve equation (1) for b:

$$b = \frac{3.50 \text{ in.}}{\tan 32.4°} \tag{2}$$

Substituting for tan 32.4° = 0.6346 (from Table 16) gives

$$b = \frac{3.50 \text{ in.}}{0.6346} \tag{3}$$

which yields

$$b = 5.52 \text{ in.} \tag{4}$$

Slide rule solution. Set H over the shorter leg (3.50) on D and move slide to place 32.4° under H on T. This setting places the right index of T over the longer leg b on D. Read $b = 5.52$ in. under right index of T on D (*Note:* The setting is the same as if you had known that $b = 5.52$ in. at the beginning.)

example   *One leg and an acute angle known.* A triangular-shaped orifice is to be broached in a circular forging as shown in Figure 8-51. (In this example the

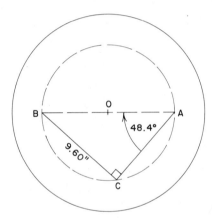

Figure 8-51

known angle is greater than 45°, making the known side the longer leg.) Proceed as directed in Case IV and set up the proportion

$$\tan 48.4° = \frac{BC}{AC} = \frac{9.60 \text{ in.}}{AC} \tag{1}$$

Solve equation (1) for AC:

$$AC = \frac{9.60 \text{ in.}}{\tan 48.4°} \tag{2}$$

Substituting tan 48.4° = 1.126 (from Table 16), you obtain

$$AC = \frac{9.60 \text{ in.}}{1.126} \tag{3}$$

which yields

$$AC = 8.52 \text{ in.} \tag{4}$$

Slide rule solution. Set the right index of *T* over the longer leg (9.60) on *D* and *H* over 48.4° on *T*. Read the shorter leg, *AC* = 8.52 in., under *H* on *D*.

Summary of Basic Procedure Using S or T Scales

All right triangle problems can be solved by the basic procedure used in the previous examples.

1. Establish a sin θ or tan θ ratio in terms of two sides:

$$\sin \theta = \frac{\text{side opposite } \theta}{\text{hypotenuse}} \qquad \tan \theta = \frac{\text{side opposite } \theta}{\text{side adjacent } \theta}$$

2. Always set *H* over the shorter side (involved in the ratio) on *D* and over the angle θ* on *S* or *T* (depending on the ratio). If you do not know the value of the shorter side, you know the angle and vice-versa.

3. An index of *S* and *T* is always placed (or ends) over the longer side (involved in the ratio) on *D*.

4. The setting(s) over known side(s) always precedes the angle setting (or reading).

EXERCISE 8.8

For each of the Problems 1–12, draw to scale and dimension an appropriate right triangle, ABC, as shown in Figure 8-52. Determine the designated part(s) by measuring and then solve using trigonometry.

1. $a = 25.6$ in., $b = 14.2$ in.; find $\angle A$ and $\angle B$.
2. $a = 14.4$ ft, $b = 26.6$ ft; find $\angle A$ and $\angle B$.
3. $a = 7.44$ ft, $c = 12.3$ ft; find $\angle A$ and $\angle B$.
4. $b = 14.2$ cm, $c = 20.4$ cm; find $\angle A$ and $\angle B$.
5. $c = 41.4$ in., $a = 11.2$ in.; find $\angle A$ and $\angle B$.
6. $a = 17.5$ in., $\angle A = 24.4°$; find side c.
7. $a = 42.4$ ft, $\angle A = 55.4°$; find side c.
8. $b = 12.6$ in., $\angle A = 32.8°$; find $\angle B$ and side c.

* If either acute angle is known, it is simple to determine its complement.

9. $b = 48.2$ ft, $\angle B = 40.2°$; find $\angle A$ and side c.
10. $b = 126$ m, $\angle A = 35.8°$; find $\angle B$ and side a.
11. Side $c = 18.7$ in., $\angle A = 42.7°$; find side a.
12. $c = 59.7$ yd, $\angle B = 35.3°$; find side b.

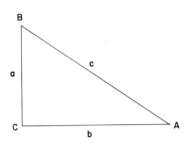

Figure 8-52

In Problems 13–16, draw to scale an appropriate right triangle as shown in Figure 8-53 and dimension the given parts. This triangle is called an impedance triangle in electrical work. Determine the required information by measuring and by using trigonometry.

13. Side $X_L = 24.6$ units, side $R = 31.2$ units; find $\angle \theta$ and side Z.
14. Side $X_L = 32.9$ units, side $R = 28.6$ units; find $\angle \theta$ and side Z.
15. Side $R = 52.6$ ohms, $\angle \theta = 41.2°$; find side X_L and side Z.
16. Side $Z = 38.2$ ohms, $\angle \theta = 33.5°$; find side X_L and side R.

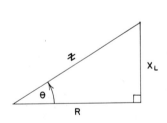

Figure 8-53

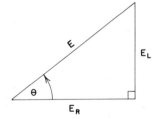

Figure 8-54

In Problems 17–20, draw an appropriate right triangle to scale as shown in Figure 8-54. Dimension the given parts and solve for the designated part(s) by measuring and by trigonometry. This triangle is called a voltage triangle in electrical work.

17. Side $E_L = 52.3$ units, side $E_R = 42.4$ units; find $\angle \theta$ and E.
18. Side $E_L = 138$ units, side $E_R = 245$ units; find $\angle \theta$ and E.
19. Side $E = 120$ volts, $\angle \theta = 25.2°$; find E_L and E_R.
20. Side $E_L = 405$ volts, $\angle \theta = 54.3°$; find E_R and E.

In Problems 21–24, draw to scale an appropriate figure showing dimensions for the given data and determine the designated part(s) by measuring and by trigonometry. See Figure 8-55.

21. $OP = 1560$ ft, $\theta = 32.6°$; find distances OB and PB.
22. $OP = 2180$ ft, $\theta = 44.2°$; find distances PB and OB.
23. Surveyors call the distance PB (Figure 8-55) the *latitude* of point P from point O, and the distance OB the *departure* of point P from point O. Find the latitude and departure of point P (from O) in Figure 8-55 if the course distance $OP = 465$ ft and $\theta = 28.2°$.
24. Determine $\sphericalangle \theta$ (Figure 8-55) if the latitude $PB = 462$ ft and the departure $OB = 512$ ft. Then determine the course distance OP.

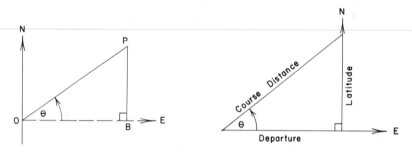

Figure 8-55   **Figure 8-56**

25. Using Figure 8-56 and in terms of "course distance," "latitude," and "departure," write ratios for the following:
    (a) $\sin \theta$. Then solve for the latitude.
    (b) $\cos \theta$. Then solve for the departure.
    (c) $\tan \theta$.

In Problems 26–30, use the statements given below and the slide rule to calculate the required information. (Refer to Figure 8-56.)

$$\text{latitude} = (\text{course distance})(\sin \theta)$$

$$\text{departure} = (\text{course distance})(\cos \theta)$$

$$\tan \theta = \frac{\text{latitude}}{\text{departure}}$$

26. Course distance $= 652$ ft, $\theta = 31.4°$; determine latitude and departure.
27. Course distance $= 1650$ ft, $\theta = 22.5°$; determine latitude and departure.
28. Course distance $= 755$ ft, $\theta = 48.3°$; determine latitude and departure.
29. Latitude $= 157$ ft, departure $= 206$ ft; determine $\sphericalangle \theta$ and course distance.
30. Course distance $= 1250$ ft, latitude $= 486$ ft; determine $\sphericalangle \theta$ and departure.

8.9   FUNCTIONS OF ANGLES OF ANY MAGNITUDE

An angle in standard position may be of any magnitude (positive or negative). Consider a directed line segment running from the origin O to a point P with coordinates (x, y) and forming the terminal side of $\angle\,\theta$ (in standard position). See Figure 8-57. The line OP is called a *radius vector* because the segment OP has both magnitude (represented by its length) and direction (represented by the arrow pointing away from the origin along the terminal side of $\angle\,\theta$). For our present purpose, we shall represent the distance from O to P by the letter r (for radius); i.e., $r =$ length of radius vector OP.

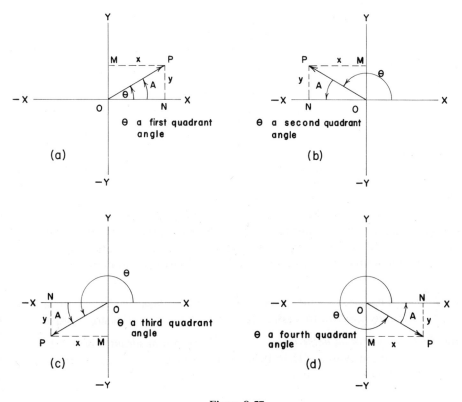

Figure 8-57

When θ is an acute first quadrant angle (Figure 8-57(a)), you will have no difficulty working with the functions of $\angle\,\theta$. When θ becomes equal to or greater than 90°, you will need additional information to make such ratios as sin 90°, tan 240°, or sin 315° meaningful.

Definitions of Trigonometric Functions

The definitions of trigonometric functions for an angle* of any magnitude are:

$$\sin \theta = \frac{\text{ordinate value}}{\text{radius}} = \frac{y}{r}$$

$$\cos \theta = \frac{\text{abscissa value}}{\text{radius}} = \frac{x}{r}$$

$$\tan \theta = \frac{\text{ordinate value}}{\text{abscissa value}} = \frac{y}{x}$$

$$\cot \theta = \frac{\text{abscissa value}}{\text{ordinate value}} = \frac{x}{y}$$

$$\sec \theta = \frac{\text{radius}}{\text{abscissa value}} = \frac{r}{x}$$

$$\csc \theta = \frac{\text{radius}}{\text{ordinate value}} = \frac{r}{y}$$

Referring again to Figures 8-57(a), (b), (c), and (d), you see that $\angle A$ is the positive acute angle between the terminal side OP and the x axis. Angle A is called *the working angle or reference angle* for $\angle \theta$.

Also note from Figure 8-57 and the definitions given above that in some cases the trigonometric ratios are positive and in other cases they are negative. We shall discuss the signs of the functions for a first and a second quadrant angle in standard position and leave the discussion of a third and a fourth quadrant angle as a part of the next exercise.

case I Functions of an angle terminating in the first quadrant. The x, y, and r values are positive; therefore, all the trigonometric ratios are positive.

case II Functions of an angle terminating in the second quadrant. The value for x is negative in quadrant II, but y and r are positive. Therefore, those functions defined by the ratio involving x will be negative:

$$\sin \theta = \frac{y}{r} \qquad \text{positive}$$

$$\cos \theta = \frac{x}{r} \qquad \text{negative}$$

---

* The angle must be in standard position and have any point $P(x, y)$ on its terminal side a distance r from the origin.

$$\tan \theta = \frac{y}{x} \quad \text{negative}$$

$$\cot \theta = \frac{x}{y} \quad \text{negative}$$

$$\sec \theta = \frac{r}{x} \quad \text{negative}$$

$$\csc \theta = \frac{r}{y} \quad \text{positive}$$

The following ASTC code proves helpful in remembering those functions with positive ratios (Figure 8-58).

1. *A* (in quadrant I) stands for *all* functions are positive.
2. *S* (in quadrant II) stands for the *sine* function being positive. (*Note:* The reciprocal of the sine, i.e. cosecant, is also positive.)
3. *T* (in quadrant III) stands for the *tangent* (and its reciprocal, cotangent) ratio being positive.
4. *C* (in quadrant IV) stands for the *cosine* (and its reciprocal, secant) ratio being positive.

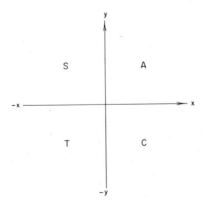

Figure 8-58

Once you determine the two functions of an angle which are positive in quadrant II, III, or IV, you know that the remaining four functions are negative.

Determining the Functional Value of a Specific Angle

The specific angle may terminate in any quadrant. This determination is done in three steps:

1. Determine the *working angle A*. It is helpful to sketch (approximately to scale) the given angle in standard position.
2. Use a set of trigonometric tables (or slide rule) to determine the value for the function of the working angle.
3. Determine the proper sign (+ or −) for the ratio.

example   Determine sin 210°.

1. Sketch the angle 210° in standard position. See Figure 8-59. The working angle $A = 30°$.
2. From Table 16 (or slide rule), sin 30° = 0.5000.
3. 210° is a third quadrant angle. In quadrant III the tangent and cotangent ratios are positive. Therefore, the sine ratio (y/r) is negative.

$$\sin 210° = -\sin 30° = -0.50$$

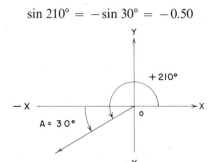

Figure 8-59

example   Determine csc 675°.

1. Sketch 675° in standard position. See Figure 8-60. The working angle $A = 45°$.
2. From the slide rule,* csc 45° = 1.414.
3. 675° is a fourth quadrant angle and csc in this quadrant is negative.

$$\csc 675° = -\csc 45° = -1.414$$

example   Determine tan −600°.

1. Sketch −600° in standard position. See Figure 8-61. The working angle is 60°.
2. From the slide rule,† tan 60° = 1.732.

---

* *H* over 45° on *S*, read *CI* (or *DI* with rule closed).
† *H* over 60° on *T*, read *CI* (or *DI* with rule closed).

3. $-600°$ is a second quadrant angle, and the tangent of a second quadrant angle is negative.

$$\tan -600° = -\tan 60° = -1.732$$

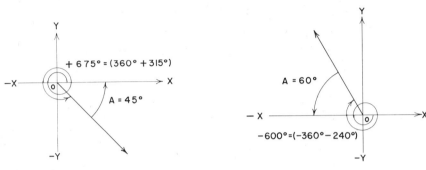

| Figure 8-60 | Figure 8-61 |

Quadrantal Angles

A special group of angles called quadrantal angles have not yet been discussed specifically. An angle in standard position which has its terminal side on one of the coordinate axes is called a *quadrantal angle*. Examples of quadrantal angles are $+90°$, $-180°$, $+270°$, $+360°$, and $-450°$.

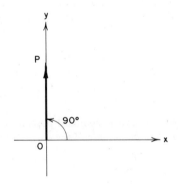

Figure 8-62

We need to discuss briefly the functions of quadrantal angles because we shall have occassion to use ratios such as sin 90°, cos 270°, and tan 180°. For example, let us determine the six trigonometric functions for a 90° angle. Figure 8-62 shows a radius vector OP in the direction of 90°. For point P on the y axis, $y =$ the length of $OP = r$, and $x = 0$. From the definitions:

$$\sin 90° = \frac{y}{r} = \frac{y}{y} = 1 \qquad (y = r)$$

$$\cos 90° = \frac{x}{r} = \frac{0}{r} = 0$$

$$\tan 90° = \frac{y}{x} = \frac{y}{0} \qquad \text{(Undefined, i.e., } \tan 90° \text{ does not exist.)}$$

$$\cot 90° = \frac{x}{y} = \frac{0}{y} = 0$$

$$\sec 90° = \frac{r}{x} = \frac{r}{0} \qquad \text{(Undefined, i.e., } \sec 90° \text{ does not exist.)}$$

$$\csc 90° = \frac{r}{y} = \frac{y}{y} = 1 \qquad (y = r)$$

Functions of 0°, 180°, 270°, and 360° can be determined in a similar manner.

EXERCISE 8.9

1. Use the definitions of the functions and the coordinates x and y along with the radius r to determine the sign (positive or negative) of the following ratios for $\sphericalangle \theta$ terminating in the given quadrant:

    (a) $\sin \theta$, quadrant III.   (e) $\tan \theta$, quadrant II.
    (b) $\tan \theta$, quadrant IV.   (f) $\cot \theta$, quadrant III.
    (c) $\sec \theta$, quadrant II.   (g) $\sin \theta$, quadrant IV.
    (d) $\cos \theta$, quadrant III.

2. Sketch each of the following angles in standard position. Label the initial side and the terminal side and indicate the magnitude of the working angle.

    (a) 315°   (c) 200°   (e) −405°   (g) −720°
    (b) 715°   (d) −315°   (f) −240°   (h) 750°

    In Problems 3–6, follow the three-step method to determine the following:

3. $\sin \theta$ and $\tan \theta$ if $\theta = 150°$.   5. $\sin \theta$ and $\sec \theta$ if $\theta = 315°$.
4. $\cos \theta$ and $\cot \theta$ if $\theta = -120°$.   6. $\cos \theta$ and $\csc \theta$ if $\theta = -330°$.

7. Determine the exact values for the six functions of θ if θ is in standard position and its terminal side passes through the point $(-5, 6)$.

8. If θ is a third quadrant angle and $\tan \theta = 0.25$, determine the exact values for the other five functions of θ.

9. Using a mental sketch only, determine the working angles for the following:
   (a) 315°   (c) −150°   (e) −210°   (g) −330°
   (b) 300°   (d) 240°   (f) −240°   (h) 800°

10. From memory, write the names of those functions which are positive ratios in each of the four quadrants.

In Problems 11–14, use the procedure discussed for the functions of 90° to determine the trigonometric functions of the given angle.

11. 180°   12. 270°   13. 360°   14. 0°

In Problems 15–20, determine the smallest positive ∡ θ for the given functional values.

15. $\sin \theta = -0.5$ (Hint: $-0.5 = -\frac{1}{2}$.)   18. $\sec \theta = -1.480$
16. $\tan \theta = -0.75$ 19. $\cot \theta = -0.4522$
17. $\cos \theta = -0.8124$ 20. $\csc \theta = -1.064$

21. Discuss the signs of the six functions for a third quadrant angle, using a method similar to that used in Case II.

22. Discuss the signs of the six trigonometric functions for a fourth quadrant angle, using a method similar to that used in Case II.

8.10 SOLUTIONS OF OBLIQUE TRIANGLES

The modern technician frequently has to solve problems involving oblique triangles as well as right triangles. For this reason, we shall now consider two commonly used methods of solving oblique triangles.

Law of Sines

If we develop this law from our present knowledge of the right triangle, it will add to our understanding of this law. Start with the oblique triangle *ABC* (Figure 8-63) and drop a perpendicular from point *B* forming two right triangles. In $\triangle AOB$, $BO = c \sin A$, and in $\triangle COB$, $BO = a \sin C$. Because $BO = BO$, you can equate $a \sin C = c \sin A$. If you divide both members by $\sin A$ and by $\sin C$,

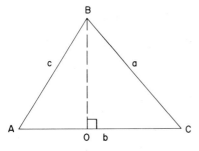

Figure 8-63

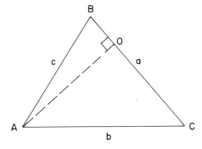

Figure 8-64

you obtain the proportion

$$\frac{a}{\sin A} = \frac{c}{\sin C} \tag{1}$$

If you had dropped the altitude from point A (Figure 8-64) instead of from point B, you would have obtained $AO = b \sin C$ and $AO = c \sin B$. By equating right members, $b \sin C = c \sin B$. Dividing both members by $\sin B$ and by $\sin C$, you obtain the proportion

$$\frac{b}{\sin B} = \frac{c}{\sin C} \tag{2}$$

From equations (1) and (2), you can state

$$\frac{a}{\sin A} = \frac{b}{\sin B} = \frac{c}{\sin C}$$

This important result is known as the *Law of Sines*. Stated in words it says: *In any triangle the sides are proportional to the sine of the opposite angles.*

The Law of Sines lends itself to calculations with logarithms or with the slide rule, because the operations involved are multiplication and division. To apply the Law of Sines, you must have sufficient known data to allow you to set up one of the following proportions:

$$\frac{a}{\sin A} = \frac{b}{\sin B} \qquad \text{(You need any three of the four factors.)}$$

or

$$\frac{b}{\sin B} = \frac{c}{\sin C}$$

or

$$\frac{a}{\sin A} = \frac{c}{\sin C}$$

example Solve the triangle shown in Figure 8-65 for $\sphericalangle B$, $\sphericalangle C$, and side c. To find $\sphericalangle B$, use the Law of Sines to obtain

$$\frac{27.0}{\sin 14.7°} = \frac{51.5}{\sin B}$$

Recalling from the algebra section on ratio and proportion that you can state the equality of the reciprocals of equal ratios, write

$$\frac{\sin 14.7°}{27.0} = \frac{\sin B}{51.5}$$

This change is unnecessary, but it may at first be easier to follow the slide rule procedure with the proportion written in this form. Set 14.7° on the *S* scale over 27.0 on *D*, move *H* to 51.5 on *D*, and read ∢ *B* = 28.9° under *H* on *S*. Keep this setting for the next part of the problem. To find ∢ *C*:

$$∢ C = 180° - (A + B) = 180° - 14.7° - 28.9° = \underline{136.4°}$$

To find side *c*, without changing the slide move *H* to ∢ *C* = 136.4° on the *S* scale. The *S* scale only goes to 90°, so make use of the working angle obtaining sin 136.4° = sin 43.6°. (Note that *A* + *B* = 43.6°.) Move *H* to 43.6° on the *S* scale and read side *c* = $\underline{73.4 \text{ ft}}$ under *H* on *D*.

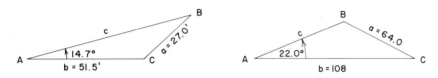

| Figure 8-65 | Figure 8-66 |

example   Solve the triangle shown in Figure 8-66 for ∢ *B*. Using

$$\frac{a}{\sin A} = \frac{b}{\sin B}, \quad \text{with } A = 22.0°, b = 108, a = 64.0$$

you have

$$\frac{64.0}{\sin 22.0°} = \frac{108}{\sin B}$$

Set 22.0° on *S* over 64.0 on *D*; you run out of scale when moving *H* to 108 on *D*. You need only reverse indices on the *S* scale and then move *H* to 108 on *D*. Read 39.2° under *H* on *S*. From Figure 8-66, you see that *B* is obtuse. Therefore,

$$∢ B = (180° - 39.2°) = 140.8°$$

Law of Cosines

The Law of Cosines states that *in any triangle the square of any side is equal to the sum of the squares of the other two sides diminished by twice the product of the other two sides and the cosine of their included angle.* Written algebraically for △*ABC* with sides *a*, *b*, and *c* (Figure 8-63), the law states $c^2 = a^2 + b^2 - 2ab \cos C$. As a special case, when ∢ *C* = 90°, side *c* becomes an hypotenuse and cos 90° = 0. The statement then becomes $c^2 = a^2 + b^2$ which you recognize as the Pythagorean theorem.

Solving triangles by the Law of Cosines requires more labor than does the Law of Sines because it involves addition and subtraction of terms which eliminate the continuous use of logarithms and also the slide rule. However, it is an important method for solving an oblique triangle whenever you are given three sides of a triangle or two sides and the included angle (cases where the Law of Sines does not work).

example Solve for $\sphericalangle A$ in $\triangle ABC$ with sides $a = 4.0$, $b = 5.0$, and $c = 6.0$. Start with $a^2 = b^2 + c^2 - 2bc \cos A$ and solve for $\cos A$:

$$\cos A = \frac{a^2 - b^2 - c^2}{-2bc} = \frac{b^2 + c^2 - a^2}{2bc}$$

Substituting for a, b, and c, you have

$$\cos A = \frac{5^2 + 6^2 - 4^2}{2 \cdot 5 \cdot 6} = \frac{45}{60} = 0.75$$

From Table 16, Appendix D (or slide rule),

$$\sphericalangle A = 41° \, 24' \approx 41° \qquad \text{(two-figure accuracy)}$$

example Solve for side c of $\triangle ABC$, given side $a = 8.0$, side $b = 5.0$, and $\sphericalangle C = 26°$. Start with $c^2 = a^2 + b^2 - 2ab \cos C$ and substitute given data:

$$c^2 = 8^2 + 5^2 - 2(8)(5)(\cos 26°)$$

$$c^2 = 64 + 25 - 80(0.899)$$

$$c^2 = 89 - 71.9 = 17.1$$

$$c = \sqrt{17.1} \approx 4.1$$

EXERCISE 8.10A

In Problems 1–6, use the slide rule and the Law of Sines to solve for the designated part of $\triangle ABC$.

1. $a = 5.00$, $b = 6.00$, and $A = 26.0°$; solve for $\sphericalangle B$. Is this triangle unique?
2. $a = 15.0$, $b = 21.0$, and $B = 38.0°$; solve for $\sphericalangle A$. Is this triangle unique?
3. $a = 12.0$, $b = 15.0$, and $A = 41.0°$; solve for $\sphericalangle B$. Is this triangle unique?
4. $a = 15.0$, $A = 35.0°$, and $B = 42.0°$; solve for side b. Is this triangle unique?
5. $a = 10.0$, $A = 40.0°$, and $B = 50.0°$; solve for side c. Is this triangle unique?
6. $a = 12.0$, $c = 12.0$, and $B = 68.0°$; solve for side b. Is this triangle unique?

In Problems 7–10, use the Law of Cosines to find the designated part of △*ABC.*

7. $a = 18.0$, $b = 11.0$, and $c = 12.0$; find ∡ A.
8. Find ∡ B in Problem 7.
9. $a = 10.0$, $b = 20.0$, and $C = 25.0°$; find side c.
10. $b = 15.0$, $c = 10.0$, and $A = 45.0°$; find side a.
11. A 212-ft tower is designed to lean 20.0° from vertical (Figure 8-67).
    (a) How long a guy wire is needed to reach from P to T?
    (b) What angle does the wire form with the ground at point P?
12. A triangular metal template has sides of 8.0, 11, and 14 in. Find the angle included between the 8.0-in. and 14-in. sides.

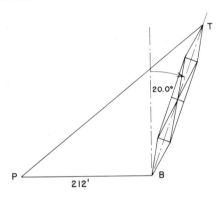

Figure 8-67

Calculating with Logarithms*

To review the use of logarithms and also to become acquainted with Table 19, Appendix D, let us use logarithms to solve the triangle shown in Figure 8-66. Solving

$$\frac{64.0}{\sin 22.0°} = \frac{108}{\sin B}$$

for sin B, you have

$$\sin B = \frac{(108)(\sin 22.0°)}{64.0}$$

Then take the log of both members:

$$\log (\sin B) = \log \frac{(108)(\sin 22.0°)}{64.0} = \log 108 + \log \sin 22.0° - \log 64.0$$

---

* This section may be omitted until the section on logarithms, Appendix C, has been covered.

Log sin 22.0° means the logarithm of the number which equals sin 22.0°. You could look up sin 22.0° in Table 16 and then look up the log of that number in Table 15. However, Table 19 gives directly the logarithms of functions of angles from 0° to 90° for each tenth of a degree or each 6 minutes.

Using Table 15:

$$\log 108 = 2.0334$$

Using Table 19:

$$\log \sin 22.0° = \quad 9.5736 - 10*$$
$$\text{Adding} \quad \overline{11.6070 - 10}$$
$$\log 64.0 = \quad 1.8062$$
$$\text{Subtracting, } \log \sin B = \quad \overline{9.8008 - 10}$$

Using Table 19, $\angle B = 39.2°$ or its supplement 140.8°. You can see from the figure that $\angle B$ is obtuse; therefore, you know that $\angle B = 140.8°$.

EXERCISE 8.10B

In Problems 1–8, use Table 19 to determine the following.

1. log sin 31.2°
2. log cos 18° 12′
3. log tan 12° 42′
4. log tan 73.7°
5. $\angle A$ if log sin $A = 9.4482 - 10$.
6. $\angle B$ if log tan $B = 10.5147 - 10$.
7. $\angle A$ if $\angle A$ is a second quadrant angle such that log sin $A = 9.6480 - 10$.
8. $\angle B$ such that $\angle B$ is a fourth quadrant angle with log cos $B = 9.9275 - 10$.
9. Using logarithms from Tables 15 and 19 and the Law of Sines, calculate side a given $\triangle ABC$ with side $b = 82.41$ ft, $\angle A = 39° 24'$, and $\angle B = 62° 6'$.
10. Using logarithms from Tables 15 and 19 and the Law of Sines, calculate $\angle B$ given $\triangle ABC$ with side $b = 173.2$ ft, $\angle A = 31° 57'$, and side $a = 156.5$ ft.

8.11   VECTORS

When a quantity such as mass, length, area, volume, or temperature is completely described when its magnitude is given, it is said to be a *scalar quantity* or simply a *scalar*. Quantities that possess both magnitude and direction are called *vector quantities* or simply *vectors*. For example, a force of 40 lb acting vertically

---

* The -10 part of the characteristic is understood.

downward is a vector quantity with a magnitude of 40 lb and a specified direction of vertically downward.

To distinguish the symbol for a vector quantity from a scalar quantity, we write the symbol for the vector in bold-face type, **V**, or use a small arrow over the letter, $\vec{V}$. Marks on the scale of a thermometer illustrate a method of representing a scalar quantity. In general, scalar quantities can be represented by marks or positions on appropriate fixed scales. Vector quantities may be represented in several ways, but the best for our purpose is the use of directed line segments. Figure 8-68 illustrates a vector $\mathbf{V}_1$ acting vertically downward. The length of the vector (to some convenient scale) represents the magnitude of the vector. The position of the line segment along with the arrow tip represents the direction in which the magnitude acts or exists. The position of a vector has meaning only when it is related to a frame of reference such as vertical and horizontal planes or lines, north-south lines, and x-y coordinate axes.

The magnitude of a vector is frequently referred to as the absolute value of the vector. Symbolically the magnitude or absolute value is represented by writing the vector symbol between vertical bars, $|\mathbf{V}|$, or by writing the vector symbol in ordinary type, thus V. For the magnitude of the vector in the previous example, you can write $|\mathbf{V}_1| = 40$ lb.

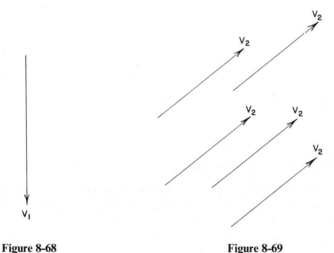

Figure 8-68   Figure 8-69

The graphic representation of a vector has the property that it may be moved to any other position as long as its length (magnitude) and direction remain the same. Figure 8-69 shows the vector $\mathbf{V}_2$ represented in several positions. You may draw (or move) a vector in any position you wish as long as you preserve the two properties of magnitude and direction.

Addition of Vectors

The following example illustrates how vector addition differs from regular (scalar) addition. Figure 8-70 shows a partial sketch of a school campus. Students may walk from *A* to *B* and then from *B* to *C* or, as many do, they may take a short cut and walk from *A* to *C*. Regardless of the route taken, the resulting position is the same. By scalar addition the first group of students walks distances *AB* + *BC* or 150 ft + 200 ft = 350 ft. The second group walks the distance *AC* = 250 ft.

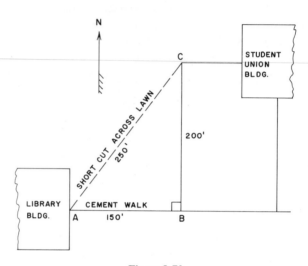

Figure 8-70

By vector addition, you have **AB** + **BC** = **AC**. This particular statement does not say that the length of *AB* plus the length of *BC* equals the length of *AC*. The statement does imply the following:

1. The easterly distance or component of **AB** (150 ft) added to the easterly distance or component of **BC** (0 ft) equals the easterly distance or component of **AC** (150 ft).

2. The northerly component of **AB** (0 ft) added to the northerly component of **BC** (200 ft) equals the northerly component of **AC** (200 ft).

A student who starts at point *A* and reaches point *C* is 150 ft east of point *A* and 200 ft north of point *A*, regardless of his route from point *A* to point *C*.

Addition by Graphic Methods—Parallelogram Law

The sum of vectors V_1 and V_2, called the *resultant*, can be obtained graphically by constructing a parallelogram using V_1 for one pair of parallel sides and V_2 for the other pair of parallel sides. The diagonal of the parallelogram is the vector sum or resultant. See Figure 8-71.

An alternate graphic method makes use of the fact that the diagonal of a parallelogram divides the parallelogram into two equal triangles. Note that each triangle (Figure 8-71) has two sides formed by vectors V_1 and V_2 and the third side is the resultant. By constructing V_2 with its initial point at the terminal point of V_1, the vector joining the initial point of V_1 to the terminal point of V_2 is the vector sum of V_1 and V_2 (Figure 8-72). To add several vectors, place the initial

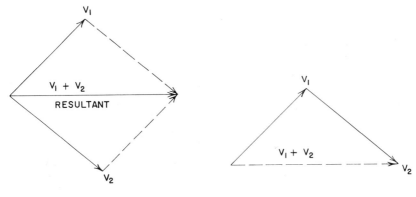

Figure 8-71   **Figure 8-72**

point of the second vector at the terminal point of the first. Then place the initial point of the third vector at the terminal point of the second vector. Continue in this manner until the last vector has been drawn. The vector joining the initial point of the first vector to the terminal point of the last vector is the required sum or resultant. Figure 8-73 illustrates the addition of five vectors, $V_1, V_2, V_3, V_4,$ and V_5.

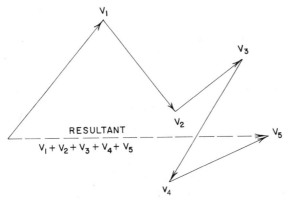

Figure 8-73

A vector may be subtracted graphically by constructing the vector to be subtracted in a direction 180° in advance of its original direction (i.e., by reversing the direction of the vector).

Vector Notation

It is frequently advantageous to represent a vector in a rectangular co-ordinate system because it is easier to specify the property of direction by means of an angle in standard position. Use the notation $V_1 = 75\underline{/20°}$ to mean a vector

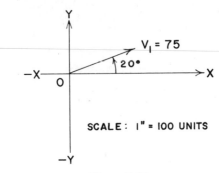

Figure 8-74

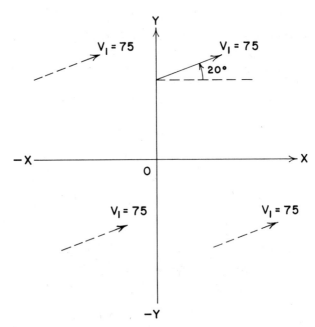

Figure 8-75

V_1 having a magnitude of 75 units in the direction of 20° (standard position). Figure 8-74 shows this vector. Figure 8-75 shows the same vector in several positions.

EXERCISE 8.11A

Problems 1–10 list some scalar quantities and some vector quantities. Write the word scalar or vector to describe each quantity; explain your answer. (Remember that direction is a necessary property of a vector quantity.)

1. The weight of a car.
2. The speed of a car.
3. The path of a car along a straight highway.
4. The force causing a car to slide off a road.
5. The area of a farm.
6. The cost of producing bricks.
7. The force exerted by a stretched spring.
8. The volume of air in a room.
9. The flow of water in a pipe.
10. The freezing point of water.
11. Represent the following vector quantities on the same coordinate system, each having the origin as an initial point:

$$V_1 = 100\,\text{lb}\;\underline{/75°} \qquad V_3 = 200\,\text{lb}\;\underline{/135°}$$
$$V_2 = 150\,\text{lb}\;\underline{/0°} \qquad V_4 = 225\,\text{lb}\;\underline{/225°}$$

12. Represent the vectors in Problem 11 on a rectangular coordinate system, using initial point (*IP*) as specified:

$$V_1 \text{ with } IP = (50, 50) \qquad V_3 \text{ with } IP = (50, 0)$$
$$V_2 \text{ with } IP = (0, -50) \qquad V_4 \text{ with } IP = (50, 0)$$

13. Represent the following vector quantities, using the origin as an initial point for each vector:

$$V_1 = 250\,\text{mph}\;\underline{/0°} \qquad V_3 = 450\,\text{mph}\;\underline{/135°}$$
$$V_2 = 325\,\text{mph}\;\underline{/45°} \qquad V_4 = 400\,\text{mph}\;\underline{/315°}$$

14. Represent the vectors in Problem 13 on a rectangular coordinate system, each having its initial point at the point (400, 400).
15. Using the parallelogram law, draw vectors V_1 and V_2 to scale and find their sum. $V_1 = 100$ units and forms an angle of 45° with the direction of V_2. $V_2 = 200$ units $\underline{/0°}$.
16. Find the sum of V_1 and V_2 in Problem 15 if the angle between the directions of the two vectors is 60°.
17. Change the angle in Problem 15 to 90° and find the sum of V_1 and V_2.
18. Double the magnitude of V_1 in Problem 15; then determine the sum of V_1 and V_2. What change occurred in the resultant?

19. Double the magnitudes of V_1 and V_2 in Problem 15; then calculate the sum of V_1 and V_2. What change occurred in the resultant?

20. A force of 500 lb acts horizontally on a point at the same time that another force of 1000 lb acts vertically upward. Using the parallelogram method, find the resultant of these two forces.

21. Use the alternate graphic method to determine the vector sum of the following vectors:

$$V_1 = 100 \underline{/0°}, \qquad V_2 = 50 \underline{/90°}, \qquad V_3 = 150 \underline{/30°}, \qquad V_4 = 200 \underline{/90°}$$

22. Determine the vector sum for the following vectors:

$$V_1 = 80 \underline{/-30°} \qquad V_2 = 40 \underline{/45°} \qquad V_3 = 50 \underline{/180°} \qquad V_4 = 35 \underline{/270°}$$

23. Subtract vector $V_1 = 75/135°$ from vector $V_2 = 90/75°$.

24. Subtract vector V_2 from vector V_1 in Problem 23.

Vector Components

To develop a systematic and analytic method of operating with vector quantities, vectors are resolved into two components which are usually perpendicular. *Resolution* of a vector means the determination of specified components of the vector. Figure 8-76 shows the vector **OC** with components **OA** and **OB**. A vector

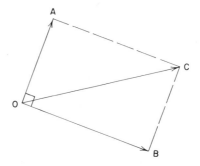

Figure 8-76

may be the sum of any number of different pairs of components. Figure 8-77 shows a vector **OC** with components of V_1 and V_2 or V_3 and V_4. Figure 8-78 shows the mutually perpendicular components obtained by projecting vector **OC** onto lines l_1 and l_2.

The *rectangular component method* of working with vector quantities is the method most frequently used by engineers and technicians. This method utilizes the following procedures.

1. All vectors are oriented with a rectangular coordinate system by noting their direction as an angle in standard position.

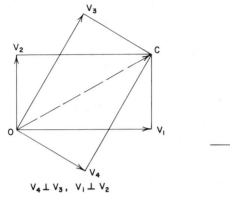

$V_4 \perp V_3, \ V_1 \perp V_2$

Figure 8-77

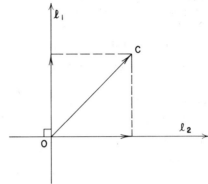

Figure 8-78

2. All vectors are resolved into mutually perpendicular components, specifically into components in the direction of the *x* axis, called *x* components, and in the direction of the *y* axis, called *y* components.

Because the projection of a vector always involves a right triangle, an important relationship is that *the magnitude of the projection or component of a vector in any specified direction is equal to the product of the magnitude of the vector and the cosine of the angle between the vector and the specified direction.*

Figure 8-79 shows a vector **OA** with **OB** as its *x* component and **OC** as its *y* component. If you apply the trigonometry of the right triangle to $\triangle ABO$ in Figure 8-79, you can state that the length of the *x* component $= |\mathbf{OA}|(\cos \theta)$ and the length of the *y* component $= |\mathbf{OA}| \cos (90° - \theta)$ or $|\mathbf{OA}|(\sin \theta)$.

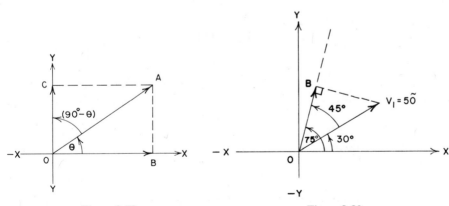

Figure 8-79   **Figure 8-80**

example Calculate the component of $V_1 = 5\tilde{0}/30°$ in the direction of 75°. See Figure 8-80.

$$|\mathbf{OB}| = |V_1|\cos(75° - 30°) = 50\cos 45° = 35$$
$$\mathbf{OB} = 35/75°$$

example Calculate the x and y components of the vector $V_1 = 82/65°$. See Figure 8-81.

$$|x\ \text{component}| = |\mathbf{OA}| = |V_1|\cos 65° = (82)(\cos 65°) = 35$$
$$\mathbf{x}_1^* = \mathbf{OA} = 35/0°$$
$$|y\ \text{component}| = |\mathbf{OB}| = |V_1|\cos 25° = (82)(\cos 25°) = 74$$
$$\mathbf{y}_1^\dagger = \mathbf{OB} = 74/90°$$

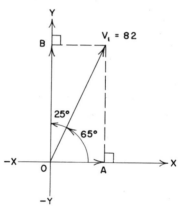

Figure 8-81

The notations $/0°$ and $/90°$ for the directions of the x component and y component are really unnecessary. When a vector has a component in the direction 0° or 90°, express its x or y direction as positive ($+$). When the component is in the direction of 180° or 270°, express its x or y direction as negative ($-$). Hence a vector having a component of 60 in the direction of 180° would be expressed as x component $= -60$.

---

* x_1 and x_2 denote the x-components of V_1 and V_2, respectively.
† x_1 and y_1 denote the x and y components of V_1.

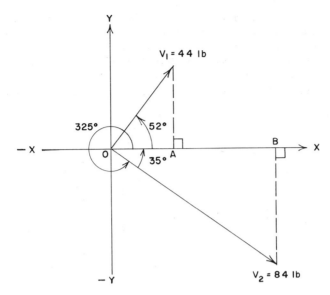

Figure 8-82

example Given $V_1 = 44\,\text{lb}\,\underline{/52°}$ and $V_2 = 84\,\text{lb}\,\underline{/325°}$, calculate the sum of the x components for these two vectors. Vectors V_1 and V_2 are shown in Figure 8-82.

$$x_1 = 44\,\text{lb}\,(\cos 52°) = +27\,\text{lb}$$
$$x_2 = 84\,\text{lb}\,(\cos 35°) = +69\,\text{lb}$$
$$\overline{(x_1 + x_2) = x \qquad\qquad = +96\,\text{lb}}$$

EXERCISE 8-11B

Sketch an appropriate diagram for each of the following problems and solve. (Watch significant figures.)

1. Find the component of $V_1 = 75.0\,\underline{/40.0°}$ in the direction 60.0°.
2. Find the component of $V_2 = 67\,\underline{/0°}$ in the direction 25°.
3. Find the x and y components of $V_1 = 150\,\underline{/38.0°}$.
4. Find the x and y components of the force vector $F_1 = 285\,\text{lb}\,\underline{/15.0°}$.
5. Find the x and y components of the force vector $F_2 = 554\,\text{lb}\,\underline{/120°}$.
6. Find the x and y components of the velocity vector $V_1 = 750\,\text{mph}\,\underline{/315°}$.
7. Find the resultant of $V_1 + V_2$ graphically and then find x and y of the resultant, given:

$$V_1 = 80.0\,\underline{/15.0°} \qquad V_2 = 10\tilde{0}\,\underline{/80.0°}$$

8. Check Problem 7 by calculating the x and y components of $\mathbf{V}_1$ and $\mathbf{V}_2$ and then summing the x components and summing the y components.

9. Three weights in a gyroscopic compass exert forces as shown in Figure 8-83. Determine $\mathbf{C}$ to exactly balance the system at O. (*Hint:* The magnitude of $\mathbf{C}$ must be equal to the magnitude of the sum of the x components for $\mathbf{A}$ and $\mathbf{B}$.)

10. If the 30° angles shown in Problem 9 were changed to 60°, what force must $\mathbf{C}$ exert?

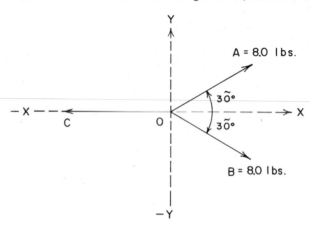

Figure 8-83

Applications Involving Components

Force of gravity. The force acting to cause an object to slide or roll down a ramp is a component of the force of gravity. The force acting perpendicular to the surface between the two objects is called the force *normal* and is also a component of gravity. Figure 8-84 shows a block with weight $\mathbf{W}$ (force of gravity); vectors

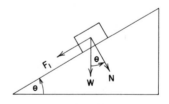

Figure 8-84

$\mathbf{F}_1$ and $\mathbf{N}$ (normal) are components of $\mathbf{W}$. The magnitudes of these components may be calculated by

$$|\mathbf{F}_1| \text{ (force acting down the ramp)} = |\mathbf{W}| \cos(90° - \theta) = |\mathbf{W}| \sin\theta$$

$$|\mathbf{N}| \text{ (force normal)} = |\mathbf{W}| \cos\theta$$

Friction. Experimental evidence shows that the force of friction, acting to prevent or resist one object from sliding along another, is directly proportional to the perpendicular force holding one object against the other. This relationship is stated algebraically as $|\mathbf{F}| = \mu|\mathbf{N}|$. Where $|\mathbf{F}|$ is the force of friction, μ (mu) is the proportionality constant called the *coefficient of friction** and $|\mathbf{N}|$ is the perpendicular or *normal* force between the two objects.

If a block is placed on an inclined plane and the angle of inclination ($\sphericalangle \theta$ in Figure 8-84) is increased to the place where the block will continue to slide when set in motion and will remain at rest when stopped, this specific $\sphericalangle \theta$ is called the *angle of friction* (denoted by θ_f). The coefficient of friction μ for this particular block and plane turns out to equal $\tan \theta_f$, i.e., $\mu = \tan \theta_f$.

example For a particular block of brake lining placed on a cast iron surface, the angle of friction is 21.8°. What is the coefficient of friction for these two surfaces?

$$\mu = \tan 21.8° = 0.40$$

example When the two surfaces in the example above become wet, the angle of friction is 11.3°. What is the coefficient of friction for these wet surfaces?

$$\mu - \tan 11.3° - 0.20$$

example At what angle would an object just start to slide on an inclined plane if the two surfaces have a coefficient of friction $\mu = 0.421$?

$$\tan \theta_f = 0.421$$

From Table 16 (or the slide rule),

$$\theta_f = 22° \, 50' \qquad (22.8°)$$

example What would be the force of friction for the system shown in Figure 8-84 with $\theta = 25°$, $\mathbf{W} = 260$ lb, and θ_f for the surfaces $= 33°$?

1. $$\mu = \tan \theta_f = \tan 33.0° = 0.649$$

2. $\quad |\mathbf{N}|$ (force normal) $= |\mathbf{W}| \, (\cos \theta)$

$$|\mathbf{N}| = (260)(\cos 25°) = 236 \, lb$$

---

* The coefficient of friction involves both surfaces. If either surface is changed, the value of the proportionality constant μ will also change.

3. $\qquad$ $|\mathbf{F}|$ (friction) $= \mu|\mathbf{N}| = (0.649)(236)$ lb

$$|\mathbf{F}| = 150\,\text{lb}$$

$$\mathbf{F} = 150\,\text{lb}\;\underline{/25^\circ}\;\text{or*}\;150\,\text{lb}\;\underline{/205^\circ}$$

Projections of Areas

The method of determining a component or projection of a vector is also applicable to the projection of any plane area onto a given plane. The projected area is equal to the given area multiplied by the cosine of the angle between the two planes (Figure 8-85):

$$\text{area } \triangle ABC' = (\text{area } \triangle ABC)(\cos\theta) \tag{1}$$

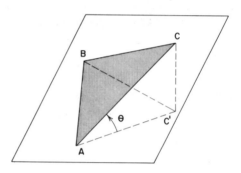

Figure 8-85

example   The roof of a house is inclined at an angle of 24° with the horizontal and has an area of 480 sq ft (not counting the eaves). What is the area of the floor covered by the roof?

$$\text{Projected area (floor area)} = 480\,(\cos 24^\circ) = 440\,\text{sq ft}$$

To determine the original vector or area from the projection, divide the magnitude of the component (or projection) by the cosine of the included angle. In short, when finding a component, multiply by $\cos\theta$. When determining a vector from a component, divide by $\cos\theta$. For example, solving equation (1) for area $\triangle ABC$, you have

$$\text{area } \triangle ABC = \frac{\text{area } \triangle ABC'}{\cos\theta}$$

* The force of friction always acts in a direction opposing the direction of motion. In this example, there is no motion; however, force $\mathbf{F}_1$ is acting so as to pull the block down the ramp. With $\mathbf{F}_1 = (260)$ $(\sin 25^\circ) = 110$ lb, the actual friction (acting up the ramp) is $\mathbf{F} = 110\,\text{lb}\;\underline{/25^\circ}$. In order that $|\mathbf{F}| = 150$ lb, it is necessary that an additional force be added to $\mathbf{F}_1$ so that the total force acting in the direction of $\mathbf{F}_1$ is equal to or greater than 150 lb.

example A moist-air supply duct in an air-conditioning system is to be made from a duct with a square cross section measuring 4.00 ft on a side (Figure 8-86). Determine the cross-sectional area at the joint. Let

$$A_j = \text{cross-sectional area of joint}$$

$$A_d = \text{cross-sectional area of duct}$$

Then:

$$A_d = A_j \cos 22.5° \text{ and}$$

$$A_j = \frac{A_d}{\cos 22.5°}$$

$$A_j = \frac{(4.00 \text{ ft})^2}{\cos 22.5°} = 17.3 \text{ sq ft}$$

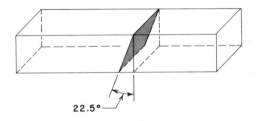

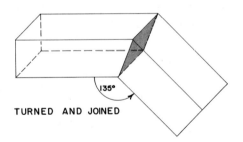

TURNED AND JOINED

Figure 8-86

example After projection through 37°, a component of a vector has a magnitude of 24 units. What is the magnitude of the original vector?

$$|\mathbf{V}| = \frac{24}{\cos 37°} = 3\tilde{0} \text{ units}$$

Not only can it be proved mathematically but experiments also show that the effect that a vector quantity has in *any specified direction* is exactly the effect of the component of the vector in *the specified direction*.

Vector Components in Surveying

Surveyors work with vector quantities and their components both in the field and in making calculations from their notes. A *course* in a field survey has the properties of distance and direction and is therefore a vector quantity. Courses are resolved into two components called latitude and departure. Latitude is the component in the north-south direction and departure is the component in the east-west direction.

example A course *OB* forms an angle of 38.6° with true north (Figure 8-87). Determine the latitude and departure of point *B* with respect to point *O*.

$$\text{Latitude of } B = (\text{course distance})(\cos 38.6°)$$

$$\text{Latitude of } B = (728 \text{ ft})(\cos 38.6°) = 569 \text{ ft}$$

$$\text{Departure of } B = (\text{course distance})(\sin 38.6°)$$

$$\text{Departure of } B = (728 \text{ ft})(\sin 38.6°) = 454 \text{ ft}$$

Figure 8-88 shows a course running between two stations of different elevations. Measurements made up or down a slope are always resolved into horizontal components for recording. *Course distances are always horizontal distances.* The vertical component of the slope distance *AB* gives the change in elevation from Station *A* to Station *B*. The horizontal component of the slope distance gives the course distance from *A* to *B*.

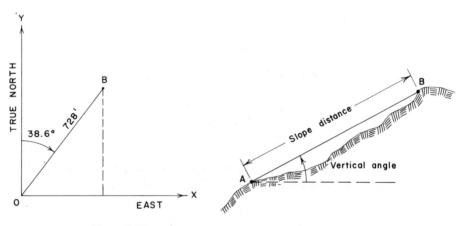

Figure 8-87   **Figure 8-88**

Voltage and Current Vectors

Figure 8-89(a) shows a voltage triangle for an *RL* circuit with vector $\mathbf{E}_{line}$ (line voltage) and its two components $\mathbf{E}_R$ (voltage drop across the resistor) and $\mathbf{E}_L$ (voltage drop across the inductor). Angle θ, called the phase angle, shows the phase relationship between the line voltage and the current **I** (shown dotted for a reference). Figure 8-89(b) shows a voltage triangle for an *RC* circuit with vector $\mathbf{E}_{line}$

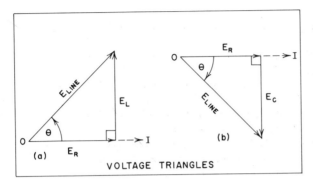

Figure 8-89

(line voltage) and its two components $\mathbf{E}_R$ (voltage drop across the resistor) and $\mathbf{E}_C$ (voltage drop across the capacitor). The phase relationship between $\mathbf{E}_{line}$ and the current **I** (shown dotted for a reference) is given by $\sphericalangle\, \theta$. Note that θ is a positive angle in (a) and a negative angle in (b).

Figure 8-90(a) shows a current triangle for an *RC* circuit with vector $\mathbf{I}_{line}$ (line current) and its components $\mathbf{I}_R$ (current through the resistor) and $\mathbf{I}_C$ (current through the capacitor). Angle θ shows the phase relationship between $\mathbf{I}_{line}$ and the voltage **V** (shown dotted for a reference). Figure 8-90(b) shows a current

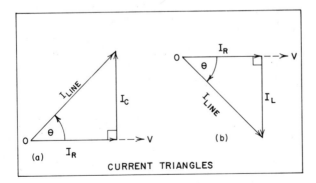

Figure 8-90

triangle for an *RL* circuit with vector $\mathbf{I}_{\text{line}}$ (line current) and its components $\mathbf{I}_R$ (current through the resistor) and $\mathbf{I}_L$ (current through the inductor). Phase ∢ θ shows the phase relationship between $\mathbf{I}_{\text{line}}$ and the voltage **V** (shown dotted for reference).

EXERCISE 8.11C

1. Refer to Figure 8-84 and compute the magnitudes of vectors $\mathbf{F}_1$ and **N** when **W** = 52.8 tons and θ = 14.6°.
2. Refer to Figure 8-84 and compute **W** and $\mathbf{F}_1$ if **N** = 73.6 g and θ = 26.8°. (*Hint:* **N** is a component of **W**.)
3. Refer to Figure 8-85 and compute the area of △ABC' if the area of △ABC is 15.7 sq yd and ∢ θ = 57° 20′.
4. Refer to Figure 8-85 and compute the area of △ABC if △ABC' has a base AB of 4.72 cm, an altitude of 3.08 cm, and ∢ θ = 60.0°.
5. Refer to Figure 8-87 and compute the latitude and departure of point B (from O) if the 38.6° angle is changed to 43.2° and the course distance is changed to 1728 ft.
6. Determine the course distance and direction of course AB if point B has a latitude of 512 yd (from A) and a departure of 314 yd (from A).
7. Refer to Figure 8-88 and compute the course distance for AB if the slope distance is 512 ft and the vertical angle (angle of elevation) is 22.3°.

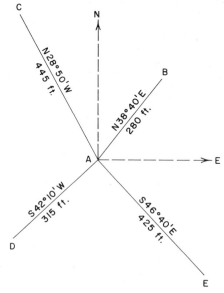

Figure 8-91

8. Determine the difference in elevations between point B and point A in Problem 7.
9. Determine the latitude and departure (from transit Station A) for Stations B, C, D, and E in Figure 8-91. Bearings for AB and AC are measured from the north. Bearings for AD and AE are measured from the south.
10. Refer to Figure 8-92 and compute the latitude and departure of Station B (from Station O).
11. Calculate the magnitude of $\mathbf{F}_1$ (neglecting friction) in Figure 8-93 if $\mathbf{W} = 50.0$ lb and $\theta = 38.6°$.

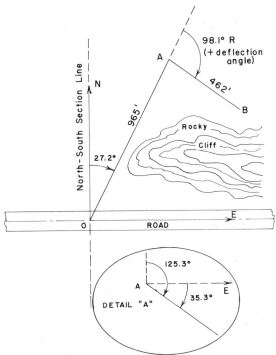

Figure 8-92

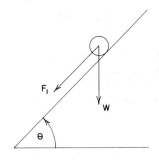

Figure 8-93

12. Calculate the force (neglecting friction) needed to pull the object in Figure 8-93 up the ramp if $W = 786$ lb and:
    (a) $\theta = 30.0°$.
    (b) $\theta = 15.6°$.
    (c) $\theta = 42.4°$.
    (d) $\theta = 65.0°$.

13. Two steel rollers are connected by a cable running over a pulley* at P (Figure 8-94).
    (a) Which way will the rollers move?
    (b) How much should the weight of the 390-lb roller be changed for the system to be balanced?

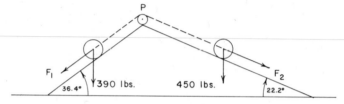

Figure 8-94

14. Four steel rollers are connected by a cable running over pulleys at points P as shown in Figure 8-95. Neglecting friction, which way will the system move?

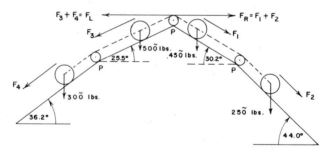

Figure 8-95

15. (a) Compute the projected area $ABC'D'$ (Figure 8-96) if $AB = 41.0$ ft, $AD = 12.0$ ft, and the angle between the planes is $36.5°$.
    (b) Compute the area of $ABCD$ if the area of $ABC'D'$ is $42\tilde{0}$ cm^2.

16. Compute the cross-sectional area of the cylinder shown in Figure 8-97 if the area of the ellipse is 48.2 sq in.

---

* Neglecting friction, a pulley allows a change in the direction of a force without diminishing its magnitude.

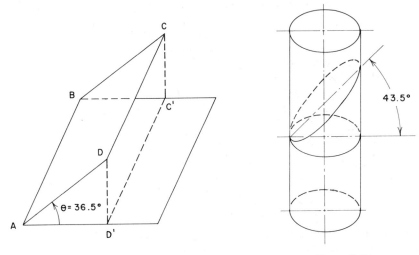

Figure 8-96   **Figure 8-97**

17. A pipe with a radius of 6.0 in. is to be cut at 45° and the two pieces welded to form a 90° angle as shown in Figure 8-98.
    (a) Compute the area of the elliptical cross section.
    (b) Using the fact that the area of an ellipse = πab (a is the semimajor axis and b is the semiminor axis), compute the major axis of the ellipse. (*Hint:* The minor axis = $2b$ = the diameter of the pipe.)

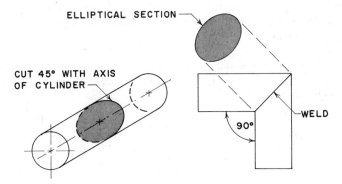

Figure 8-98

18. A steel shaft with a radius of 4.00 in. is cut at 30° and the two pieces are welded as shown in Figure 8-99. (Assume three-figure accuracy.)
    (a) Compute the area of the elliptical cross section.
    (b) Compute the major axis of the ellipse.

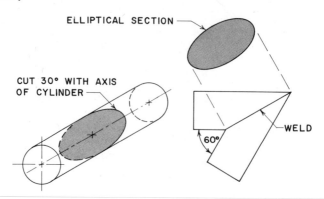

ELLIPTICAL SECTION

CUT 30° WITH AXIS
OF CYLINDER

WELD

60°

Figure 8-99

19. An air duct has a square cross section measuring 14.0 in. on a side. It is to be cut from square at a 17.0° angle, turned and joined similar to Figure 8-86.
    (a) What is the area of the cross section of the joint?
    (b) What are the dimensions of the cross section of the joint?

20. An air duct with a rectangular section of 12.0 in. by 18.0 in. is to be cut, turned, and joined so that the area of the cross section of the joint is 50.0% larger than the cross section of the duct. At what $\angle \theta$ should the duct be cut? (*Hint:* Calculate the required area of the joint and resolve it through some $\angle \theta$ to yield the cross-sectional area of the duct.) Make a sketch similar to Figure 8-86.

21. An ore-conveyor belt runs on rollers as shown in Figure 8-100. Assuming a uniform load of 60 lb/lineal ft of belt and neglecting friction, how much force must be continually applied to the belt at point C to keep the belt moving? Assume dimensions correct to nearest foot. (*Hint:* The load for any section can be treated as if it acted at a single point along the section.)

22. An ore-conveyor belt used to move molybdenum ore has a section inclined at an angle of 24.5° with the horizontal. If the loaded belt has an average weight of 96.5 lb/running ft and the total tensional force on the belt is not to exceed 4.0×10^2 lb, what is the maximum allowable distance between the driving rollers?

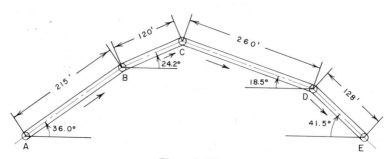

120'

260'

C

24.2°

215'

B

18.5°

D

128'

36.0°

41.5°

A

E

Figure 8-100

For Problems 23–26, refer to Figure 8-89.

23. Compute the magnitude of E_L and E_R if $E_{line} = 21\tilde{0}$ volts and $\theta = 21.5°$.
24. Compute the magnitude of E_{line} and the phase $\sphericalangle\,\theta$ if $E_R = 65.4$ volts and $E_L = 43.2$ volts.
25. Compute the magnitude of E_{line} and E_R if $\sphericalangle\,\theta = -24.0°$ and $E_C = 48.5$ volts.
26. Compute the magnitude of E_R and E_C if $E_{line} = 115$ volts and $\sphericalangle\,\theta = -28.2°$.

For Problems 27–30, refer to Figure 8-90.

27. Compute the magnitude of I_{line} and $\sphericalangle\,\theta$ if $I_C = 16.2$ amp and $I_R = 12.6$ amp.
28. Compute the magnitude of I_{line} and I_C if $\sphericalangle\,\theta = 41.0°$ and $I_R = 4.05$ amp.
29. Compute the magnitude of I_{line} and I_R if $\sphericalangle\,\theta = -17.6°$ and $I_L = 0.0125$ amp.
30. Compute the magnitude of I_{line} and $\sphericalangle\,\theta$ if $I_R = 6.06$ amp and $I_L = 2.15$ amp.

Challenge Problems

31. Refer to Figure 8-84 and determine the required force (acting up the ramp) to: (a) hold the block on the ramp and (b) pull the block up the ramp when $\theta = 57.4°$, $W = 524$ lb, and $\theta_f = 31.6°$ for the system.
32. Show that the coefficient of friction $\mu = \tan\theta_f$. (Hint: $\tan\theta = \sin\theta/\cos\theta$.)

A Review of Arithmetic

Proficiency in basic arithmetic skills is a necessity for all technicians. To improve your proficiency, you should first discover the areas in which you are weak. This appendix is designed to serve as a review and also to point out the specific skills that need additional study.

You will have difficulty with arithmetic operations if you have forgotten the basic addition and multiplication combinations. The combinations shown below are to be added or multiplied in 90 sec. Practice on them for a few minutes each day until you can answer all 90 combinations in the time allowed.

Inventory Test I begins on page 359. After taking this test, check and correct your work, using the solutions in the answer section on page 458. Study the solutions of the problems missed.

BASIC COMBINATIONS

10	12	9	11	8	7	12	11	6	1	4	3	11	9
10	0	2	11	0	5	11	1	6	0	4	3	10	3

12	7	8	10	5	12	6	11	9	10	7	11	12	2
10	4	1	9	5	1	0	9	4	8	3	0	9	8

11	5	10	9	11	12	7	10	5	9	12	8	7	9
3	4	7	5	8	8	2	0	3	1	2	3	1	6

12	6	0	11	5	10	8	10	7	2	5	6	12	9
7	1	9	7	2	6	4	1	0	2	1	2	6	7

4	10	11	8	3	2	11	1	4	6	10	12	8	10
3	5	6	5	2	1	5	1	2	3	4	5	6	2

3	12	7	9	5	6	11	4	8	2	7	3	12	4
1	4	6	8	0	4	4	1	7	0	7	0	3	0

9	12	6	10	11	8
9	12	5	3	2	8

Inventory Test II begins on page 361. After taking this test, check and correct your work, using the solutions in the answer section on page 460. If, for example, you miss Problems 1, 7, 11, 15, 17, and 19 on Test II, first study the solutions to these problems and then work additional problems from Practice Sets 1, 7, 11, 15, 17, and 19 (pages 362–367). These Practice Sets are numbered (by sets) to agree with the problems on Inventory Test II. Practice Set answers on page 465.

Inventory Test III begins on page 367. After taking this test, check and correct your work, using the solutions in the answer section on page 466. Study the solutions of the problems missed. Then work additional problems from the Practice Sets whose numbers correspond to the numbers of the problems missed.

INVENTORY TEST I

Answer each problem by providing the proper word or words. Check your work.

1. The ($+$), called a "plus" sign, is the symbol for what operation? What is meant when the ($+$) is used between two numbers? What does $7 + 12$ mean?
2. The ($-$), called a "minus" sign, is the symbol for what operation? What does the statement $16 - 7$ mean?
3. What is meant when the equal sign, ($=$), appears between two quantities? Give an example using the equal sign.
4. What is the symbol for multiplication?
5. ($\div$) is the symbol for what operation? What does $42 : 7$ mean?
6. In the problem $7\overline{)42}$, what is the symbol $\overline{)}$ called and what does it mean?
7. Show the indicated operation in Problem 6 by using a fraction.
8. What are the symbols "( )" called? What are they used for?
9. What other symbols are used for grouping?
10. Use parentheses to indicate the multiplication of 8×12.
11. What is the symbol $\sqrt{}$ called?
12. Use the proper inequality symbol to show the relationship between 7 and 10.
13. What does the symbol ($\approx$) mean?
14. Write the first 10 Arabic numerals that are used as digits.
15. Write the number $7\,2\,0\,5\,.\,6\,0\,3$ and then circle each digit in the number.
16. Which of the following represent a unit?
    (a) $\frac{3}{4}$ (c) 1 yd (e) 1 hr (g) 100
    (b) $\frac{5}{5}$ (d) 0.5 (f) 1 ton
17. Which of the following are integers or represent an integer?
    (a) 0.75 (c) $3\frac{5}{9}$ (e) 0.1
    (b) 126 (d) $\frac{125}{25}$ (f) 2,365,487
18. What is an integer? Give four examples of integers.
19. Write four consecutive integers.
20. Write four consecutive even integers.
21. Write four consecutive odd integers.
22. What is a prime number?
23. List all prime numbers from 2 to 53.

24. How does a composite number differ from a prime number? Give an example of a composite number.
25. List three multiples of 7.
26. List three multiples of 3.
27. List the smallest common multiple of 3 and 7.
28. A number can be divided evenly by what number if the last digit of the number is even? Give an example of such a number.
29. A number can be divided evenly by what number if the sum of its digits can be divided evenly by 3? Give an example having at least 5 digits.
30. A number can be divided evenly by 5 if the last digit of the number is what digit?
31. What is the test for divisibility by 9? Add a fifth digit to the number 7923 so that the resulting number is divisible by 9.
32. What does $17 < 19$ mean? What does $17 \neq 19$ mean?
33. What are numbers being added together called?
34. What does "find the 'sum' of 21, 23, and 25" mean?
35. What is the number written to show a part of something called?
36. What are the terms of a fraction?
37. In the fraction $\frac{7}{10}$, what is the 7 called? What is the 10 called? What are the 7 and the 10 together called?
38. Which of the following terms apply to the fraction $\frac{7}{10}$? Explain each choice and give reasons for skipping the others.
    (a) Prime fraction   (f) Improper fraction
    (b) Common fraction   (g) Integer
    (c) Decimal fraction   (h) Ratio
    (d) Proper fraction   (i) Digit
    (e) Mixed number
39. Explain the meaning and use of the word cancellation when applied to reducing a fraction to lower terms.
40. Can a fraction be reduced to higher terms? Explain.
41. If a fraction can be reduced to both lower and higher terms, are the resulting fractions and the initial fraction all equivalent? Explain.
42. Give an example of three equivalent fractions.
43. What is a mixed number? Give two examples.
44. If any of the following numbers have reciprocals, write its reciprocal.
    (a) 7   (c) 0.25
    (b) $\frac{2}{5}$ (d) $(6 + 8)$
45. Give an example of a mixed decimal.
46. Does multiplying by the reciprocal of a number correspond to dividing by the number?
47. Fractions must have *common* denominators before they can be (added, subtracted, multiplied, divided). Choose the correct answer.
48. Which of the following fractions are similar?
    (a) $\frac{6}{8}$ (b) $\frac{1}{8}$ (c) $\frac{3}{4}$ (d) $\frac{3}{24}$ (e) $\frac{12}{16}$
49. Give an example of three prime fractions having a common denominator of 24.
50. How is the denominator of a decimal fraction determined?
51. What is the denominator of 0.312?

INVENTORY TEST II

Check all work carefully.

1. (a) Write the number "forty-one thousand and forty one-thousandths."
   (b) Write the following number in words: 0.68135.

2. Add:

   66,815
   2,864
   385
   97
   6,489

3. Subtract:

   60,819,032
   3,002,854

4. Multiply:

   89,654
   3,728

5. Divide and check: 5,122,533 ÷ 589.

6. Add: 889.030 + 46.977 + 3.0851 + 0.0205 + 4765.890.

7. Subtract 86.69 from 423.5025.

8. Multiply 897.3 by 1.876.

9. Divide 33.603 by 4.87.

10. (a) Reduce $\frac{132}{204}$ to lowest terms.
    (b) Change $\frac{5}{12}$ to 84ths.
    (c) Supply the missing numerators for the following:

    $$7\frac{5}{9} - 7\frac{(?)}{36} - 6\frac{(?)}{36}$$

11. (a) Write $\frac{89}{7}$ as a mixed number.
    (b) Write $10\frac{8}{15}$ as an improper fraction.

12. Find the lowest common denominator (LCD) for the following fractions:

    $$\frac{2}{5} \qquad \frac{5}{6} \qquad \frac{7}{8} \qquad \frac{1}{3} \qquad \frac{5}{12}$$

13. Add: (b) Subtract $17\frac{3}{8}$ from $46\frac{5}{9}$.
    (a) $12\frac{2}{5}$
    $67\frac{5}{6}$
    $38\frac{7}{8}$
    $49\frac{1}{3}$
    $17\frac{5}{12}$

14. Multiply $\frac{9}{16}$ by $\frac{7}{12}$.

15. Divide $\frac{7}{8}$ by $\frac{5}{16}$.

16. Multiply: $3\frac{3}{4} \times 7\frac{3}{5} \times 8\frac{2}{3}$.

17. Divide $4\frac{5}{9}$ by $7\frac{3}{5}$.

18. (a) Multiply by the short method: $63.005 \times 10,000$.
    (b) Divide by the short method: $6.05 \div 100$.

19. (a) Change $\frac{137}{200}$ to a decimal fraction.
    (b) Change 0.0275 to a common fraction and simplify.

20. Determine the average of the following numbers: 925, 631, 402, 89, 173, 632, and 81.

PRACTICE PROBLEMS

Set 1   Reading and Writing Numbers

In Problems 1–15, use commas (when necessary) to form proper groups of digits and then write the number in words.

1. 138547
2. 8623546
3. 170016
4. 28900
5. 1029.5
6. 31.04
7. 0.01205
8. 0.000315
9. 45.045
10. 31000031
11. $\frac{47}{1000}$
12. $\frac{40}{7000}$
13. $\frac{52}{100000}$
14. $\frac{50}{200000}$
15. 81.81

In Problems 16–30, write the numbers in standard form, using commas (when necessary) to form proper groups.

16. Forty-seven thousandths
17. Forty and seven thousandths
18. Twenty-one hundredths
19. Twenty one-hundredths
20. Eight million, four hundred forty-four
21. Sixteen ten-thousandths
22. Thirty-eight thousand, four hundred fifty-six millionths
23. Three and twenty-one hundredths
24. Three and twenty one-hundredths
25. Four hundred-ten thousandths
26. Four hundred ten-thousandths
27. Fifty-four hundred-thousandths
28. Eighty-seven million, three hundred nineteen thousand, five hundred.
29. Four million, four
30. One hundred ten-millionths

Set 2   Addition of Whole Numbers

1. 28 + 456 + 39 + 7285 + 274
2. 755 + 391 + 746 + 8652 + 45
3. 8976 + 4567 + 3000 + 4500
4. 7256 + 3951 + 7858 + 3711 + 9888
5. 7235 + 29,787 + 411,509 + 285
6. 86,413 + 29,755 + 86 + 9 + 73,595
7. 61 + 485 + 5000 + 90,000 + 8951
8. 23,145 + 86,988 + 37,129 + 86,141
9. 88,619 + 89,726 + 38,717 + 91,400
10. 8291 + 7345 + 76 + 884 + 80,808

Set 3   Subtraction of Whole Numbers

1. 2381 from 6485
2. 78 from 295
3. 485 from 6235
4. 2888 from 2999
5. 78 from 5000
6. 198 from 3001
7. 785,142 from 86,000,000
8. 614,295,387 from 989,121,000
9. 73,009 from 84,900
10. 4005 from 9123

Set 4   Multiplication of Whole Numbers

1. 732×87
2. 456×28
3. 816×49
4. 980×97
5. 2138×725
6. 7819×2907
7. 856×35
8. 288×41
9. 707×53
10. 476×235

Set 5   Division of Whole Numbers

Divide and check.

1. $8\overline{)3145}$
2. $7\overline{)6543}$
3. $9\overline{)29,773}$
4. $12\overline{)985}$
5. $15\overline{)3651}$
6. $22\overline{)83,145}$
7. $326\overline{)82,179}$
8. $415\overline{)98,387}$
9. $782\overline{)147,382}$
10. $727\overline{)90,082}$

Set 6   Addition of Decimals

Add and check.

1. $3.7215 + 68.4 + 98.75 + 30.0071$
2. $48.073 + 295.775 + 88.080 + 3.1005$
3. $2.973 + 897.05 + 31.727 + 4.0755 + 38.88$
4. $31.46 + 283.095 + 36.215 + 3.007$
5. $65.03 + 128.6 + 3.0921 + 45.9656$
6. $165.091 + 38.385 + 101.292 + 363.0983$
7. $45.402 + 123.6 + 483.92 + 161.41$
8. $141.305 + 29.606 + 1.0029 + 38.0515$
9. $47.392 + 89.717 + 129.867 + 3.05$
10. $109.307 + 287.0313 + 64.297 + 3.1818$

Set 7   Subtraction of Decimals

Subtract and check.

1. 68.407 from 395.2
2. 7.07 from 61.0392
3. 867.359 from 2345.6789
4. 70.010 from 385.0007
5. 619.0132 from 6,000
6. 0.000135 from 0.001212
7. 0.01001 from 0.1
8. 739.27 from 10,000
9. 8.00023 from 9.01010
10. 2.0030 from 7.01001

Set 8   Multiplication of Decimals

Multiply and check.

1. 68.41 by 8.09
2. 7.865 by 2.39
3. 86.72 by 9.08
4. 2.005 by 0.0012

5. 273.2 by 0.987
6. 0.857 by 0.198
7. 798.87 by 68.097

8. 31.029 by 0.9883
9. 55.55 by 88.88
10. 213.027 by 98.765

Set 9 Division of Decimals

Carry out division as indicated and check.

1. $73.5 \div 6.07$ (nearest tenth)
2. $8.08 \div 9.9$ (nearest hundredth)
3. $175.1 \div 53.24$ (nearest ten-thousandth)
4. $66.7 \div 89.315$ (nearest thousandth)
5. $2.005 \div 8.7$ (nearest hundredth)
6. $31.4 \div 7.28$ (nearest hundredth)
7. $12 \div 0.413$ (nearest one)
8. $29.7 \div 6.78$ (nearest hundredth)
9. $18.3 \div 0.023$ (nearest ten)
10. $77.05 \div 0.093$ (nearest ten)

Set 10 Equivalent Fractions

In Problems 1–6, change to equivalent fractions with the specified denominators.

1. $\frac{3}{4}$ to 28ths
2. $\frac{7}{9}$ to 81sts
3. $\frac{28}{35}$ to 5ths

4. $\frac{1}{7}$ to 21sts
5. $\frac{3}{5}$ to hundredths
6. $\frac{66}{78}$ to 13ths

In Problems 7–8, supply the missing terms.

7. $3\frac{2}{5} = 3\frac{?}{25} = 2\frac{?}{25}$
8. $16\frac{3}{8} = 16\frac{?}{24} = 15\frac{?}{24}$

In Problems 9–10 reduce to lowest terms.

9. $\frac{84}{144}$
10. $\frac{153}{187}$

Set 11 Mixed Numbers and Improper Fractions

In Problems 1–5 change to mixed numbers.

1. $\frac{28}{5}$
2. $\frac{144}{13}$
3. $\frac{100}{7}$

4. $\frac{88}{25}$
5. $\frac{1000}{625}$

In Problems 6–10 change to improper fractions.

6. $3\frac{7}{8}$
7. $12\frac{4}{5}$
8. $6\frac{9}{16}$

9. $44\frac{1}{4}$
10. $23\frac{1}{3}$

In Problems 11–12 change to improper fractions having indicated denominators.

11. $7\frac{3}{4}$ (8ths)

12. $9\frac{5}{8}$ (16ths)

Set 12 Lowest Common Denominators

Find the LCD for the following groups of fractions by proper selection of prime factors.

1. $\frac{2}{3}, \frac{3}{4}, \frac{5}{6}$
2. $\frac{7}{8}, \frac{5}{12}$
3. $\frac{5}{6}, \frac{4}{5}, \frac{3}{10}, \frac{7}{12}$
4. $\frac{1}{2}, \frac{1}{3}, \frac{1}{4}, \frac{1}{5}, \frac{1}{6}, \frac{1}{12}$
5. $\frac{3}{14}, \frac{5}{28}, \frac{9}{16}$

6. $\frac{7}{8}, \frac{5}{12}, \frac{3}{14}, \frac{5}{16}$
7. $\frac{5}{12}, \frac{4}{3}, \frac{3}{5}, \frac{9}{16}$
8. $\frac{5}{42}, \frac{9}{56}, \frac{17}{24}$
9. $\frac{8}{9}, \frac{7}{18}, \frac{3}{16}, \frac{5}{24}$
10. $\frac{2}{19}, \frac{3}{25}, \frac{4}{9}, \frac{3}{10}$

Set 13 Addition and Subtraction of Fractions

1. $\frac{2}{3} + \frac{3}{4} + \frac{5}{6}$
2. $\frac{7}{8} - \frac{5}{12}$
3. $\frac{5}{6} + \frac{4}{5} + \frac{3}{10} + \frac{7}{12}$
4. $\frac{1}{2} + \frac{1}{3} + \frac{1}{4} + \frac{1}{5} + \frac{1}{6} + \frac{1}{12}$
5. $\frac{3}{14} + \frac{5}{28} + \frac{9}{16}$

6. $5\frac{2}{3} + 3\frac{3}{4}$
7. $16\frac{7}{9} - 12\frac{14}{15}$
8. $5\frac{1}{5} + 3\frac{2}{3} + 4\frac{4}{7} + 6\frac{5}{12}$
9. $67\frac{1}{2} - 55\frac{9}{16}$
10. $88 - 35\frac{3}{4}$

Set 14 Multiplication of Fractions

Multiply and simplify whenever possible.

1. $\frac{3}{4} \times \frac{4}{5} \times \frac{5}{6}$
2. $\frac{7}{8} \times \frac{9}{16} \times \frac{24}{27}$
3. $\frac{7}{12} \times \frac{4}{14} \times \frac{8}{42}$
4. $\frac{5}{16} \times \frac{8}{25} \times \frac{9}{20} \times \frac{15}{30}$
5. $\frac{56}{66} \times \frac{11}{12} \times \frac{3}{4}$

6. $\frac{16}{144} \times \frac{12}{64} \times \frac{27}{81}$
7. $\frac{3}{4} \times \frac{4}{5}$
8. $17 \times \frac{6}{34}$
9. $\frac{1}{5} \times 15$
10. $\frac{5}{8} \times 12 \times \frac{4}{15}$

Set 15 Division of Fractions

1. $\frac{3}{4} \div \frac{4}{5}$
2. $\frac{8}{9} \div \frac{3}{4}$
3. $\frac{7}{8} \div \frac{1}{5}$
4. $\frac{6}{7} \div \frac{3}{14}$
5. $\frac{3}{8} \div \frac{7}{9}$

6. $\frac{4}{54} \div \frac{8}{9}$
7. $\frac{21}{22} \div \frac{7}{88}$
8. $\frac{1}{4} \div \frac{1}{4}$
9. $7 \div \frac{3}{4}$
10. $\frac{2}{3} \div 5$

Set 16 Multiplication of Mixed Numbers

Multiply and simplify.

1. $3\frac{1}{4} \times 7\frac{2}{3}$
2. $5\frac{1}{6} \times 6\frac{3}{8}$
3. $6\frac{3}{5} \times 12\frac{2}{3}$
4. $8 \times 5\frac{1}{4}$
5. $6\frac{1}{3} \times 12$

6. $10\frac{1}{2} \times 3\frac{1}{7}$
7. $4\frac{4}{5} \times 8\frac{1}{3}$
8. $8\frac{1}{8} \times 6\frac{2}{3} \times 3\frac{3}{13}$
9. $120 \times 2\frac{1}{3}$
10. $6\frac{1}{3} \times 4\frac{1}{5}$

Set 17   Division of Mixed Numbers

1. $2\frac{1}{3} \div 3\frac{1}{4}$
2. $4\frac{1}{2} \div 5\frac{1}{4}$
3. $8\frac{1}{8} \div 7\frac{5}{6}$
4. $6\frac{2}{3} \div 4\frac{4}{5}$
5. $2\frac{1}{16} \div 3\frac{3}{8}$

6. $7\frac{1}{4} \div 5$
7. $16 \div 2\frac{3}{8}$
8. $12 \div 3\frac{3}{5}$
9. $8\frac{2}{5} \div 7$
10. $14\frac{4}{5} \div 16\frac{3}{9}$

Set 18   Multiplication and Division by 10, 100, 1000, etc.

1. 8.615×100
2. $38.45 \div 10$
3. $681.4164 \div 1000$
4. $0.0123 \times 1,000,000$
5. $68.68 \div 1000$

6. 40×100
7. $5 \div 100,000$
8. 27.023×1000
9. $45 \times 10 \times 100$
10. $5.6 \div 10$

Set 19   Changing to Equivalent Fractions

Change the given common fractions to equivalent decimals (nearest 0.0001) and the given decimal fractions to equivalent common fractions.

1. $\frac{7}{8}$
2. $\frac{3}{4}$
3. $\frac{5}{9}$

4. 0.0612
5. $\frac{5}{16}$
6. $\frac{1}{5}$

7. $\frac{1}{12}$
8. 0.016
9. 0.125

10. 0.40
11. 0.55
12. 0.024

Set 20   Averages

Find the average for Problems 1–5.

1. 28, 32, 72, 44, 19
2. 91, 82, 75, 78, 82, 86, 94
3. 0.290, 0.288, 0.299, 0.286, 0.283
4. $1.98, $2.05, $1.91, $2.12, $2.03, $2.09
5. 15.22 in., 12.07 in., 20.05 in.
6. A certain city had rainfall recorded as follows:

Jan.	0.52 in.	July	0.48 in.
Feb.	0.44 in.	Aug.	0.22 in.
Mar.	1.22 in.	Sept.	0.95 in.
Apr.	3.01 in.	Oct.	0.81 in.
May	2.49 in.	Nov.	0.71 in.
June	1.09 in.	Dec.	0.43 in.

What was the average monthly rainfall for this city? (Leave answer to the nearest 0.01 in.)

7. A sample of 10 bolts had the following breaking strengths:

18,400 lb	17,592 lb	18,100 lb	18,152 lb	17,988 lb
18,524 lb	18,429 lb	18,312 lb	18,475 lb	18,304 lb

What is the average breaking strength for these bolts? (Leave answer to nearest pound.)

8. Find the average cost of the items below:

$3.20 $5.12 $8.31 $4.56 $8.04

(Leave answer to nearest cent.)

9. Twelve students measured the same cylinder and recorded the following diameters. What was the average measurement? (Leave answer to the nearest 0.001 in.)

3.505 in.	3.498 in.	3.499 in.	3.500 in.
3.499 in.	3.501 in.	3.502 in.	3.500 in.
3.501 in.	3.504 in.	3.503 in.	3.500 in.

10. A student has scores of 80, 83, 86, 90, 72, 76, and 79. What must he make on the next test to give him an average of 83?

INVENTORY TEST III

Check all work carefully.

1. (a) Write the following in words: 83,685,435.68135.
   (b) Write the following as a number in conventional form: Six billion, eight hundred thirty-one million, five hundred forty-two thousand, nine hundred sixty-seven.

2. Add:
   48
   725
   396
   87
   9

3. Subtract:
   3925
   1748

4. Multiply:
   253
   87

5. Divide:
   $7\overline{)2695}$

6. Add: $3.86 + 29.145 + 0.705 + 385.1$
7. Subtract 68.456 from 83.9.
8. Multiply 3.27×2.8.
9. Divide 738.45 by 6.7.
10. (a) Reduce $\frac{18}{78}$ to lowest terms.
    (b) Change $\frac{4}{7}$ to 35ths.
    (c) Supply the missing terms for the following:

$$6\frac{3}{5} = 6\frac{?}{10} = 5\frac{?}{20}$$

11. (a) Write $6\frac{3}{5}$ as an improper fraction.
    (b) Write $\frac{30}{9}$ as a mixed number.
12. Find the LCD for the following fractions:

$$\frac{2}{3}, \quad \frac{4}{5}, \quad \frac{3}{4}, \quad \frac{7}{10}$$

13. (a) Add the following mixed numbers:

$$42\frac{3}{4}, \quad 26\frac{1}{3}, \quad 15\frac{5}{16}$$

   (b) Subtract $15\frac{3}{5}$ from $27\frac{3}{8}$.

14. Multiply $\frac{3}{5}$ by $\frac{4}{7}$.
15. Divide $\frac{3}{4}$ by $\frac{7}{8}$.
16. Multiply $3\frac{3}{4}$ by $7\frac{1}{5}$.
17. Divide $8\frac{5}{8}$ by $2\frac{2}{3}$.
18. (a) Multiply by the short method: 6.054×100.
     (b) Divide by the short method: $6.035 \div 10{,}000$
19. (a) Change $\frac{5}{8}$ to a decimal fraction correct to three decimal places.
     (b) Change 0.2135 to a common fraction and simplify.
20. Find the average of the following scores: 92, 83, 61, 47, 77, 82, and 100.

B · Geometric Facts

LINES AND LINE SEGMENTS

[1] One and only one straight line can be drawn through two given points; any two points determine a straight line.

[2] A line segment may be extended indefinitely or it may be terminated at any point.

[3] The shortest path (or distance) between any two points is the line segment joining them.

ANGLES

[4] If two straight lines intersect, the vertical (opposite) angles are equal.

[5] Two angles having the same vertex and a common side are called adjacent angles.

[6] Every point on the bisector of an angle is equidistant from the sides of the angle.

[7] If the sides of two angles are parallel, right to right and left to left, the angles are equal. If the sides are parallel right to left and left to right, the angles are supplementary.

[8] If the sides of one angle are perpendicular to the sides of another angle, right to right and left to left, the angles are equal. If the sides are perpendicular right to left and left to right, the angles are supplementary.

PARALLEL LINES AND TRANSVERSALS

[9] If a line intersects one of two parallel lines, it intersects the other also.

[10] A line that intersects two or more lines is called a transversal.

[11] If two parallel lines are cut by a transversal, the alternate interior angles are equal, and the alternate exterior angles are equal.

[12] If two parallel lines are cut by a transversal, the exterior-interior (corresponding) angles on the same side of the transversal are equal.

[13] If two or more lines are cut by a number of parallel transversals, the corresponding angles are equal.

[14] If three or more parallel lines intercept equal parts on one transversal, they intercept equal parts on every transversal.

[15] If three or more lines pass through the same point and intersect two parallel lines, they intersect proportional segments on the parallel lines. (Visualize a third parallel through the point of intersection of the several lines.)

PERPENDICULAR LINES

[16] If a line is perpendicular to one of two parallel lines, it is perpendicular to the other also.

[17] Two points, each equidistant from the extremities of a line segment, determine the perpendicular bisector of the segment.

[18] Two lines in the same plane that are perpendicular to the same line are parallel.

TRIANGLES

[19] The sum of the angles of any triangle is 180°.

[20] The acute angles of a right triangle are complementary. (Their sum = 90°.)

[21] The altitude of any triangle is the perpendicular drawn from any vertex to the opposite side.

[22] The altitude to the hypotenuse of a right triangle forms two right triangles which are similar to each other and to the whole right triangle and is a mean proportional between the segments of the hypotenuse. (The mean proportional of two numbers is the square root of their product.)

[23] In an isosceles triangle the angles opposite the equal sides are equal.

[24] If two angles of a triangle are equal, the sides opposite these angles are equal and the triangle is isoscles.

[25] If two angles of a triangle are unequal, the sides opposite these angles are unequal, and the side opposite the larger angle is the longer side.

[26] If one acute angle of a right triangle is 30°, the side opposite this angle is equal to one-half the length of the hypotenuse.

[27] If the corresponding angles of two triangles are equal, the triangles are called similar.

[28] If two triangles are similar, the corresponding sides are proportional.

[29] If two angles of one triangle are equal to two angles of another triangle, their third angles are equal and the triangles are similar.

[30] An exterior angle of a triangle is formed by extending any side and is equal to the sum of the two opposite interior angles.

[31] In an isosceles or equilateral triangle, the bisector of the vertex angle is the ·perpendicular bisector of the base. (It is also the altitude of the triangle.)

[32] The line joining the midpoints of two sides of a triangle is parallel to the third side and is equal to one-half the length of the third side.

[33] A line parallel to one side of a triangle and meeting the other two sides divides these sides proportionally.

[34] The midpoint of the hypotenuse of a right triangle is equidistant from the three vertices and is the center of the circumscribed circle.

[35] The perpendicular bisectors of the sides of any triangle meet at a common point. (This point is the center of the circumscribed circle.)

[36] The bisectors of the angles of any triangle meet at a common point. (This point is the center of the inscribed circle.)

[37] In any triangle the bisector of an angle divides the opposite side into segments which are proportional to the other two sides.

[38] The three altitudes of any triangle meet at a common point.

[39] A median of a triangle is a line segment drawn from a vertex to the midpoint of the side opposite the angle.

[40] The medians of a triangle meet at a common point which is two-thirds of the distance from any vertex to the midpoint of the opposite side.

[41] The area of a triangle is equal to one-half the product of its base and its altitude.

[42] In any right triangle the square of the hypotenuse equals the sum of the squares of the other two sides.

CONGRUENT TRIANGLES

[43] If three sides of one triangle are equal, respectively, to three sides of another triangle, the triangles are congruent (exactly equal in all respects).

[44] Two right triangles are congruent if the hypotenuse and an acute angle of the first triangle are equal to the hypotenuse and the corresponding acute angle of the second triangle.

[45] Two right triangles are congruent if the hypotenuse and a leg of one are equal to the hypotenuse and leg of the other.

[46] Two triangles are congruent if two angles and the included side (common side) of one triangle are equal, respectively, to two angles and the included side of the other triangle.

[47] Two triangles are congruent if two sides and the included angle of one are equal, respectively, to two sides and the included angle of the other.

[48] Two triangles are congruent if two angles and a side of one are equal to two angles and the corresponding side of the other.

[49] Corresponding parts of congruent triangles are equal.

QUADRILATERALS—ANGLES, DIAGONALS, AND AREAS

[50] The sum of the angles of any quadrilateral is 360°.

[51] A parallelogram is a quadrilateral with opposite sides parallel and equal.

[52] The diagonals of a parallelogram bisect each other, and the point of intersection is the geometric center of the parallelogram.

[53] The opposite angles of a parallelogram are equal.

[54] The line joining the midpoints of the legs of a trapezoid is parallel to the bases and equal to their arithmetic mean (average) length.

[55] The area of a trapezoid is equal to the altitude multiplied by the average of the bases.

POLYGONS

[56] The sum of the interior angles of a polygon of n sides is equal to $(n - 2)(180°)$.

[57] Each interior angle of a regular polygon of n sides is equal to

$$\frac{(n - 2)}{n}(180°).$$

[58] Each exterior angle of a regular polygon of n sides is equal to $360°/n$.

[59] The sum of the exterior angles of a polygon formed by extending the sides in succession is 360°.

[60] In any two similar polygons, corresponding parts (sides, diagonals, perimeters) are proportional.

[61] The areas of two similar polygons form a ratio which is equal to the ratio formed by the squares of any two corresponding dimensions.

[62] The area of a regular polygon is equal to one-half the product of its perimeter and its apothem (the radius of the inscribed circle).

CIRCLES, ARCS, CHORDS, AND ANGLES

[63] One and only one circle can be drawn through three given points not lying on a straight line.

[64] A central angle has its vertex at the center of a circle and is measured by its intercepted arc.

[65] An inscribed angle has its vertex on the circle and is measured by one-half its intercepted arc.

[66] An inscribed angle having the same (or equal) intercepted arc as a central angle is equal to one-half the central angle.

[67] An angle inscribed in a semicircle is a right angle.

[68] A radius drawn to the point of contact of a tangent line is perpendicular to the tangent line.

[69] An angle formed by two chords intersecting within a circle is measured by one-half the sum of their intercepted arcs.

[70] If two circles intersect, the line of centers is the perpendicular bisector of their common chord.

[71] In the same circle or equal circles, equal central angles intercept equal arcs and equal chords. (Equal inscribed angles intercept equal arcs and chords also.)

[72] An angle formed by a tangent and a chord drawn from the point of contact is measured by one-half the intercepted arc.

[73] An angle formed by two secants, by two tangents, or by a secant and a tangent meeting outside the circle is measured by one-half the difference between the intercepted arcs.

[74] If two chords of a circle intersect, the product of the segments of one chord is equal to the product of the segments of the other chord.

[75] A circle may be circumscribed about, or inscribed within, any regular polygon.

TANGENCY

[76] If two circles are tangent to the same line at the same point, the line connecting their centers (line of centers) passes through the point of tangency and forms right angles with the common tangent.

[77] Two tangents drawn to a circle from a common point are equal in length and make equal angles with the line drawn from the point to the center of the circle.

[78] If two tangents are drawn to a circle from a common point, the line joining the point to the center of the circle is the perpendicular bisector of the chord joining the points of tangency. This line also bisects the angle formed by the two tangents.

MISCELLANEOUS

[79] The areas of two similar plane figures are proportional to the squares of corresponding dimensions.

[80] The areas of two similar solid figures are proportional to the squares of corresponding dimensions.

[81] The volumes of two similar solid figures are proportional to the cubes of corresponding dimensions.

[82] The intersection of a plane with the surface of a sphere is a circle. If the plane passes through the center of the sphere, the intersection is called a great circle.

C Logarithms

C.1 WHAT ARE LOGARITHMS

Let us begin our study of logarithms by preparing a small table of powers of 2:

$2^0 = 1$	$2^6 = 64$	$2^{12} = 4{,}096$
$2^1 = 2$	$2^7 = 128$	$2^{13} = 8{,}192$
$2^2 = 4$	$2^8 = 256$	$2^{14} = 16{,}384$
$2^3 = 8$	$2^9 = 512$	$2^{15} = 32{,}768$
$2^4 = 16$	$2^{10} = 1{,}024$	$2^{16} = 65{,}536$
$2^5 = 32$	$2^{11} = 2{,}048$	$2^{17} = 131{,}072$

To see how to multiply or divide using powers of 2, consider the following examples.

example   To multiply 128 by 512, write the problem as $2^7 \times 2^9$. Recall from the section on exponents that $2^7 \times 2^9 = 2^{16}$. Returning to the table, you see that $2^{16} = 65{,}536$. Therefore, $128 \times 512 = 65{,}536$.

example   To divide 32,768 by 2048, represent the problem as $2^{15} \div 2^{11}$. You then obtain $2^4 = 16$ as the quotient.

Recalling that the square root of a power can be obtained by dividing the exponent by 2, you can find the square root of some numbers by using the table:

$$\sqrt{16{,}384} = \sqrt{(2^{14})} = 2^{14 \div 2} = 2^7 = 128$$

$$\sqrt{65{,}536} = \sqrt{(2^{16})} = 2^{16 \div 2} = 2^8 = 256$$

To cube a number written in power form, you only need to multiply the exponent by 3.

For example, to cube a number such as 32:

$$32^3 = (2^5)^3 = 2^{15} = 32{,}768$$

Note that all of the arithmetic operations in these examples are performed on the exponents and not on the base numbers. Because arithmetic operations on exponents can facilitate certain calculations, the basic rules and conditions which apply to these operations come under the subject of *logarithms*. *Logarithms are exponents*.

To emphasize this last statement, let us pick from our table $2^6 = 64$ and really digest it. The base number 2 has an exponent of 6 which is called the logarithm of 64 to the base 2; i.e., $\log_2 64 = 6$. Thus, $2^6 = 64$ and $\log_2 64 = 6$ are equivalent statements.

Using the table of the powers of 2, what is the $\log_2$ of 128? Because $2^7 = 128$, $\log_2 128 = 7$. The $\log_2 8192 = 13$ because $2^{13} = 8192$. What number has a log of 10? Because $2^{10} = 1024$, 1024 has a log of 10.

Because logarithms are special exponents, we present here in abbreviated form the following laws of exponents and definitions:

1. $b^n = b \cdot b \cdot b \cdot b \cdot b \cdot b \cdot b \cdot b \cdots n$ times
2. $b^n \cdot b^m = b^{n+m}$
3. $b^n \div b^m = b^{n-m}$
4. $(ab)^n = a^n b^n$

5. $\left(\dfrac{a}{b}\right)^n = \dfrac{a^n}{b^n}$

6. $(b^n)^m = b^{nm}$

7. Definition: $b^0 = 1 \qquad (b \neq 0)$

8. Definition: $b^{-n} = \dfrac{1}{b^n} \qquad (b \neq 0)$

The logarithm L of a number N to the base b ($b \neq 1$) is the exponent which, applied to the base b, makes $b^L = N$. This definition is commonly written in two equivalent algebraic forms:

$$L = \log_b N \qquad \text{(the logarithmic form)}$$

$$b^L = N \qquad \text{(the exponential form)}$$

We restrict the base to positive values to avoid imaginary numbers. This limitation, in turn, restricts N to positive values because a positive base raised to any power p is positive. Further, $b \neq 1$, because $1^p = 1$ for all values of the number p. It is impossible to find a value of p such that $1^p \neq 1$.

examples

1. $\text{Log}_3 81 = 4$; in the exponential form is $3^4 = 81$.
2. $5^3 = 125$; in the logarithmic form is $\log_5 125 = 3$.
3. $\text{Log}_{12} 144 = 2$; in the exponential form is $12^2 = 144$.
4. $7^3 = 343$; in the logarithmic form is $\log_7 343 = 3$.

EXERCISE C.1

Use the table of powers of 2 in this section to answer Problems 1–10.

1. $2^7 = ?$
2. $\log_2 4096 = ?$
3. $\log_2 16,384 = ?$
4. $\log_2 1 = ?$
5. $\log_2 N = 15, N = ?$
6. $\log_2 N = 1, N = ?$
7. $\log_2 64 + \log_2 128 = ?$
8. The sum of the logs of two numbers is 11. What is the product of these two numbers?
9. $\sqrt{4096} = ?$
10. $\sqrt[3]{32,768} = ?$
11. Write each of the following in exponential form:
    (a) $\log_5 25 = 2$ (e) $\log_4 1 = 0$
    (b) $\log_{10} 1000 = 3$ (f) $\log_{10} 100 = 2$
    (c) $\log_2 64 = 6$ (g) $\log_4 64 = 3$
    (d) $\log_5 5 = 1$
12. Write each of the following in logarithmic form:
    (a) $2^3 = 8$ (e) $10^4 = 10,000$
    (b) $5^3 = 125$ (f) $6^1 = 6$
    (c) $6^0 = 1$ (g) $2^8 = 256$
    (d) $7^2 = 49$
13. By using the table of powers of 2 and the laws of exponents, calculate the following:
    (a) $(32)(256)$ (c) $(64)^2$
    (b) $16,384 \div 1024$ (d) $\sqrt[5]{32,768}$
14. Prepare a table of powers of 3 from 3^0 to 3^{10} and use it to find the following:
    (a) $\log_3 6561$ (d) $19,683 \div 2187$
    (b) $\log_3 243$ (e) $\sqrt{59,049}$
    (c) $(81)(729)$ (f) $(27)^3$

C.2   LAWS OF LOGARITHMS

By using the definition of a logarithm, we can modify the laws of exponents to obtain some operating rules for working with logarithms.

rule 1 $\text{Log}_b (m \cdot n) = \log_b m + \log_b n$

This rule states that the log of a product of two factors is equal to the sum of the logs of the factors. (This rule is not limited to the product of just two factors but includes any integral number of factors.)

example

$$\text{Log}_5 (25)(125) = \log_5 25 + \log_5 125$$
$$= 2 + 3 = 5$$

Therefore,

$$(25)(125) = 5^5 = 3125$$

rule 2 $\operatorname{Log}_b (m \div n) = \log_b m - \log_b n$

This rule states that the log of a quotient of two factors equals the difference between the logs of the two factors. (The log of the divisor is subtracted from the log of the dividend.)

example

$$\operatorname{Log}_2 \frac{32{,}768}{4096} = \log_2 32{,}768 - \log_2 4096$$

$$= 15 - 12 = 3$$

Therefore,

$$\frac{32{,}768}{4096} = 2^3 = 8$$

rule 3 $\operatorname{Log}_b M^p = p \cdot \log_b M$

This rule states that the log of a number raised to a power p is equal to the power times the log of the number.

example $\operatorname{Log}_2 (32)^3 = 3 \cdot \log_2 32$

From the table of powers of two in Section C.1, $\log_2 32 = 5$. Therefore,

$$3 \cdot \log_2 32 = 3 \cdot 5 = 15$$

$$32^3 = 2^{15} = 32{,}768$$

rule 4 $\operatorname{Log}_b \sqrt[q]{M} = 1/q \log_b M$

This rule is really a case of Rule 3 with a fractional exponent.

example $\operatorname{Log}_2 \sqrt[5]{1024} = \log_2 (1024)^{1/5} = (\tfrac{1}{5}) \cdot (\log_2 1024)$

From the table in Section C.1,

$$\log_2 1024 = 10$$

$$\frac{1}{5} \log_2 1024 = \frac{1}{5} \cdot 10 = 2$$

Therefore,

$$\sqrt[5]{1024} = 2^2 = 4$$

rule 5 $\operatorname{Log}_b 1 = 0$

This rule is really the logarithmic form for the definition $b^0 = 1$. It states that the logarithm of 1 to any suitable base b is equal to zero.

examples

$$\text{Log}_6 \; 1 = 0, \quad \text{because } 6^0 = 1$$

$$\text{Log}_{100} \; 1 = 0, \quad \text{because } 100^0 = 1$$

$$\text{Log}_{688} \; 1 = 0, \quad \text{because } 688^0 = 1$$

rule 6 $\text{Log}_b \, b = 1$

This rule states that the log of any number b to the base b is equal to 1.

examples

$$\text{Log}_{12} \; 12 = 1, \quad \text{because } 12^1 = 12$$

$$\text{Log}_{88} \; 88 = 1, \quad \text{because } 88^1 = 88$$

C.3 COMMON LOGARITHMS AND NATURAL LOGARITHMS

Because our number system has a base of 10, it is not surprising to find that 10 is used as a base for a logarithmic system. Logarithms having the base 10 are referred to as *common logarithms* or *Briggs Logarithms* (named for the English mathematician H. Briggs). Another system of logarithms, called *natural logarithms,* have as a base a theoretically derived number e ($e = 2.71828+$). This system was developed by the Scotch mathematician J. Napier.

We shall use common logarithms. A brief table of powers of 10 appears below for convenient reference:

$10^0 = 1$	$10^{-1} = 0.1$
$10^1 = 10$	$10^{-2} = 0.01$
$10^2 = 100$	$10^{-3} = 0.001$
$10^3 = 1000$	$10^{-4} = 0.0001$
$10^4 = 10,000$	$10^{-5} = 0.00001$
$10^5 = 100,000$	$10^{-6} = 0.000001$
$10^6 = 1,000,000$	

C.4 THE CHARACTERISTIC AND MANTISSA OF A LOGARITHM

Referring to the table in Section C.3, you see that log* 10,000 = 4 (equivalent to $10^4 = 10,000$) and log* 100,000 = 5 (equivalent to $10^5 = 100,000$). Now

---

* We henceforth drop the writing of the base number 10; it should be understood. Log 100 means the log of 100 to the base 10. Whenever we wish to indicate a base other than 10, we shall write it in.

consider the logarithm of a number such as 55,000. Since 55,000 falls halfway between 10,000 and 100,000, it seems natural to assume that 10 should be raised to a power halfway between 4 and 5 to equal 55,000. A good guess would be $10^{4.5} = 55{,}000$. However, the logarithmic function is not a linear function; the exponent needed is 4.74036. That is, $10^{4.74036} = 55{,}000$.

We are not interested at this point in how the fraction .74036 was obtained. What we are interested in is the fact that the logarithm of most numbers consists of a whole number and a fraction. In this case the log of 55,000 is 4.74036. The 4 is the whole number part and the .74036 is the fractional part. The whole number part is called the *characteristic* of the logarithm. The fractional part of the logarithm is called the *mantissa*.

When a number is written in the power of 10 form, the exponent of the 10 is equal to the characteristic* of the number. For example:

$$210 = 2.10 \times 10^{2}, \qquad \text{characteristic* of 210 is 2}$$

$$0.035 = 3.5 \times 10^{-2}, \qquad \text{characteristic of 0.035 is } -2$$

$$520{,}000 = 5.2 \times 10^{5}, \qquad \text{characteristic of 520,000 is 5}$$

Numbers greater than 1 have characteristics that are one less than the number of digits in the whole number part of the number. For example, a whole number with 6 digits has a characteristic of 5. Numbers less than 1 have negative characteristics equal to the number of places the decimal point must be moved to write the number in scientific (power of 10) form. Thus, 0.012 has a characteristic of -2 because in scientific form it would be 1.2×10^{-2}.

examples

Number	Characteristic*
256	2
18,935.4	4
88.0	1
6.5	0
0.882	-1
0.0235	-2
0.00089	-4

* When we refer to the characteristic of a number, we mean the characteristic of the logarithm of the number.

EXERCISE C.4

In Problems 1–10, determine the characteristic of the logarithm of each number.

1.	68.3	6.	6,000,000
2.	30.05	7.	6.3×10^5
3.	88.88	8.	8.2×10^{-6}
4.	888.8	9.	3.01×10^7
5.	0.00000691	10.	3.00013

In Problems 11–20, place the decimal point in the proper position so that the logarithm of the resulting number will have the indicated characteristic.

11.	38125, characteristic of -2	16.	38125, characteristic of -4
12.	38125, characteristic of 0	17.	38125, characteristic of 7
13.	38125, characteristic of 5	18.	38125, characteristic of -1
14.	38125, characteristic of 1	19.	3333, characteristic of 2
15.	38125, characteristic of 3	20.	3333, characteristic of -2

C.5   THE MANTISSA AND THE USE OF TABLE 15

Consider the complete logarithm of the number 563. By inspection you can determine that the characteristic is 2, so write $\log 563 = 2.----$. To get the mantissa, turn to Table 15, Appendix D. Look down the left-hand column (headed N for number) to find the first two digits of the number, 56. Because the third digit is a 3, you find the mantissa at the intersection of line 56 with the column headed 3.

N	0	1	2	3
56				7505

The mantissa is 0.7505. Combining the characteristic and mantissa, you have $\log 563 = 2.7505$. This means that $10^{2.7505} = 563$.

example   Determine the logarithm of 30.7. By inspection, the characteristic is 1. To find the mantissa, locate the first two digits, 30, in the column headed N.

N	0	1	2	3	4	5	6	7	8	9
30								4871		

The third digit is a seven, so look across to Column 7 (on the line with 30) where you find the mantissa 4871. Log $30.7 = 1.4871$. This means that $10^{1.4871} = 30.7$.

As previously mentioned, a logarithm has two parts, a characteristic and a mantissa. The characteristic is always an integer and may be positive or negative. The mantissa is a decimal fraction and is always positive. This leads to confusion when you need to carry out arithmetic operations with logs such as -2.7114.

To avoid this confusion, two methods of writing a logarithm are used. One method has the negative sign written over the characteristic (e.g., $\bar{2}.7114$) as a reminder that only the characteristic is negative. The second method (and the one which we shall use) amounts to adding 10 to the characteristic and subtracting 10 from the whole log. Therefore, $\bar{2}.7114$ would be written $8.7114 - 10$.

example   Determine the log of 0.00244 and write it in the $10 - 10$ form. The characteristic $= -3$. From Table 15, the mantissa is 0.3874. Adding 10 to -3, you get 7; subtracting 10 from this result, you obtain

$$\log 0.00244 = 7.3874 - 10$$

If we have the logarithm of a number, we can determine the number itself. It is common practice to call the number so determined the *antilog*. The *antilog* is the number whose log you have at the start.

example   Find the antilog of $8.3541 - 10$. By inspection, you know that the characteristic is -2. Table 15 shows the digits of the number whose mantissa is 0.3541 to be 226. Locating the decimal point, you have 0.0226 as the number whose log is $8.3541 - 10$. (Thus, 0.0226 is the antilog of $8.3541 - 10$.)

example   Find the antilog of $6.6325 - 10$. First, determine the characteristic to be -4. Then from Table 15, you see that the digits of the number with a mantissa of .6325 are 429. Next, use the -4 characteristic to locate the decimal point, obtaining 0.000429 as the antilog of $6.6325 - 10$.

The list below contains several numbers and their logarithms. Check the given characteristics by inspection and use Table 15 to check the mantissas.

Number	Characteristic	Mantissa
812	2	0.9096
0.0052	-3	0.7160
23.8	1	0.3766
0.241	-1	0.3820
38,500	4	0.5855
629,000	5	0.7987
2.88	0	0.4594
331.0	2	0.5198

EXERCISE C.5

In Problems 1–10, determine the logarithm of the given number.

1.	36	6.	0.000310
2.	300	7.	34.6
3.	0.0256	8.	999
4.	0.025	9.	977.0
5.	38,400	10.	1.380

In Problems 11–20, determine the antilog of the given logarithm.

11.	2.1492	16.	1.8102
12.	3.3979	17.	8.8704 − 10
13.	0.5263	18.	5.8407 − 10
14.	9.6503 − 10	19.	3.9000 − 10
15.	4.7275	20.	1.9773

C.6 LINEAR INTERPOLATION

The interpolation procedure used with tables of mantissas is the same as that used with tables of trigonometric ratios. See Section 8.5 for review.

example   Determine the logarithm of 8065. By inspection, you see that the log has a characteristic of 3. In Table 15, you find that 8060 has a mantissa of 0.9063 and that 8070 has a mantissa of 0.9069. The number 8065 is halfway between 8060 and 8070, so you can expect the mantissa of 8065 to have a value halfway between 0.9063 and 0.9069, which would be 0.9066. Therefore, the log of 8065 = 3.9066.

The following form presents a more systematic method for determining a logarithm when interpolation becomes necessary:

$$\left.\begin{array}{c}\left.\begin{array}{c}8060 \\ 8065 \\ 8070\end{array}\right]{-}5\end{array}\right]{-}10 \qquad \left.\begin{array}{c}9063 \\ \\ 9069\end{array}\right]\text{difference is 6}$$

The number falls 0.5 of the way between the listed numbers. The mantissa you want then is one that falls halfway between the corresponding listed mantissas. Since 0.5 of 6 = 3, add 3 to the 9063, obtaining 9066. The log then is 3.9066.

example   Determine the logarithm of 84,540. By inspection, the characteristic is 4. From Table 15, determine

$$\left.\begin{array}{c}\left.\begin{array}{c}8450 \\ 8454 \\ 8460\end{array}\right]{-}4\end{array}\right]{-}10 \qquad \left.\begin{array}{c}9269 \\ \\ 9274\end{array}\right]\text{difference is 5}$$

8454 is 0.4 of the way between 8450 and 8460; 0.4 × 5 = 2. Adding 2 to 9269 gives 9271 for the required mantissa. The complete log is 4.9271.

example   Determine the antilogarithm of 2.8354. Looking in Table 15, you find that 8354 falls between the mantissas 8351 and 8357. In tabular form, you have

$$
\left.\begin{array}{c} 6840 \\ \\ 6850 \end{array}\right\}\!\!-10 \qquad \left.\left.\begin{array}{c} 8351 \\ 8354 \\ 8357 \end{array}\right\}\!\!-3\ \right\}\!\!-6, \tfrac{3}{6}\ \text{or}\ \tfrac{1}{2}
$$

$\frac{1}{2}$ of 10 = 5 and 5 added to the 6840 gives 6845 for the digits of the antilog. Using the characteristic of 2, the number is 684.5.

Note: To facilitate the interpolation process, zeros were annexed to the digits making up the numbers corresponding to the listed mantissas. Because the characteristic is the sole indicator of the decimal point location, you are free to annex as many zeros as needed. Specifically, in the last example, it is easier to think of adding $\frac{1}{2}$ of 10 to 6840 than adding $\frac{1}{2}$ of 1 to 684.

EXERCISE C.6

Determine the logarithms of the numbers in Problems 1–10, employing the tabular form used in the examples.

1.	3.857	6.	0.0003008
2.	21.73	7.	38,450
3.	987.2	8.	69.88
4.	0.6077	9.	456,390
5.	0.01234	10.	2,873,852

In Problems 11–20, find the antilog for the logarithm.

11.	2.8672	16.	7.5203 − 10
12.	3.7208	17.	9.3506 − 10
13.	1.6766	18.	3.0513
14.	0.6390	19.	10.1507 − 10 (characteristic is zero)
15.	8.4546 − 10	20.	6.3600 − 6

C.7   THE USE OF LOGARITHMS IN MULTIPLICATION AND DIVISION PROBLEMS

In this section, we shall apply Rules 1 and 2.

rule 1   To multiply two factors, add their logarithms.

rule 2   To divide two factors, subtract the log of the divisor from the log of the dividend.

When calculating with logs, it is advisable to divide the problem into five parts:

1. Diagram the problem.
2. Make a data table consisting of the numerical factors and their corresponding logs.
3. Carry out the arithmetic involved in Step 1.
4. Look up the antilog for the result of Step 3.
5. Check each intermediate step; i.e., check each step as you complete it rather than attempting to check the finished problem.

The following examples illustrate these steps.

example   Calculate the indicated product of the following integers:

$$(685)(491)(38,200) = P*$$

Step 1. Diagram:

$$\log P = \log 685 + \log 491 + \log 38,200$$

Step 2. Make data table:

$$\log 685 = 2.8357$$

$$\log 491 = 2.6911$$

$$\log 38,200 = 4.5821$$

Step 3. Carry out operations indicated in Step 1. In this case, add the logs:

$$
\begin{aligned}
\log 685 &= 2.8357 \\
\log 491 &= 2.6911 \\
\log 38,200 &= 4.5821 \\
\hline
\log P &= 10.1089
\end{aligned}
$$

Step 4. Determine the antilog of 10.1089. Whenever interpolation is necessary, it is good practice to show the interpolation procedure:

$$
\left.\begin{array}{c} 1280 \\ P \\ 1290 \end{array}\right] {\scriptstyle -10} \qquad \left.\left.\begin{array}{c} 1072 \\ 1089 \\ 1106 \end{array}\right] {\scriptstyle -17}\right] {\scriptstyle -34}
$$

$$\tfrac{17}{34} = \tfrac{1}{2}, \qquad \tfrac{1}{2} \times 10 = 5, \qquad 1280 + 5 = 1285$$

---

* It is good practice to indicate a product or quotient by the letter P or the letter Q, respectively, so that the logarithm of the product or quotient can be referred to as $\log P$ or $\log Q$.

Locate the decimal point corresponding to the characteristic of 10, obtaining $P = 12,850,000,000$.

example   Calculate the indicated quotient:

$$68,400 \div 396 = Q$$

Step 1. Diagram:

$$\log Q = \log 68,400 - \log 396$$

Step 2. Make table of data:

$$\log 68,400 = 4.8351$$
$$\log 396 = 2.5977$$

Step 3. Carry out arithmetic operation indicated in Step 1:

$$\log 68,400 = 4.8351$$
$$\underline{\log 396 = 2.5977}$$
$$\log Q = 2.2374$$

Step 4. Determine antilog of 2.2374. Because the 396 factor has only three significant figures, you need not interpolate to get the antilog of 2.2374. Just determine the number whose mantissa is closest to .2374. In this case, 173 has a mantissa nearest to 2374. Locating the decimal point corresponding to a +2 characteristic, you obtain $Q = 173$.

example   Carry out the indicated operations:

$$\frac{(68,200)(46,300)}{(29,800)} = Q$$

Step 1. Diagram:

$$\log Q = \log 68,200 + \log 46,300 - \log 29,800$$

Step 2. Make data table:

$$\log 68,200 = 4.8338$$
$$\log 46,300 = 4.6656$$
$$\log 29,800 = 4.4742$$

Step 3. Carry out the arithmetic operations indicated in Step 1:

$$\begin{aligned}
\log 68{,}200 &= 4.8338 \\
\log 46{,}300 &= 4.6656 \\
\hline
\text{sum} &= 9.4994 \\
\log 29{,}800 &= 4.4742 \\
\hline
\log Q &= 5.0252
\end{aligned}$$

Step 4. Determine antilog of 5.0252:

$$\log Q = 5.0252$$

$$Q = 106{,}000 \qquad \text{(No interpolation is necessary.)}$$

EXERCISE C.7

Follow the procedure used in the above examples in calculating the indicated product or quotient. Interpolate* only when necessary.

1. (86.30)(290.5)(36.4)
2. (645.2)(88.61)(395)
3. Find the volume in cubic feet of a rectangular solid which measures 682 in. by 423 in. by 406 in.
4. Calculate the volume in cubic feet of a rectangular solid measuring 12.75 ft by 68.61 ft by 37.4 ft.
5. $63.44 \div 2.92$
6. $785.40 \div 423$

7. $\dfrac{(6.83)(49.5)(7.26)(886)}{(51.9)(38.2)(7.88)}$

8. $\dfrac{(682.1)(49.52)(713.50)}{(498.2)(71.05)(912)}$

9. $\dfrac{(0.00213)(0.01285)(0.08213)}{(0.000515)(0.00188)}$

10. $\dfrac{(0.00688)(0.002135)(0.000304)}{(0.0812)(0.000215)}$

* When the least accurate factor contains three or fewer significant digits, the other factors can be rounded off to three significant figures, thus avoiding the need to interpolate.

C.8   CALCULATING ROOTS AND POWERS BY USING LOGARITHMS

In this section, we take up Rules 3 and 4.

rule 3   In calculating with a power of a number, multiply the log of the number by the power.

rule 4   In calculating the root of a number, divide the log of the number by the index of the root.

example   Calculate the indicated power:

$$(27.32)^6 = N$$

Step 1. Diagram:

$$\log N = \log(27.32)^6 = 6 \cdot \log 27.32$$

Step 2. Make table of data:

$$\log 27.32 = 1.4365$$

Step 3. Carry out the arithmetic operation indicated in Step 1:

$$\log N = (6)(1.4365) = 8.6190$$

Step 4. Determine antilog of 8.6190: ·

$$N = 415,900,000$$

example   Calculate the indicated root:

$$\sqrt[5]{0.2913} = R$$

Step 1. Diagram:

$$\log R = \log(0.2913)^{1/5} = \tfrac{1}{5}\log 0.2913 \text{ or } \frac{\log 0.2913}{5}$$

Step 2. Make table of data:

$$\log 0.2913 = 9.4643 - 10$$

Step 3. Carry out arithmetic operations indicated in Step 1:

$$\tfrac{1}{5}\log 0.2913 = (9.4643 - 10) \div 5$$

$$\frac{1.8929 - 2}{5)\overline{9.4643 - 10}} \quad (\text{divide } -10 \text{ by } 5 = -2)$$

Step 4. Determine antilog of $1.8929 - 2$:

$$R = 0.7814$$

example   Calculate the indicated root:

$$\sqrt[3]{0.01235} = R$$

Step 1. Diagram:

$$\log R = \log(0.01235)^{1/3} = \tfrac{1}{3} \log 0.01235$$

Step 2. Make table of data:

$$\log 0.01235 = 8.0917 - 10$$

Step 3. Carry out the arithmetic operation indicated in Step 1. To carry out the division by 3, add and subtract a number to the characteristic $(8 - 10)$ so that the division of the negative part of the characteristic is exact. In this case, add 20 and subtract 20, obtaining $28.0917 - 30$. Then divide by 3, obtaining

$$\log R = 9.3639 - 10$$

Step 4. Determine the antilog of $9.3639 - 10$:

$$\log R = 9.3639 - 10$$
$$R = 0.2312$$

example   Carry out the indicated operations:

$$N = \frac{(28.31)^{1/4}(688.2)^3}{\sqrt{6,823,000}}$$

Step 1. Diagram:

$$\log N = (\tfrac{1}{4})(\log 28.31) + (3)(\log 688.2) - (\tfrac{1}{2})(\log 6,823,000)$$

Step 2. Make table of data:

$$\log 28.31 = 1.4519, \qquad \tfrac{1}{4}\log 28.31 = 0.3630$$
$$\log 688.2 = 2.8377, \qquad 3\log 688.2 = 8.5131$$
$$\log 6,823,000 = 6.8340, \qquad \tfrac{1}{2}\log 6,823,000 = 3.4170$$

Step 3. Carry out the arithmetic operations indicated in Step 1:

$$\tfrac{1}{4}\log 28.31 = 0.3630$$
$$3\log 688.2 = 8.5131$$
$$\text{Sum} = \overline{8.8761}$$
$$\tfrac{1}{2}\log 6,823,000 = 3.4170$$
$$\text{Difference} = \overline{5.4591}$$

Step 4. Determine the antilog of 5.4591:

$$N = \text{antilog } 5.4591$$
$$N = 287,800$$

EXERCISE C.8

Use logarithms to calculate the indicated product or quotient. Interpolate only when necessary.

1. $(68.2)^{1/2}$
2. $(293.0)^{1/5}$
3. $(738.3)^{1/4}$
4. $(0.002154)^{1/3}$
5. $(0.00888)^5$
6. $(0.0101)^3$

7. $(38.4)^2 \cdot (261)^{1/3}$
8. $(561)^{1/4} \cdot (3.8650)^3$
9. $(989)^{1/2} \cdot (1.46)^5$
10. $\dfrac{(68.3)^2(3.8450)^3}{\sqrt[3]{6954.0}}$

D Tables

Table 1 Greek Alphabet of Capital and Lower-Case Letters

A	α	Alpha	N	ν	Nu
B	β	Beta	Ξ	ξ	Xi
Γ	γ	Gamma	O	o	Omicron
Δ	δ	Delta	Π	π	Pi
E	ε	Epsilon	P	ρ	Rho
Z	ζ	Zeta	Σ	σ	Sigma
H	η	Eta	T	τ	Tau
Θ	θ	Theta	Υ	υ	Upsilon
I	ι	Iota	Φ	ϕ	Phi
K	κ	Kappa	X	χ	Chi
Λ	λ	Lambda	Ψ	ψ	Psi
M	μ	Mu	Ω	ω	Omega

Table 2 Dimensional Prefixes

Mega $= \times 10^{6}$	Centi $= \times 10^{-2}$
Kilo $= \times 10^{3}$	Milli $= \times 10^{-3}$
Hecto $= \times 10^{2}$	Micro $= \times 10^{-6}$
Deca $= \times 10^{1}$	Millimicro $= \times 10^{-9}$
Deci $= \times 10^{-1}$	Micromicro $= \times 10^{-12}$

Table 3 Symbols and Abbreviations

Sign	Meaning	Sign	Meaning
$+$	Plus (sign of addition)	g	Acceleration due to gravity
$+$	Positive		(32.16 ft per sec per sec)
$-$	Minus (sign of subtraction)	i (or j)	Imaginary unit ($\sqrt{-1}$)
$-$	Negative	sin	Sine
$\pm(\mp)$	Plus or minus (minus or plus)	cos	Cosine
$\times$	Multiplied by (multiplication sign)	tan	Tangent
		cot	Cotangent
$\cdot$	Multiplied by (multiplication sign)	sec	Secant
		csc	Cosecant
$\div$	Divided by (division sign)	Δ	Delta (increment of)
$:$	Is to (in proportion)	$\angle$	Angle
$=$	Equals	$\llcorner$	Right angle
$\neq$	Is not equal to	$\perp$	Perpendicular to
$\equiv$	Is identical to	$\llcorner$	Rt $\angle$s
$::$	Equals (in proportion)	$\parallel$	Parallel to
$\approx$	Approximately equals	$\triangle$	Triangle
$\sim$	Similar	$\odot$	Circle
$>$	Greater than	$\square$	Parallelogram
$<$	Less than	$°$	Degree (circular arc, or temperature)
$\geqslant$	Greater than or equal to		
$\leqslant$	Less than or equal to	$'$	Minutes or feet
$\propto$	Varies directly as	$''$	Seconds or inches
$\sqrt{}$	Square root	a'	a prime
$\sqrt[3]{}$	Cube root	a''	a double prime
$\sqrt[4]{}$	4th root	a_1	a sub one
$\sqrt[n]{}$	nth root	a_2	a sub two
$\dfrac{1}{n}$	Reciprocal value of n	a_n	a sub n
		$(\ )$	Parentheses
$\log n$	Common logarithm of n	$[\ ]$	Brackets
μ	Mu (coefficient of friction)	$\{\ \}$	Braces
π	Pi (3.1416)	$\overline{}$	vinculum
ω	Omega (angles measured in radians)	$P(x, y)$	Rectangular coordinates of point P

Table 4 Frequently Used Constants

$\pi = 3.141593$ 1 micron $= 10^{-4}$ cm
$\sqrt{2} = 1.414214$ 1 Angstrom $= 10^{-8}$ cm
$\sqrt{3} = 1.732051$

Speed of light $= 186,272$ miles/sec $\approx 3 \times 10^{10}$ cm/sec
Acceleration of gravity $= 980$ cm/sec^2 $= 32.16$ ft/sec^2
Radius of earth:
               mean polar $= 3950$ miles
           mean equatorial $= 3963$ miles

Density of water $= 62.4$ lb/cu ft (approximately)
1 gal $= 231$ in^3 (exact) $= 0.134$ cu ft
1 cu ft $= 7.48$ gal
1 m $= 39.37$ in. (exact) $= 0.000621$ mile
1 mile $= 1760$ yd $= 5280$ ft (exact)
1 knot $= 1.151$ mph
1 horsepower $= 33,000$ ft-lb of work/min (exact)
              $= 550$ ft-lb of work/sec (exact)
              $= 746$ watts (exact)

Table 5 Decimal Equivalents of Fractions

Fractions	Decimals	Fractions	Decimals	Fractions	Decimals	Fractions	Decimals
$\frac{1}{64}$	0.015625	$\frac{17}{64}$	0.265625	$\frac{33}{64}$	0.515625	$\frac{49}{64}$	0.765625
$\frac{1}{32}$	0.03125	$\frac{9}{32}$	0.28125	$\frac{17}{32}$	0.53125	$\frac{25}{32}$	0.78125
$\frac{3}{64}$	0.046875	$\frac{19}{64}$	0.296875	$\frac{35}{64}$	0.546875	$\frac{51}{64}$	0.796875
$\frac{1}{16}$	0.0625	$\frac{5}{16}$	0.3125	$\frac{9}{16}$	0.5625	$\frac{13}{16}$	0.8125
$\frac{5}{64}$	0.078125	$\frac{21}{64}$	0.328125	$\frac{37}{64}$	0.578125	$\frac{53}{64}$	0.828125
$\frac{3}{32}$	0.09375	$\frac{11}{32}$	0.34375	$\frac{19}{32}$	0.59375	$\frac{27}{32}$	0.84375
$\frac{7}{64}$	0.109375	$\frac{23}{64}$	0.359375	$\frac{39}{64}$	0.609375	$\frac{55}{64}$	0.859375
$\frac{1}{8}$	0.125	$\frac{3}{8}$	0.375	$\frac{5}{8}$	0.625	$\frac{7}{8}$	0.875
$\frac{9}{64}$	0.140625	$\frac{25}{64}$	0.390625	$\frac{41}{64}$	0.640625	$\frac{57}{64}$	0.890625
$\frac{5}{32}$	0.15625	$\frac{13}{32}$	0.40625	$\frac{21}{32}$	0.65625	$\frac{29}{32}$	0.90625
$\frac{11}{64}$	0.171875	$\frac{27}{64}$	0.421875	$\frac{43}{64}$	0.671875	$\frac{59}{64}$	0.921875
$\frac{3}{16}$	0.1875	$\frac{7}{16}$	0.4375	$\frac{11}{16}$	0.6875	$\frac{15}{16}$	0.9375
$\frac{13}{64}$	0.203125	$\frac{29}{64}$	0.453125	$\frac{45}{64}$	0.703125	$\frac{61}{64}$	0.953125
$\frac{7}{32}$	0.21875	$\frac{15}{32}$	0.46875	$\frac{23}{32}$	0.71875	$\frac{31}{32}$	0.96875
$\frac{15}{64}$	0.234375	$\frac{31}{64}$	0.484375	$\frac{47}{64}$	0.734375	$\frac{63}{64}$	0.984375
$\frac{1}{4}$	0.25	$\frac{1}{2}$	0.5	$\frac{3}{4}$	0.75	1	1

Table 6   Fractional Equivalents: Common, Percentage, and Decimal

Common							Percentage	Decimal
2nds	3rds	4ths	6ths	8ths	12ths	16ths		
						1	$6\frac{1}{4}$	0.0625
					1		$8\frac{1}{3}$	$0.083\frac{1}{3}$
				1			$12\frac{1}{2}$	0.125
			1				$16\frac{2}{3}$	$0.166\frac{2}{3}$
						3	$18\frac{3}{4}$	0.1875
		1					25	0.250
						5	$31\frac{1}{4}$	0.3125
	1						$33\frac{1}{3}$	$0.333\frac{1}{3}$
				3			$37\frac{1}{2}$	0.375
					5		$41\frac{2}{3}$	$0.416\frac{2}{3}$
						7	$43\frac{3}{4}$	0.4375
1							50	0.500
						9	$56\frac{1}{4}$	0.5625
					7		$58\frac{1}{3}$	$0.583\frac{1}{3}$
				5			$62\frac{1}{2}$	0.625
	2						$66\frac{2}{3}$	$0.666\frac{2}{3}$
						11	$68\frac{3}{4}$	0.6875
		3					75	0.750
						13	$81\frac{1}{4}$	0.8125
			5				$83\frac{1}{3}$	$0.833\frac{1}{3}$
				7			$87\frac{1}{2}$	0.875
					11		$91\frac{2}{3}$	$0.916\frac{2}{3}$
						15	$93\frac{3}{4}$	0.9375

Table 7 Numbers (1–50), Squares, Cubes, Square Roots, Cube Roots, Reciprocals, Circumferences, and Circular Areas

N	N^2	N^3	$\sqrt{N}$	$\sqrt[3]{N}$	$\dfrac{1000}{N}$	πN	$\dfrac{\pi N^2}{4}$
1	1	1	1.0000	1.0000	1000.000	3.142	0.7854
2	4	8	1.4142	1.2599	500.000	6.283	3.1416
3	9	27	1.7321	1.4422	333.333	9.425	7.0686
4	16	64	2.0000	1.5874	250.000	12.566	12.5664
5	25	125	2.2361	1.7100	200.000	15.708	19.6350
6	36	216	2.4495	1.8171	166.667	18.850	28.2743
7	49	343	2.6458	1.9129	142.857	21.991	38.4845
8	64	512	2.8284	2.0000	125.000	25.133	50.2655
9	81	729	3.0000	2.0801	111.111	28.274	63.6173
10	100	1000	3.1623	2.1544	100.000	31.416	78.5398
11	121	1331	3.3166	2.2240	90.9091	34.558	95.0332
12	144	1728	3.4641	2.2894	83.3333	37.699	113.097
13	169	2197	3.6056	2.3513	76.9231	40.841	132.732
14	196	2744	3.7417	2.4101	71.4286	43.982	153.938
15	225	3375	3.8730	2.4662	66.6667	47.124	176.715
16	256	4096	4.0000	2.5198	62.5000	50.265	201.062
17	289	4913	4.1231	2.5713	58.8235	53.407	226.980
18	324	5832	4.2426	2.6207	55.5556	56.549	254.469
19	361	6859	4.3589	2.6684	52.6316	59.690	283.529
20	400	8000	4.4721	2.7144	50.0000	62.832	314.159
21	441	9261	4.5826	2.7589	47.6190	65.973	346.361
22	484	10648	4.6904	2.8020	45.4545	69.115	380.133
23	529	12167	4.7958	2.8439	43.4783	72.257	415.476
24	576	13824	4.8990	2.8845	41.6667	75.398	452.389
25	625	15625	5.0000	2.9240	40.0000	78.540	490.874
26	676	17576	5.0990	2.9625	38.4615	81.681	530.929
27	729	19683	5.1962	3.0000	37.0370	84.823	572.555
28	784	21952	5.2915	3.0366	35.7143	87.965	615.752
29	841	24389	5.3852	3.0723	34.4828	91.106	660.520
30	900	27000	5.4772	3.1072	33.3333	94.248	706.858
31	961	29791	5.5678	3.1414	32.2581	97.389	754.768
32	1024	32768	5.6569	3.1748	31.2500	100.531	804.248
33	1089	35937	5.7446	3.2075	30.3030	103.673	855.299
34	1156	39304	5.8310	3.2396	29.4118	106.814	907.920
35	1225	42875	5.9161	3.2711	28.5714	109.956	962.113
36	1296	46656	6.0000	3.3019	27.7778	113.097	1017.88
37	1369	50653	6.0828	3.3322	27.0270	116.239	1075.21
38	1444	54872	6.1644	3.3620	26.3158	119.381	1134.11
39	1521	59319	6.2450	3.3912	25.6410	122.522	1194.59
40	1600	64000	6.3246	3.4200	25.0000	125.66	1256.64
41	1681	68921	6.4031	3.4482	24.3902	128.81	1320.25
42	1764	74088	6.4807	3.4760	23.8095	131.95	1385.44
43	1849	79507	6.5574	3.5034	23.2558	135.09	1452.20
44	1936	85184	6.6332	3.5303	22.7273	138.23	1520.53
45	2025	91125	6.7082	3.5569	22.2222	141.37	1590.43
46	2116	97336	6.7823	3.5830	21.7391	144.51	1661.90
47	2209	103823	6.8557	3.6088	21.2766	147.65	1734.94
48	2304	110592	6.9282	3.6342	20.8333	150.80	1809.56
49	2401	117649	7.0000	3.6593	20.4082	153.94	1885.74
50	2500	125000	7.0711	3.6840	20.0000	157.08	1963.50

Table 7 (*continued*)   Numbers (51–100), Squares, Cubes, Square Roots, Cube
Roots, Reciprocals, Circumferences, and Circular Areas

N	N^2	N^3	$\sqrt{N}$	$\sqrt[3]{N}$	$\dfrac{1000}{N}$	πN	$\dfrac{\pi N^2}{4}$
51	2601	132651	7.1414	3.7084	19.6078	160.22	2042.82
52	2704	140608	7.2111	3.7325	19.2308	163.36	2123.72
53	2809	148877	7.2801	3.7563	18.8679	166.50	2206.18
54	2916	157464	7.3485	3.7798	18.5185	169.65	2290.22
55	3025	166375	7.4162	3.8030	18.1818	172.79	2375.83
56	3136	175616	7.4833	3.8259	17.8571	175.93	2463.01
57	3249	185193	7.5498	3.8485	17.5439	179.07	2551.76
58	3364	195112	7.6158	3.8709	17.2414	182.21	2642.08
59	3481	205379	7.6811	3.8930	16.9492	185.35	2733.97
60	3600	216000	7.7460	3.9149	16.6667	188.50	2827.43
61	3721	226981	7.8102	3.9365	16.3934	191.64	2922.47
62	3844	238328	7.8740	3.9579	16.1290	194.78	3019.07
63	3969	250047	7.9373	3.9791	15.8730	197.92	3117.25
64	4096	262144	8.0000	4.0000	15.6250	201.06	3216.99
65	4225	274625	8.0623	4.0207	15.3846	204.20	3318.31
66	4356	287496	8.1240	4.0412	15.1515	207.35	3421.19
67	4489	300763	8.1854	4.0615	14.9254	210.49	3525.65
68	4624	314432	8.2462	4.0817	14.7059	213.63	3631.68
69	4761	328509	8.3066	4.1016	14.4928	216.77	3739.28
70	4900	343000	8.3666	4.1213	14.2857	219.91	3848.45
71	5041	357911	8.4261	4.1408	14.0845	223.05	3959.19
72	5184	373248	8.4853	4.1602	13.8889	226.19	4071.50
73	5329	389017	8.5440	4.1793	13.6986	229.34	4185.39
74	5476	405224	8.6023	4.1983	13.5135	232.48	4300.84
75	5625	421875	8.6603	4.2172	13.3333	235.62	4417.86
76	5776	438976	8.7178	4.2358	13.1579	238.76	4536.46
77	5929	456533	8.7750	4.2543	12.9870	241.90	4656.63
78	6084	474552	8.8318	4.2727	12.8205	245.04	4778.36
79	6241	493039	8.8882	4.2908	12.6582	248.19	4901.67
80	6400	512000	8.9443	4.3089	12.5000	251.33	5026.55
81	6561	531441	9.0000	4.3267	12.3457	254.47	5153.00
82	6724	551368	9.0554	4.3445	12.1951	257.61	5281.02
83	6889	571787	9.1104	4.3621	12.0482	260.75	5410.61
84	7056	592704	9.1652	4.3795	11.9048	263.89	5541.77
85	7225	614125	9.2195	4.3968	11.7647	267.04	5674.50
86	7396	636056	9.2736	4.4140	11.6279	270.18	5808.80
87	7569	658503	9.3274	4.4310	11.4943	273.32	5944.68
88	7744	681472	9.3808	4.4480	11.3636	276.46	6082.12
89	7921	704969	9.4340	4.4647	11.2360	279.60	6221.14
90	8100	729000	9.4868	4.4814	11.1111	282.74	6361.73
91	8281	753571	9.5394	4.4979	10.9890	285.88	6503.88
92	8464	778688	9.5917	4.5144	10.8696	289.03	6647.61
93	8649	804357	9.6437	4.5307	10.7527	292.17	6792.91
94	8836	830584	9.6954	4.5468	10.6383	295.31	6939.78
95	9025	857375	9.7468	4.5629	10.5263	298.45	7088.22
96	9216	884736	9.7980	4.5789	10.4167	301.59	7238.23
97	9409	912673	9.8489	4.5947	10.3093	304.73	7389.81
98	9604	941192	9.8995	4.6104	10.2041	307.88	7542.96
99	9801	970299	9.9499	4.6261	10.1010	311.02	7697.69
100	10000	1000000	10.0000	4.6416	10.0000	314.16	7853.98

Table 8   Overall Coefficients of Heat Transfer (K) for Building Structures Expressed in B.t.u. per Hour per Square Foot per Degree Fahrenheit*

Building Structure			
Walls—Thickness in inches	8	12	16
Brick, without interior plaster	0.50	0.36	0.28
Brick, with interior plaster	0.46	0.34	0.27
Concrete, with interior plaster	0.62	0.49	0.44
Hollow tile, with interior plaster	0.38	0.29	0.24
Partitions			
4-in. hollow clay tile, plaster both sides	0.40		
4-in. common brick, plaster both sides	0.43		
Wood lath and plaster on both sides of studding	0.34		
2-in. corkboard and plaster on one side of studding	0.12		

Floors—thickness in inches	4	6	8	10
Concrete, no ceiling and no flooring	0.65	0.59	0.53	0.49
Concrete, plastered ceiling and no flooring	0.59	0.54	0.50	0.45
Concrete, on ground and terrazzo floor	0.98	0.84	0.74	0.66
Frame construction, no ceiling, maple or oak flooring on yellow pine subflooring on joists				0.34

Roofs, tar and gravel—thickness in inches	2	4	6
Concrete, no ceiling and no insulation	0.82	0.72	0.64
Concrete, no ceiling and 1-in. rigid insulation	0.24	0.23	0.22
Wood shingles, rafters exposed			0.46
Asphalt shingles, plaster board and plaster			0.32

Glass	
Single windows and skylights	1.13
Double windows and skylights	0.45
Hollow glass tile wall, 6 × 6 × 2-in. thick blocks:	
Still air inside and outside surfaces	0.48
Still air inside, 15 mph wind outside	0.60

*Correction for exposure:	North	East	South	West
Multiply K by	1.3	1.1	1.0	1.2

Table 9 Properties of Common Materials

Metal	Tensile Strength, lb/in.²	Specific Gravity	Weight/cubic foot	Melting Point, °F.	Linear Expansion unit length/°F.	Electric Conductivity*
Aluminum	19,000–48,000	2.7	168.5	1220	0.00001244	63.0
Copper	60,000–70,000	8.89	544.7	1981	0.00000900	96.61
Gold	20,000	19.3	1204.3	1945	0.00000778	76.61
Iron, cast	20,000–60,000	7.03–7.73	438.7–482.4	1990–2300	0.00000655	...
Lead	2600–3300	11.342	707.7	621	0.0000163	8.12
Magnesium	33,000	1.741	108.6	1204	0.00001444	39.44
Mercury	...	13.546	845.3	−38	...	1.75
Nickel	45,000–165,000	8.8	549.1	2651	0.00000700	12.89
Platinum	50,000	21.37	1333.5	3224	0.00000496	14.43
Silver	42,000	10.42–10.53	650.2–657.1	1761	0.00001025	100.00*
Steel	60,000–82,000	7.85	490.0	2500	0.00000633	12.00
Tin	4000–5000	7.29	454.9	449	0.00001496	14.39
Tungsten	590,000	18.6–19.1	1161–1192	6098	0.00000239	14.00
Other						
Brick	40–400	2.0–2.2	125–137	...	...	...
Concrete	200	2.2	137	...	...	...
Walnut	6000–8000	0.59	38	...	...	...

* Silver electric conductivity is 100.00.

Table 10   Copper Wire Table*

Size, American Wire Gage	Area, Circular Mils	Resistance, Ohms per 1000 ft at 25° C.	Weight, Pounds per 1000 ft	Current Capacity in Amperes		
				Rubber Insulation	Varnished Cloth	Other Insulations
18	1,620	6.51	4.92	3	. . .	5
16	2,580	4.09	7.82	6	. . .	10
14	4,110	2.58	12.4	15	18	20
12	6,530	1.62	19.8	20	25	25
10	10,400	1.02	31.4	25	30	30
8	16,500	0.641	50.0	35	40	50
6	26,300	0.403	79.5	50	60	70
5	33,100	0.319	100	55	65	80
4	41,700	0.253	126	70	85	90
3	52,600	0.205	163	80	95	100
2	66,400	0.162	205	90	110	125
1	83,700	0.129	258	100	120	150
0	106,000	0.102	326	125	150	200
00	133,000	0.0811	411	150	180	225
000	168,000	0.0642	518	175	210	275
0000	212,000	0.0509	653	225	270	325
. . .	300,000	0.0360	926	275	330	400
. . .	400,000	0.0270	1240	325	390	500
. . .	500,000	0.0216	1540	400	480	600
. . .	600,000	0.0180	1850	450	540	680
. . .	700,000	0.0154	2160	500	600	760

*For wires larger than No. 4, the values given are for standard wires. The carrying capacity of insulated aluminum wire is 84% of that given for copper.

Table 11   Weights of Materials

Material	Pounds per Cubic Foot	Material	Pounds per Cubic Foot
Air*	0.0809	Lime	53–75
Alabaster	168	Limestone	156–162
Alcohol	49–57	Loam	65–88
Amber	67		
Asbestos	125–175		
		Marble	157–177
		Masonry	100–165
Basalt	180	Mica	165–200
Brass	510–542	Molybdenum	529
Bronze	545–555	Mortar, hard	103
		Mud	80–130
Calcium	98.6		
Carbon	125–144		
Carbon dioxide	0.124	Naphtha	53
Celluloid	90		
Cement, loose	72–105	Oil, cottonseed	60.2
Chalk	119–175	Oil, linseed	58.8
Charcoal	17–35	Oil, lubricating	56.2–57.7
Chromium	368	Oil, petroleum	54.8
Clay, hard	129–133	Oil, turpentine	54.2
Clay, soft	118		
Coal, anthracite	81–106		
Coal, anthracite, loose	47–58	Paper	44–72
Coal, bituminous	78–88	Paraffin	54–57
Coal, bituminous, loose	44–54	Plaster of Paris	144
Coke	62–105	Porcelain	143–156
Coke, loose	23–32	Pumice stone	23–56
Concrete (1:2:4)	146		
Concrete (1:1½:3)	139	Quartz	165
Concrete (1:3:6)	156		
Copper, pure	554	Talc	168
Cork	15.6	Tar	62.4
		Tile	113
German silver	515–535	Tile, hollow	26–45
Glass, common	150–175	Trap rock	187–190
Glass, flint	180–280	Turf	20–30
Glycerine	78.6		
Gold	1203		
Granite	125–187	Water, maximum density	62.4
Gravel	90–147	Water, sea	64.0–64.3
Gun metal	533	Wax, bees	60.5
Gypsum	144	Wood, ash	45–47
		Wood, bamboo	22–25
Ice	55–57	Wood, cedar	37–38
Ivory	114	Wood, cherry	43–56

*At 0°C. and atmospheric pressure.

Table 11 (*continued*)   Weights of Materials

Material	Pounds per Cubic Foot	Material	Pounds per Cubic Foot
Resin	67	Wood, cypress	32–37
Rubber, pure	58.0–60.5	Wood, ebony	69–83
		Wood, fir	34–35
		Wood, hickory	53–58
		Wood, mahogany	32–53
Salt	129–131	Wood, maple	49–50
Sand	90–120	Wood, oak	37–56
Sandstone	124–200	Wood, pine	24–45
Shale	162	Wood, poplar	24–27
Slate	162–205	Wood, red wood	30–32
Snow, fresh fallen	5–12	Wood, spruce	25–32
Snow, wet compact	15–50	Wood, walnut	38–45
Soapstone	162–175		
Sulphur	120–130	Zinc	448

Table 12   Heat of Combustion (British Thermal Units)

Substance*	Per Pound	Per Gallon	Cubic Foot†
Acetylene	21,500	. . .	1480
Alcohol, ethyl, denatured	11,600	78,900	. . .
Alcohol, methyl (0.798)	9,540	63,700	. . .
Benzine (0.879)	18,500	136,000	3810
Carbon, to CO	4,400	. . .	. . .
Carbon, to CO_2	14,500	. . .	. . .
Carbon monoxide, to CO_2	4,370	. . .	323
Charcoal, wood	13,500	. . .	. . .
Coal, anthracite	11,500–14,000	. . .	. . .
Coal, bituminous	11,000–15,300	. . .	. . .
Coke	12,000–14,400	. . .	. . .
Gas, blast furnace	. . .	. . .	90–110
Gas, coal	. . .	. . .	630–680
Gas, coke oven	. . .	. . .	430–600
Gas, natural	. . .	. . .	700–2470
Gas, oil	. . .	. . .	450–950
Gasoline (0.770)	20,000	129,000	. . .
Hydrogen	62,000	. . .	326
Kerosene (0.783)	20,000	131,000	. . .
Kerosene (0.800)	20,160	136,000	. . .
Peat	3,500–10,000	. . .	. . .
Petroleum (0.785)	20,000	131,000	. . .
Petroleum (1.000)	18,300	153,000	. . .
Straw	5,100–6,700	. . .	. . .
Wood, air-dried	5,420–6,830	. . .	. . .

* Numbers indicate specific gravity.
† At 60°F. and atmospheric pressure.

Table 13   Conversion Factors*

Multiply	By	To Obtain
Abamperes	10	Amperes
Acres	43,560	Square feet
Acres	4047	Square meters
Acres	1.562×10^{-3}	Square miles
Acres	4840	Square yards
Acre-feet	43,560	Cubic feet
Acre-feet	3.259×10^5	Gallons
Amperes	$\frac{1}{10}$	Abamperes
Ares	0.02471	Acres
Ares	100	Square meters
Atmospheres	76	Centimeters of mercury
Atmospheres	29.92	Inches of mercury
Atmospheres	33.90	Feet of water
Atmospheres	10,332	Kilograms per square meter
Atmospheres	14.70	Pounds per square inch
Atmospheres	1.058	Tons per square foot
Bars	0.9869	Atmospheres
Bars	14.50	Pounds per square inch
Board-feet	144 sq in. $\times$ 1 in.	Cubic inches
British thermal units	778.2	Foot-pounds
British thermal units	3.930×10^{-4}	Horsepower-hours
British thermal units	2.930×10^{-4}	Kilowatt hours
British thermal units per minute	12.97	Foot-pounds per second
British thermal units per minute	0.02358	Horsepower
British thermal units per minute	0.01758	Kilowatts
British thermal units per minute	17.58	Watts
Centimeters	3.281×10^{-2}	Feet
Centimeters	0.3937	Inches
Centimeters of mercury	0.01316	Atmospheres
Centimeters of mercury	0.1934	Pounds per square inch
Centimeters per second	1.968	Feet per minute
Centimeters per second	0.03281	Feet per second
Circular mils	5.067×10^{-6}	Square centimeters
Circular mils	7.854×10^{-7}	Square inches
Cubic centimeters	3.531×10^{-5}	Cubic feet
Cubic centimeters	6.102×10^{-2}	Cubic inches
Cubic feet	2.832×10^4	Cubic centimeters
Cubic feet	7.481	Gallons
Cubic feet	28.32	Liters
Cubic feet per minute	472.0	Cubic centimeters per second
Cubic feet per minute	0.1247	Gallons per second
Cubic inches	16.39	Cubic centimeters
Cubic yards	0.7646	Cubic meters
Cubic yards	202.0	Gallons

*The customary units of weight and mass are avoirdupois units unless designated otherwise.

Table 13 (*continued*)   Conversion Factors

Multiply	By	To Obtain
Degrees per second	0.01745	Radians per second
Drams	1.772	Grams
Feet of water	0.02950	Atmospheres
Feet of water	0.8826	Inches of mercury
Feet per second	30.48	Centimeters per second
Foot-pounds	1.285×10^{-3}	British thermal units
Gallons	3785	Cubic centimeters
Gallons	0.1337	Cubic feet
Gallons	4.951×10^{-3}	Cubic yards
Gallons	3.785	Liters
Gallons per minute	2.228×10^{-3}	Cubic feet per second
Gallons per minute	0.06308	Liters per second
Grams	2.205×10^{-3}	Pounds
Horsepower	42.40	British thermal units per minute
Horsepower	33,000	Foot-pounds per minute
Horsepower	550	Foot-pounds per second
Horsepower	1.014	Horsepower (metric)
Horsepower	10.68	Kilograms-calories per minute
Horsepower	0.7457	Kilowatts
Horsepower	745.7	Watts
Horsepower (boiler)	33.520	British thermal units per hour
Inches of mercury	0.03342	Atmospheres
Kilograms	980,665	Dynes
Kilograms	10^3	Grams
Kilograms	70.93	Poundals
Kilograms	2.205	Pounds
Kilograms	1.102×10^{-3}	Tons (short)
Kilometers	3281	Feet
Kilometers	0.6214	Miles
Kilometers per hour	27.78	Centimeters per second
Kilometers per hour	54.68	Feet per minute
Kilometers per hour	0.9113	Feet per second
Kilometers per hour	0.5396	Knots
Kilometers per hour	16.67	Meters per minute
Kilometers per hour	0.6214	Miles per hour
Links (engineer's)	12	Inches
Links (surveyor's)	7.92	Inches
Miles	1.609×10^5	Centimeters
Miles	5280	Feet
Miles	6.336×10^4	Inches
Miles	1.609	Kilometers
Miles	1760	Yards
Miles per hour	44.70	Centimeters per second
Miles per hour	88	Feet per minute
Miles per hour	1.467	Feet per second
Miles per hour	1.609	Kilometers per hour
Miles per hour	0.8684	Knots
Miles per hour	26.82	Meters per minute

Table 13 (*continued*)   Conversion Factors

Multiply	By	To Obtain
Miles per hour per second	44.70	Centimeters per second per second
Miles per hour per second	1.467	Feet per second per second
Ohms	10^9	Abohms
Ohms	10^{-6}	Megohms
Ohms	10^6	Microhms
Perches (masonry)	24.75	Cubic feet
Pounds	444,823	Dynes
Pounds	7000	Grains
Pounds	453.6	Grams
Pounds per square inch	0.06804	Atmospheres
Pounds per square inch	2.307	Feet of water
Pounds per square inch	2.036	Inches of mercury
Radians	57.30	Degrees
Radians	3438	Minutes
Radians	0.6366	Quadrants
Radians per second	57.30	Degrees per second
Radians per second	9.549	Revolutions per minute
Radians per second	0.1592	Revolutions per second
Reams	500	Sheets
Revolutions	360	Degrees
Revolutions	4	Quadrants
Revolutions	6.283	Radians
Revolutions per minute	6	Degrees per second
Revolutions per minute	0.1047	Radians per second
Square centimeters	1.973×10^5	Circular mils
Square centimeters	1.076×10^{-3}	Square feet
Square centimeters	0.1550	Square inches
Square centimeters	10^{-4}	Square meters
Square meters	2.471×10^{-4}	Acres
Square meters	10.76	Square feet
Square meters	1550	Square inches
Square meters	3.861×10^{-7}	Squares miles
Square meters	1.196	Square yards
Square miles	640	Acres
Square miles	27.88×10^6	Square feet
Square miles	2.590	Square kilometers
Square miles	3.098×10^6	Square yards
Square millimeters	1.973×10^3	Circular mils
Statcoulombs	$\frac{1}{3} \times 10^{-9}$	Coulombs
Temperature (°C.) + 273	1	Absolute temperature (°C.)
Temperature (°C.) + 17.8	1.8	Temperature (°F.)
Temperature (°F.) + 460	1	Absolute temperature (°F.)
Temperature (°F.) − 32	$\frac{5}{9}$	Temperature (°C.)
Watts	0.05688	British thermal units per minute
Watts	10^7	Ergs per second
Watts	44.27	Foot-pounds per minute
Watts	0.7378	Foot-pounds per second
Watts	1.341×10^{-3}	Horsepower

Table 14   Compound Interest

Years or Periods					Showing How Much $1 Will Amount to at Various Rates					Years or Periods
n	1%	1½%	2%	2½%	3%	3½%	4%	5%	6%	*n*
1	1.0100	1.0150	1.0200	1.0250	1.0300	1.0350	1.0400	1.0500	1.0600	1
2	1.0201	1.0302	1.0404	1.0506	1.0609	1.0712	1.0816	1.1025	1.1236	2
3	1.0303	1.0457	1.0612	1.0769	1.0927	1.1087	1.1249	1.1576	1.1910	3
4	1.0406	1.0614	1.0824	1.1038	1.1255	1.1475	1.1699	1.2155	1.2625	4
5	1.0510	1.0773	1.1041	1.1314	1.1593	1.1877	1.2167	1.2763	1.3382	5
6	1.0615	1.0934	1.1262	1.1597	1.1941	1.2293	1.2653	1.3401	1.4185	6
7	1.0721	1.1098	1.1487	1.1887	1.2299	1.2723	1.3159	1.4071	1.5036	7
8	1.0829	1.1265	1.1717	1.2184	1.2668	1.3168	1.3686	1.4775	1.5938	8
9	1.0937	1.1434	1.1951	1.2489	1.3048	1.3629	1.4233	1.5513	1.6895	9
10	1.1046	1.1605	1.2190	1.2801	1.3439	1.4106	1.4802	1.6289	1.7908	10
11	1.1157	1.1779	1.2434	1.3121	1.3842	1.4600	1.5395	1.7103	1.8983	11
12	1.1268	1.1956	1.2682	1.3449	1.4258	1.5111	1.6010	1.7959	2.0122	12
13	1.1381	1.2136	1.2936	1.3785	1.4685	1.5640	1.6651	1.8856	2.1329	13
14	1.1495	1.2318	1.3195	1.4130	1.5126	1.6187	1.7317	1.9799	2.2609	14
15	1.1610	1.2502	1.3459	1.4483	1.5580	1.6753	1.8009	2.0789	2.3966	15
16	1.1726	1.2690	1.3728	1.4845	1.6047	1.7340	1.8730	2.1829	2.5404	16
17	1.1843	1.2880	1.4002	1.5216	1.6528	1.7947	1.9479	2.2920	2.6928	17
18	1.1961	1.3073	1.4282	1.5597	1.7024	1.8575	2.0258	2.4066	2.8543	18
19	1.2081	1.3270	1.4568	1.5987	1.7535	1.9225	2.1068	2.5270	3.0256	19
20	1.2202	1.3469	1.4859	1.6386	1.8061	1.9898	2.1911	2.6533	3.2071	20
21	1.2324	1.3671	1.5157	1.6796	1.8603	2.0594	2.2788	2.7860	3.3996	21
22	1.2447	1.3876	1.5460	1.7216	1.9161	2.1315	2.3699	2.9253	3.6035	22
23	1.2572	1.4084	1.5769	1.7646	1.9736	2.2061	2.4647	3.0715	3.8197	23
24	1.2697	1.4295	1.6084	1.8087	2.0328	2.2833	2.5633	3.2251	4.0489	24
25	1.2824	1.4509	1.6406	1.8539	2.0938	2.3632	2.6658	3.3864	4.2919	25
26	1.2953	1.4727	1.6734	1.9003	2.1566	2.4460	2.7725	3.5557	4.5494	26
27	1.3082	1.4948	1.7069	1.9478	2.2213	2.5316	2.8834	3.7335	4.8223	27
28	1.3213	1.5172	1.7410	1.9965	2.2879	2.6202	2.9987	3.9201	5.1117	28
29	1.3345	1.5400	1.7758	2.0464	2.3566	2.7119	3.1187	4.1161	5.4184	29
30	1.3478	1.5631	1.8114	2.0976	2.4273	2.8068	3.2434	4.3219	5.7435	30
31	1.3613	1.5865	1.8476	2.1500	2.5001	2.9050	3.3731	4.5380	6.0881	31
32	1.3749	1.6103	1.8845	2.2038	2.5751	3.0067	3.5081	4.7649	6.4534	32
33	1.3887	1.6345	1.9222	2.2589	2.6523	3.1119	3.6484	5.0032	6.8406	33
34	1.4026	1.6590	1.9607	2.3153	2.7319	3.2209	3.7943	5.2533	7.2510	34
35	1.4166	1.6839	1.9999	2.3732	2.8139	3.3336	3.9461	5.5160	7.6861	35
36	1.4308	1.7091	2.0399	2.4325	2.8983	3.4503	4.1039	5.7918	8.1473	36
37	1.4451	1.7348	2.0807	2.4933	2.9852	3.5710	4.2681	6.0814	8.6361	37
38	1.4595	1.7608	2.1223	2.5557	3.0748	3.6960	4.4388	6.3855	9.1543	38
39	1.4741	1.7872	2.1647	2.6196	3.1670	3.8254	4.6164	6.7048	9.7035	39
40	1.4889	1.8140	2.2080	2.6851	3.2620	3.9593	4.8010	7.0400	10.2857	40
41	1.5038	1.8412	2.2522	2.7522	3.3599	4.0978	4.9931	7.3920	10.9029	41
42	1.5188	1.8688	2.2972	2.8210	3.4607	4.2413	5.1928	7.7616	11.5570	42
43	1.5340	1.8969	2.3432	2.8915	3.5645	4.3897	5.4005	8.1497	12.2505	43
44	1.5493	1.9253	2.3901	2.9638	3.6715	4.5433	5.6165	8.5572	12.9855	44
45	1.5648	1.9542	2.4379	3.0379	3.7816	4.7024	5.8412	8.9850	13.7646	45
46	1.5805	1.9835	2.4866	3.1139	3.8950	4.8669	6.0748	9.4343	14.5905	46
47	1.5963	2.0133	2.5363	3.1917	4.0119	5.0373	6.3178	9.9060	15.4659	47
48	1.6122	2.0435	2.5871	3.2715	4.1323	5.2136	6.5705	10.4013	16.3939	48
49	1.6283	2.0741	2.6388	3.3533	4.2562	5.3961	6.8333	10.9213	17.3775	49
50	1.6446	2.1052	2.6916	3.4371	4.3839	5.5849	7.1067	11.4674	18.4202	50
n	1%	1½%	2%	2½%	3%	3½%	4%	5%	6%	*n*

Table 15 Common Logarithms

N	0	1	2	3	4	5	6	7	8	9
0	...	0000	3010	4771	6021	6990	7782	8451	9031	9542
1	0000	0414	0792	1139	1461	1761	2041	2304	2553	2788
2	3010	3222	3424	3617	3802	3979	4150	4314	4472	4624
3	4771	4914	5051	5185	5315	5441	5563	5682	5798	5911
4	6021	6128	6232	6335	6435	6532	6628	6721	6812	6902
5	6990	7076	7160	7243	7324	7404	7482	7559	7634	7709
6	7782	7853	7924	7993	8062	8129	8195	8261	8325	8388
7	8451	8513	8573	8633	8692	8751	8808	8865	8921	8976
8	9031	9085	9138	9191	9243	9294	9345	9395	9445	9494
9	9542	9590	9638	9685	9731	9777	9823	9868	9912	9956
10	0000	0043	0086	0128	0170	0212	0253	0294	0334	0374
11	0414	0453	0492	0531	0569	0607	0645	0682	0719	0755
12	0792	0828	0864	0899	0934	0969	1004	1038	1072	1106
13	1139	1173	1206	1239	1271	1303	1335	1367	1399	1430
14	1461	1492	1523	1553	1584	1614	1644	1673	1703	1732
15	1761	1790	1818	1847	1875	1903	1931	1959	1987	2014
16	2041	2068	2095	2122	2148	2175	2201	2227	2253	2279
17	2304	2330	2355	2380	2405	2430	2455	2480	2504	2529
18	2553	2577	2601	2625	2648	2672	2695	2718	2742	2765
19	2788	2810	2833	2856	2878	2900	2923	2945	2967	2989
20	3010	3032	3054	3075	3096	3118	3139	3160	3181	3201
21	3222	3243	3263	3284	3304	3324	3345	3365	3385	3404
22	3424	3444	3464	3483	3502	3522	3541	3560	3579	3598
23	3617	3636	3655	3674	3692	3711	3729	3747	3766	3784
24	3802	3820	3838	3856	3874	3892	3909	3927	3945	3962
25	3979	3997	4014	4031	4048	4065	4082	4099	4116	4133
26	4150	4166	4183	4200	4216	4232	4249	4265	4281	4298
27	4314	4330	4346	4362	4378	4393	4409	4425	4440	4456
28	4472	4487	4502	4518	4533	4548	4564	4579	4594	4609
29	4624	4639	4654	4669	4683	4698	4713	4728	4742	4757
30	4771	4786	4800	4814	4829	4843	4857	4871	4886	4900
31	4914	4928	4942	4955	4969	4983	4997	5011	5024	5038
32	5051	5065	5079	5092	5105	5119	5132	5145	5159	5172
33	5185	5198	5211	5224	5237	5250	5263	5276	5289	5302
34	5315	5328	5340	5353	5366	5378	5391	5403	5416	5428
35	5441	5453	5465	5478	5490	5502	5514	5527	5539	5551
36	5563	5575	5587	5599	5611	5623	5635	5647	5658	5670
37	5682	5694	5705	5717	5729	5740	5752	5763	5775	5786
38	5798	5809	5821	5832	5843	5855	5866	5877	5888	5899
39	5911	5922	5933	5944	5955	5966	5977	5988	5999	6010
40	6021	6031	6042	6053	6064	6075	6085	6096	6107	6117
41	6128	6138	6149	6160	6170	6180	6191	6201	6212	6222
42	6232	6243	6253	6263	6274	6284	6294	6304	6314	6325
43	6335	6345	6355	6365	6375	6385	6395	6405	6415	6425
44	6435	6444	6454	6464	6474	6484	6493	6503	6513	6522
45	6532	6542	6551	6561	6571	6580	6590	6599	6609	6618
46	6628	6637	6646	6656	6665	6675	6684	6693	6702	6712
47	6721	6730	6739	6749	6758	6767	6776	6785	6794	6803
48	6812	6821	6830	6839	6848	6857	6866	6875	6884	6893
49	6902	6911	6920	6928	6937	6946	6955	6964	6972	6981
50	6990	6998	7007	7016	7024	7033	7042	7050	7059	7067
N	0	1	2	3	4	5	6	7	8	9

Table 15 (*continued*)   Common Logarithms

N	0	1	2	3	4	5	6	7	8	9
50	6990	6998	7007	7016	7024	7033	7042	7050	7059	7067
51	7076	7084	7093	7101	7110	7118	7126	7135	7143	7152
52	7160	7168	7177	7185	7193	7202	7210	7218	7226	7235
53	7243	7251	7259	7267	7275	7284	7292	7300	7308	7316
54	7324	7332	7340	7348	7356	7364	7372	7380	7388	7396
55	7404	7412	7419	7427	7435	7443	7451	7459	7466	7474
56	7482	7490	7497	7505	7513	7520	7528	7536	7543	7551
57	7559	7566	7574	7582	7589	7597	7604	7612	7619	7627
58	7634	7642	7649	7657	7664	7672	7679	7686	7694	7701
59	7709	7716	7723	7731	7738	7745	7752	7760	7767	7774
60	7782	7789	7796	7803	7810	7818	7825	7832	7839	7846
61	7853	7860	7868	7875	7882	7889	7896	7903	7910	7917
62	7924	7931	7938	7945	7952	7959	7966	7973	7980	7987
63	7993	8000	8007	8014	8021	8028	8035	8041	8048	8055
64	8062	8069	8075	8082	8089	8096	8102	8109	8116	8122
65	8129	8136	8142	8149	8156	8162	8169	8176	8182	8189
66	8195	8202	8209	8215	8222	8228	8235	8241	8248	8254
67	8261	8267	8274	8280	8287	8293	8299	8306	8312	8319
68	8325	8331	8338	8344	8351	8357	8363	8370	8376	8382
69	8388	8395	8401	8407	8414	8420	8426	8432	8439	8445
70	8451	8457	8463	8470	8476	8482	8488	8494	8500	8506
71	8513	8519	8525	8531	8537	8543	8549	8555	8561	8567
72	8573	8579	8585	8591	8597	8603	8609	8615	8621	8627
73	8633	8639	8645	8651	8657	8663	8669	8675	8681	8686
74	8692	8698	8704	8710	8716	8722	8727	8733	8739	8745
75	8751	8756	8762	8768	8774	8779	8785	8791	8797	8802
76	8808	8814	8820	8825	8831	8837	8842	8848	8854	8859
77	8865	8871	8876	8882	8887	8893	8899	8904	8910	8915
78	8921	8927	8932	8938	8943	8949	8954	8960	8965	8971
79	8976	8982	8987	8993	8998	9004	9009	9015	9020	9025
80	9031	9036	9042	9047	9053	9058	9063	9069	9074	9079
81	9085	9090	9096	9101	9106	9112	9117	9122	9128	9133
82	9138	9143	9149	9154	9159	9165	9170	9175	9180	9186
83	9191	9196	9201	9206	9212	9217	9222	9227	9232	9238
84	9243	9248	9253	9258	9263	9269	9274	9279	9284	9289
85	9294	9299	9304	9309	9315	9320	9325	9330	9335	9340
86	9345	9350	9355	9360	9365	9370	9375	9380	9385	9390
87	9395	9400	9405	9410	9415	9420	9425	9430	9435	9440
88	9445	9450	9455	9460	9465	9469	9474	9479	9484	9489
89	9494	9499	9504	9509	9513	9518	9523	9528	9533	9538
90	9542	9547	9552	9557	9562	9566	9571	9576	9581	9586
91	9590	9595	9600	9605	9609	9614	9619	9624	9628	9633
92	9638	9643	9647	9652	9657	9661	9666	9671	9675	9680
93	9685	9689	9694	9699	9703	9708	9713	9717	9722	9727
94	9731	9736	9741	9745	9750	9754	9759	9763	9768	9773
95	9777	9782	9786	9791	9795	9800	9805	9809	9814	9818
96	9823	9827	9832	9836	9841	9845	9850	9854	9859	9863
97	9868	9872	9877	9881	9886	9890	9894	9899	9903	9908
98	9912	9917	9921	9926	9930	9934	9939	9943	9948	9952
99	9956	9961	9965	9969	9974	9978	9983	9987	9991	9996
100	0000	0004	0009	0013	0017	0022	0026	0030	0035	0039
N	0	1	2	3	4	5	6	7	8	9

Table 16 Natural Trigonometric Functions

Degrees	Sin	Cos	Tan	Cot	Sec	Csc	
0° 00'	.0000	1.0000	.0000		1.000		**90° 00'**
10	.0029	1.0000	.0029	343.8	1.000	343.8	50
20	.0058	1.0000	.0058	171.9	1.000	171.9	40
30	.0087	1.0000	.0087	114.6	1.000	114.6	30
40	.0116	.9999	.0116	85.94	1.000	85.95	20
50	.0145	.9999	.0145	68.75	1.000	68.76	10
1° 00'	.0175	.9998	.0175	57.29	1.000	57.30	**89° 00'**
10	.0204	.9998	.0204	49.10	1.000	49.11	50
20	.0233	.9997	.0233	42.96	1.000	42.98	40
30	.0262	.9997	.0262	38.19	1.000	38.20	30
40	.0291	.9996	.0291	34.37	1.000	34.38	20
50	.0320	.9995	.0320	31.24	1.001	31.26	10
2° 00'	.0349	.9994	.0349	28.64	1.001	28.65	**88° 00'**
10	.0378	.9993	.0378	26.43	1.001	26.45	50
20	.0407	.9992	.0407	24.54	1.001	24.56	40
30	.0436	.9990	.0437	22.90	1.001	22.93	30
40	.0465	.9989	.0466	21.47	1.001	21.49	20
50	.0494	.9988	.0495	20.21	1.001	20.23	10
3° 00'	.0523	.9986	.0524	19.08	1.001	19.11	**87° 00'**
10	.0552	.9985	.0553	18.07	1.002	18.10	50
20	.0581	.9983	.0582	17.17	1.002	17.20	40
30	.0610	.9981	.0612	16.35	1.002	16.38	30
40	.0640	.9980	.0641	15.60	1.002	15.64	20
50	.0669	.9978	.0670	14.92	1.002	14.96	10
4° 00'	.0698	.9976	.0699	14.30	1.002	14.34	**86° 00'**
10	.0727	.9974	.0729	13.73	1.003	13.76	50
20	.0756	.9971	.0758	13.20	1.003	13.23	40
30	.0785	.9969	.0787	12.71	1.003	12.75	30
40	.0814	.9967	.0816	12.25	1.003	12.29	20
50	.0843	.9964	.0846	11.83	1.004	11.87	10
5° 00'	.0872	.9962	.0875	11.43	1.004	11.47	**85° 00'**
10	.0901	.9959	.0904	11.06	1.004	11.10	50
20	.0929	.9957	.0934	10.71	1.004	10.76	40
30	.0958	.9954	.0963	10.39	1.005	10.43	30
40	.0987	.9951	.0992	10.08	1.005	10.13	20
50	.1016	.9948	.1022	9.788	1.005	9.839	10
6° 00'	.1045	.9945	.1051	9.514	1.006	9.567	**84° 00'**
	Cos	Sin	Cot	Tan	Csc	Sec	Degrees

Table 16 (*continued*)   Natural Trigonometric Functions

Degrees	Sin	Cos	Tan	Cot	Sec	Csc	
6° 00'	.1045	.9945	.1051	9.514	1.006	9.567	**84° 00'**
10	.1074	.9942	.1080	9.255	1.006	9.309	50
20	.1103	.9939	.1110	9.010	1.006	9.065	40
30	.1132	.9936	.1139	8.777	1.006	8.834	30
40	.1161	.9932	.1169	8.556	1.007	8.614	20
50	.1190	.9929	.1198	8.345	1.007	8.405	10
7° 00'	.1219	.9925	.1228	8.144	1.008	8.206	**83° 00'**
10	.1248	.9922	.1257	7.953	1.008	8.016	50
20	.1276	.9918	.1287	7.770	1.008	7.834	40
30	.1305	.9914	.1317	7.596	1.009	7.661	30
40	.1334	.9911	.1346	7.429	1.009	7.496	20
50	.1363	.9907	.1376	7.269	1.009	7.337	10
8° 00'	.1392	.9903	.1405	7.115	1.010	7.185	**82° 00'**
10	.1421	.9899	.1435	6.968	1.010	7.040	50
20	.1449	.9894	.1465	6.827	1.011	6.900	40
30	.1478	.9890	.1495	6.691	1.011	6.765	30
40	.1507	.9886	.1524	6.561	1.012	6.636	20
50	.1536	.9881	.1554	6.435	1.012	6.512	10
9° 00'	.1564	.9877	.1584	6.314	1.012	6.392	**81° 00'**
10	.1593	.9872	.1614	6.197	1.013	6.277	50
20	.1622	.9868	.1644	6.084	1.013	6.166	40
30	.1650	.9863	.1673	5.976	1.014	6.059	30
40	.1679	.9858	.1703	5.871	1.014	5.955	20
50	.1708	.9853	.1733	5.769	1.015	5.855	10
10° 00'	.1736	.9848	.1763	5.671	1.015	5.759	**80° 00'**
10	.1765	.9843	.1793	5.576	1.016	5.665	50
20	.1794	.9838	.1823	5.485	1.016	5.575	40
30	.1822	.9833	.1853	5.396	1.017	5.487	30
40	.1851	.9827	.1883	5.309	1.018	5.403	20
50	.1880	.9822	.1914	5.226	1.018	5.320	10
11° 00'	.1908	.9816	.1944	5.145	1.019	5.241	**79° 00'**
10	.1937	.9811	.1974	5.066	1.019	5.164	50
20	.1965	.9805	.2004	4.989	1.020	5.089	40
30	.1994	.9799	.2035	4.915	1.020	5.016	30
40	.2022	.9793	.2065	4.843	1.021	4.945	20
50	.2051	.9787	.2095	4.773	1.022	4.876	10
12° 00'	.2079	.9781	.2126	4.705	1.022	4.810	**78° 00'**
	Cos	Sin	Cot	Tan	Csc	Sec	Degrees

Table 16 (*continued*)   Natural Trigonometric Functions

Degrees	Sin	Cos	Tan	Cot	Sec	Csc	
12° 00′	.2079	.9781	.2126	4.705	1.022	4.810	**78° 00′**
10	.2108	.9775	.2156	4.638	1.023	4.745	50
20	.2136	.9769	.2186	4.574	1.024	4.682	40
30	.2164	.9763	.2217	4.511	1.024	4.620	30
40	.2193	.9757	.2247	4.449	1.025	4.560	20
50	.2221	.9750	.2278	4.390	1.026	4.502	10
13° 00′	.2250	.9744	.2309	4.331	1.026	4.445	**77° 00′**
10	.2278	.9737	.2339	4.275	1.027	4.390	50
20	.2306	.9730	.2370	4.219	1.028	4.336	40
30	.2334	.9724	.2401	4.165	1.028	4.284	30
40	.2363	.9717	.2432	4.113	1.029	4.232	20
50	.2391	.9710	.2462	4.061	1.030	4.182	10
14° 00′	.2419	.9703	.2493	4.011	1.031	4.134	**76° 00′**
10	.2447	.9696	.2524	3.962	1.031	4.086	50
20	.2476	.9689	.2555	3.914	1.032	4.039	40
30	.2504	.9681	.2586	3.867	1.033	3.994	30
40	.2532	.9674	.2617	3.821	1.034	3.950	20
50	.2560	.9667	.2648	3.776	1.034	3.906	10
15° 00′	.2588	.9659	.2679	3.732	1.035	3.864	**75° 00′**
10	.2616	.9652	.2711	3.689	1.036	3.822	50
20	.2644	.9644	.2742	3.647	1.037	3.782	40
30	.2672	.9636	.2773	3.606	1.038	3.742	30
40	.2700	.9628	.2805	3.566	1.039	3.703	20
50	.2728	.9621	.2836	3.526	1.039	3.665	10
16° 00′	.2756	.9613	.2867	3.487	1.040	3.628	**74° 00′**
10	.2784	.9605	.2899	3.450	1.041	3.592	50
20	.2812	.9596	.2931	3.412	1.042	3.556	40
30	.2840	.9588	.2962	3.376	1.043	3.521	30
40	.2868	.9580	.2994	3.340	1.044	3.487	20
50	.2896	.9572	.3026	3.305	1.045	3.453	10
17° 00′	.2924	.9563	.3057	3.271	1.046	3.420	**73° 00′**
10	.2952	.9555	.3089	3.237	1.047	3.388	50
20	.2979	.9546	.3121	3.204	1.048	3.356	40
30	.3007	.9537	.3153	3.172	1.049	3.326	30
40	.3035	.9528	.3185	3.140	1.049	3.295	20
50	.3062	.9520	.3217	3.108	1.050	3.265	10
18° 00′	.3090	.9511	.3249	3.078	1.051	3.236	**72° 00′**
	Cos	Sin	Cot	Tan	Csc	Sec	Degrees

Table 16 (*continued*)   Natural Trigonometric Functions

Degrees	Sin	Cos	Tan	Cot	Sec	Csc	
18° 00′	.3090	.9511	.3249	3.078	1.051	3.236	**72° 00′**
10	.3118	.9502	.3281	3.047	1.052	3.207	50
20	.3145	.9492	.3314	3.018	1.053	3.179	40
30	.3173	.9483	.3346	2.989	1.054	3.152	30
40	.3201	.9474	.3378	2.960	1.056	3.124	20
50	.3228	.9465	.3411	2.932	1.057	3.098	10
19° 00′	.3256	.9455	.3443	2.904	1.058	3.072	**71° 00′**
10	.3283	.9446	.3476	2.877	1.059	3.046	50
20	.3311	.9436	.3508	2.850	1.060	3.021	40
30	.3338	.9426	.3541	2.824	1.061	2.996	30
40	.3365	.9417	.3574	2.798	1.062	2.971	20
50	.3393	.9407	.3607	2.773	1.063	2.947	10
20° 00′	.3420	.9397	.3640	2.747	1.064	2.924	**70° 00′**
10	.3448	.9387	.3673	2.723	1.065	2.901	50
20	.3475	.9377	.3706	2.699	1.066	2.878	40
30	.3502	.9367	.3739	2.675	1.068	2.855	30
40	.3529	.9356	.3772	2.651	1.069	2.833	20
50	.3557	.9346	.3805	2.628	1.070	2.812	10
21° 00′	.3584	.9336	.3839	2.605	1.071	2.790	**69° 00′**
10	.3611	.9325	.3872	2.583	1.072	2.769	50
20	.3638	.9315	.3906	2.560	1.074	2.749	40
30	.3665	.9304	.3939	2.539	1.075	2.729	30
40	.3692	.9293	.3973	2.517	1.076	2.709	20
50	.3719	.9283	.4006	2.496	1.077	2.689	10
22° 00′	.3746	.9272	.4040	2.475	1.079	2.669	**68° 00′**
10	.3773	.9261	.4074	2.455	1.080	2.650	50
20	.3800	.9250	.4108	2.434	1.081	2.632	40
30	.3827	.9239	.4142	2.414	1.082	2.613	30
40	.3854	.9228	.4176	2.394	1.084	2.595	20
50	.3881	.9216	.4210	2.375	1.085	2.577	10
23° 00′	.3907	.9205	.4245	2.356	1.086	2.559	**67° 00′**
10	.3934	.9194	.4279	2.337	1.088	2.542	50
20	.3961	.9182	.4314	2.318	1.089	2.525	40
30	.3987	.9171	.4348	2.300	1.090	2.508	30
40	.4014	.9159	.4383	2.282	1.092	2.491	20
50	.4041	.9147	.4417	2.264	1.093	2.475	10
24° 00′	.4067	.9135	.4452	2.246	1.095	2.459	**66° 00′**
	Cos	Sin	Cot	Tan	Csc	Sec	Degrees

Table 16 (*continued*) Natural Trigonometric Functions

Degrees	Sin	Cos	Tan	Cot	Sec	Csc	
24° 00′	.4067	.9135	.4452	2.246	1.095	2.459	**66° 00′**
10	.4094	.9124	.4487	2.229	1.096	2.443	50
20	.4120	.9112	.4522	2.211	1.097	2.427	40
30	.4147	.9100	.4557	2.194	1.099	2.411	30
40	.4173	.9088	.4592	2.177	1.100	2.396	20
50	.4200	.9075	.4628	2.161	1.102	2.381	10
25° 00′	.4226	.9063	.4663	2.145	1.103	2.366	**65° 00′**
10	.4253	.9051	.4699	2.128	1.105	2.352	50
20	.4279	.9038	.4734	2.112	1.106	2.337	40
30	.4305	.9026	.4770	2.097	1.108	2.323	30
40	.4331	.9013	.4806	2.081	1.109	2.309	20
50	.4358	.9001	.4841	2.066	1.111	2.295	10
26° 00′	.4384	.8988	.4877	2.050	1.113	2.281	**64° 00′**
10	.4410	.8975	.4913	2.035	1.114	2.268	50
20	.4436	.8962	.4950	2.020	1.116	2.254	40
30	.4462	.8949	.4986	2.006	1.117	2.241	30
40	.4488	.8936	.5022	1.991	1.119	2.228	20
50	.4514	.8923	.5059	1.977	1.121	2.215	10
27° 00′	.4540	.8910	.5095	1.963	1.122	2.203	**63° 00′**
10	.4566	.8897	.5132	1.949	1.124	2.190	50
20	.4592	.8884	.5169	1.935	1.126	2.178	40
30	.4617	.8870	.5206	1.921	1.127	2.166	30
40	.4643	.8857	.5243	1.907	1.129	2.154	20
50	.4669	.8843	.5280	1.894	1.131	2.142	10
28° 00′	.4695	.8829	.5317	1.881	1.133	2.130	**62° 00′**
10	.4720	.8816	.5354	1.868	1.134	2.118	50
20	.4746	.8802	.5392	1.855	1.136	2.107	40
30	.4772	.8788	.5430	1.842	1.138	2.096	30
40	.4797	.8774	.5467	1.829	1.140	2.085	20
50	.4823	.8760	.5505	1.816	1.142	2.074	10
29° 00′	.4848	.8746	.5543	1.804	1.143	2.063	**61° 00′**
10	.4874	.8732	.5581	1.792	1.145	2.052	50
20	.4899	.8718	.5619	1.780	1.147	2.041	40
30	.4924	.8704	.5658	1.767	1.149	2.031	30
40	.4950	.8689	.5696	1.756	1.151	2.020	20
50	.4975	.8675	.5735	1.744	1.153	2.010	10
30° 00′	.5000	.8660	.5774	1.732	1.155	2.000	**60° 00′**
	Cos	Sin	Cot	Tan	Csc	Sec	Degrees

Table 16 (*continued*)   Natural Trigonometric Functions

Degrees	Sin	Cos	Tan	Cot	Sec	Csc	
30° 00′	.5000	.8660	.5774	1.732	1.155	2.000	**60° 00′**
10	.5025	.8646	.5812	1.720	1.157	1.990	50
20	.5050	.8631	.5851	1.709	1.159	1.980	40
30	.5075	.8616	.5890	1.698	1.161	1.970	30
40	.5100	.8601	.5930	1.686	1.163	1.961	20
50	.5125	.8587	.5969	1.675	1.165	1.951	10
31° 00′	.5150	.8572	.6009	1.664	1.167	1.942	**59° 00′**
10	.5175	.8557	.6048	1.653	1.169	1.932	50
20	.5200	.8542	.6088	1.643	1.171	1.923	40
30	.5225	.8526	.6128	1.632	1.173	1.914	30
40	.5250	.8511	.6168	1.621	1.175	1.905	20
50	.5275	.8496	.6208	1.611	1.177	1.896	10
32° 00′	.5299	.8480	.6249	1.600	1.179	1.887	**58° 00′**
10	.5324	.8465	.6289	1.590	1.181	1.878	50
20	.5348	.8450	.6330	1.580	1.184	1.870	40
30	.5373	.8434	.6371	1.570	1.186	1.861	30
40	.5398	.8418	.6412	1.560	1.188	1.853	20
50	.5422	.8403	.6453	1.550	1.190	1.844	10
33° 00′	.5446	.8387	.6494	1.540	1.192	1.836	**57° 00′**
10	.5471	.8371	.6536	1.530	1.195	1.828	50
20	.5495	.8355	.6577	1.520	1.197	1.820	40
30	.5519	.8339	.6619	1.511	1.199	1.812	30
40	.5544	.8323	.6661	1.501	1.202	1.804	20
50	.5568	.8307	.6703	1.492	1.204	1.796	10
34° 00′	.5592	.8290	.6745	1.483	1.206	1.788	**56° 00′**
10	.5616	.8274	.6787	1.473	1.209	1.781	50
20	.5640	.8258	.6830	1.464	1.211	1.773	40
30	.5664	.8241	.6873	1.455	1.213	1.766	30
40	.5688	.8225	.6916	1.446	1.216	1.758	20
50	.5712	.8208	.6959	1.437	1.218	1.751	10
35° 00′	.5736	.8192	.7002	1.428	1.221	1.743	**55° 00′**
10	.5760	.8175	.7046	1.419	1.223	1.736	50
20	.5783	.8158	.7089	1.411	1.226	1.729	40
30	.5807	.8141	.7133	1.402	1.228	1.722	30
40	.5831	.8124	.7177	1.393	1.231	1.715	20
50	.5854	.8107	.7221	1.385	1.233	1.708	10
36° 00′	.5878	.8090	.7265	1.376	1.236	1.701	**54° 00′**
	Cos	Sin	Cot	Tan	Csc	Sec	Degrees

Table 16 (*continued*)   Natural Trigonometric Functions

Degrees	Sin	Cos	Tan	Cot	Sec	Csc	
36° 00′	.5878	.8090	.7265	1.376	1.236	1.701	**54° 00′**
10	.5901	.8073	.7310	1.368	1.239	1.695	50
20	.5925	.8056	.7355	1.360	1.241	1.688	40
30	.5948	.8039	.7400	1.351	1.244	1.681	30
40	.5972	.8021	.7445	1.343	1.247	1.675	20
50	.5995	.8004	.7490	1.335	1.249	1.668	10
37° 00′	.6018	.7986	.7536	1.327	1.252	1.662	**53° 00′**
10	.6041	.7969	.7581	1.319	1.255	1.655	50
20	.6065	.7951	.7627	1.311	1.258	1.649	40
30	.6088	.7934	.7673	1.303	1.260	1.643	30
40	.6111	.7916	.7720	1.295	1.263	1.636	20
50	.6134	.7898	.7766	1.288	1.266	1.630	10
38° 00′	.6157	.7880	.7813	1.280	1.269	1.624	**52° 00′**
10	.6180	.7862	.7860	1.272	1.272	1.618	50
20	.6202	.7844	.7907	1.265	1.275	1.612	40
30	.6225	.7826	.7954	1.257	1.278	1.606	30
40	.6248	.7808	.8002	1.250	1.281	1.601	20
50	.6271	.7790	.8050	1.242	1.284	1.595	10
39° 00′	.6293	.7771	.8098	1.235	1.287	1.589	**51° 00′**
10	.6316	.7753	.8146	1.228	1.290	1.583	50
20	.6338	.7735	.8195	1.220	1.293	1.578	40
30	.6361	.7716	.8243	1.213	1.296	1.572	30
40	.6383	.7698	.8292	1.206	1.299	1.567	20
50	.6406	.7679	.8342	1.199	1.302	1.561	10
40° 00′	.6428	.7660	.8391	1.192	1.305	1.556	**50° 00′**
10	.6450	.7642	.8441	1.185	1.309	1.550	50
20	.6472	.7623	.8491	1.178	1.312	1.545	40
30	.6494	.7604	.8541	1.171	1.315	1.540	30
40	.6517	.7585	.8591	1.164	1.318	1.535	20
50	.6539	.7566	.8642	1.157	1.322	1.529	10
41° 00′	.6561	.7547	.8693	1.150	1.325	1.524	**49° 00′**
10	.6583	.7528	.8744	1.144	1.328	1.519	50
20	.6604	.7509	.8796	1.137	1.332	1.514	40
30	.6626	.7490	.8847	1.130	1.335	1.509	30
40	.6648	.7470	.8899	1.124	1.339	1.504	20
50	.6670	.7451	.8952	1.117	1.342	1.499	10
42° 00′	.6691	.7431	.9004	1.111	1.346	1.494	**48° 00′**
	Cos	Sin	Cot	Tan	Csc	Sec	Degrees

Table 16 (*continued*)   Natural Trigonometric Functions

Degrees	Sin	Cos	Tan	Cot	Sec	Csc	
42° 00′	.6691	.7431	.9004	1.111	1.346	1.494	**48° 00′**
10	.6713	.7412	.9057	1.104	1.349	1.490	50
20	.6734	.7392	.9110	1.098	1.353	1.485	40
30	.6756	.7373	.9163	1.091	1.356	1.480	30
40	.6777	.7353	.9217	1.085	1.360	1.476	20
50	.6799	.7333	.9271	1.079	1.364	1.471	10
43° 00′	.6820	.7314	.9325	1.072	1.367	1.466	**47° 00′**
10	.6841	.7294	.9380	1.066	1.371	1.462	50
20	.6862	.7274	.9435	1.060	1.375	1.457	40
30	.6884	.7254	.9490	1.054	1.379	1.453	30
40	.6905	.7234	.9545	1.048	1.382	1.448	20
50	.6926	.7214	.9601	1.042	1.386	1.444	10
44° 00′	.6947	.7193	.9657	1.036	1.390	1.440	**46° 00′**
10	.6967	.7173	.9713	1.030	1.394	1.435	50
20	.6988	.7153	.9770	1.024	1.398	1.431	40
30	.7009	.7133	.9827	1.018	1.402	1.427	30
40	.7030	.7112	.9884	1.012	1.406	1.423	20
50	.7050	.7092	.9942	1.006	1.410	1.418	10
45° 00′	.7071	.7071	1.000	1.000	1.414	1.414	**45° 00′**
	Cos	Sin	Cot	Tan	Csc	Sec	Degrees

Table 17* Conversion of Radians to Degrees, Minutes, and
Seconds or Fractions of Degrees

Radians	Degrees	Minutes	Seconds	Fractions of Degrees
1	57°	17′	44.8″	57.2958°
2	114°	35′	29.6″	114.5916°
3	171°	53′	14.4″	171.8873°
4	229°	10′	59.2″	229.1831°
5	286°	28′	44.0″	286.4789°
6	343°	46′	28.8″	343.7747°
7	401°	4′	13.6″	401.0705°
8	458°	21′	58.4″	458.3662°
9	515°	39′	43.3″	515.6620°
10	572°	57′	28.1″	572.9578°
.1	5°	43′	46.5″	
.2	11°	27′	33.0″	
.3	17°	11′	19.4″	
.4	22°	55′	5.9″	
.5	28°	38′	52.4″	
.6	34°	22′	38.9″	
.7	40°	6′	25.4″	
.8	45°	50′	11.8″	
.9	51°	33′	58.3″	
.01	0°	34′	22.6″	
.02	1°	8′	45.3″	
.03	1°	43′	7.9″	
.04	2°	17′	30.6″	
.05	2°	51′	53.2″	
.06	3°	26′	15.9″	
.07	4°	0′	38.5″	
.08	4°	35′	1.2″	
.09	5°	9′	23.8″	
.001	0°	3′	26.3″	
.002	0°	6′	52.5″	
.003	0°	10′	18.8″	
.004	0°	13′	45.1″	
.005	0°	17′	11.3″	
.006	0°	20′	37.6″	
.007	0°	24′	3.9″	
.008	0°	27′	30.1″	
.009	0°	30′	56.4″	
.0001	0°	0′	20.6″	
.0002	0°	0′	41.3″	
.0003	0°	1′	1.9″	
.0004	0°	1′	22.5″	
.0005	0°	1′	43.1″	
.0006	0°	2′	3.8″	
.0007	0°	2′	24.4″	
.0008	0°	2′	45.0″	
.0009	0°	3′	5.6″	

*This table from the *Mathematical Handbook of Formulas and Tables* by Murray R. Spiegel, 1968 Edition, is used with permission of the McGraw-Hill Book Co.

Table 18* Conversion of Degrees, Minutes,
and Seconds to Radians

Degrees	Radians
1°	.0174533
2°	.0349066
3°	.0523599
4°	.0698132
5°	.0872665
6°	.1047198
7°	.1221730
8°	.1396263
9°	.1570796
10°	.1745329

Minutes	Radians
1′	.00029089
2′	.00058178
3′	.00087266
4′	.00116355
5′	.00145444
6′	.00174533
7′	.00203622
8′	.00232711
9′	.00261800
10′	.00290888

Seconds	Radians
1″	.0000048481
2″	.0000096963
3″	.0000145444
4″	.0000193925
5″	.0000242407
6″	.0000290888
7″	.0000339370
8″	.0000387851
9″	.0000436332
10″	.0000484814

*This table from the *Mathematical Handbook of Formulas and Tables* by Murray R. Spiegel, 1968 Edition, is used with permission of the McGraw-Hill Book Co.

Table 19 Common Logarithms of Sines, Cosines, and Tangents

Degrees	Function	0.0°	0.1°	0.2°	0.3°	0.4°	0.5°	0.6°	0.7°	0.8°	0.9°
0	log sin	$-\infty$	7.2419	7.5429	7.7190	7.8439	7.9408	8.0200	8.0870	8.1450	8.1961
	log cos	0	0	0	0	0	0	0	0	0	9.9999
	log tan	$-\infty$	7.2419	7.5429	7.7190	7.8439	7.9409	8.0200	8.0870	8.1450	8.1962
1	log sin	8.2419	8.2832	8.3210	8.3558	8.3880	8.4179	8.4459	8.4723	8.4971	8.5206
	log cos	9.9999	9.9999	9.9999	9.9999	9.9999	9.9999	9.9998	9.9998	9.9998	9.9998
	log tan	8.2419	8.2833	8.3211	8.3559	8.3881	8.4181	8.4461	8.4725	8.4973	8.5208
2	log sin	8.5428	8.5640	8.5842	8.6035	8.6220	8.6397	8.6567	8.6731	8.6889	8.7041
	log cos	9.9997	9.9997	9.9997	9.9996	9.9996	9.9996	9.9996	9.9995	9.9995	9.9994
	log tan	8.5431	8.5643	8.5845	8.6038	8.6223	8.6401	8.6571	8.6736	8.6894	8.7046
3	log sin	8.7188	8.7330	8.7468	8.7602	8.7731	8.7857	8.7979	8.8098	8.8213	8.8326
	log cos	9.9994	9.9994	9.9993	9.9993	9.9992	9.9992	9.9991	9.9991	9.9990	9.9990
	log tan	8.7194	8.7337	8.7475	8.7609	8.7739	8.7865	8.7988	8.8107	8.8223	8.8336
4	log sin	8.8436	8.8543	8.8647	8.8749	8.8849	8.8946	8.9042	8.9135	8.9226	8.9315
	log cos	9.9989	9.9989	9.9988	9.9988	9.9987	9.9987	9.9986	9.9985	9.9985	9.9984
	log tan	8.8446	8.8554	8.8659	8.8762	8.8862	8.8960	8.9056	8.9150	8.9241	8.9331
5	log sin	8.9403	8.9489	8.9573	8.9655	8.9736	8.9816	8.9894	8.9970	9.0046	9.0120
	log cos	9.9983	9.9983	9.9982	9.9981	9.9981	9.9980	9.9979	9.9978	9.9978	9.9977
	log tan	8.9420	8.9506	8.9591	8.9674	8.9756	8.9836	8.9915	8.9992	9.0068	9.0143
6	log sin	9.0192	9.0264	9.0334	9.0403	9.0472	9.0539	9.0605	9.0670	9.0734	9.0797
	log cos	9.9976	9.9975	9.9975	9.9974	9.9973	9.9972	9.9971	9.9970	9.9969	9.9968
	log tan	9.0216	9.0289	9.0360	9.0430	9.0499	9.0567	9.0633	9.0699	9.0764	9.0828
7	log sin	9.0859	9.0920	9.0981	9.1040	9.1099	9.1157	9.1214	9.1271	9.1326	9.1381
	log cos	9.9968	9.9967	9.9966	9.9965	9.9964	9.9963	9.9962	9.9961	9.9960	9.9959
	log tan	9.0891	9.0954	9.1015	9.1076	9.1135	9.1194	9.1252	9.1310	9.1367	9.1423
8	log sin	9.1436	9.1489	9.1542	9.1594	9.1646	9.1697	9.1747	9.1797	9.1847	9.1895
	log cos	9.9958	9.9956	9.9955	9.9954	9.9953	9.9952	9.9951	9.9950	9.9949	9.9947
	log tan	9.1478	9.1533	9.1587	9.1640	9.1693	9.1745	9.1797	9.1848	9.1898	9.1948
9	log sin	9.1943	9.1991	9.2038	9.2085	9.2131	9.2176	9.2221	9.2266	9.2310	9.2353
	log cos	9.9946	9.9945	9.9944	9.9943	9.9941	9.9940	9.9939	9.9937	9.9936	9.9935
	log tan	9.1997	9.2046	9.2094	9.2142	9.2189	9.2236	9.2282	9.2328	9.2374	9.2419
10	log sin	9.2397	9.2439	9.2482	9.2524	9.2565	9.2606	9.2647	9.2687	9.2727	9.2767
	log cos	9.9934	9.9932	9.9931	9.9929	9.9928	9.9927	9.9925	9.9924	9.9922	9.9921
	log tan	9.2463	9.2507	9.2551	9.2594	9.2637	9.2680	9.2722	9.2764	9.2805	9.2846
11	log sin	9.2806	9.2845	9.2883	9.2921	9.2959	9.2997	9.3034	9.3070	9.3107	9.3143
	log cos	9.9919	9.9918	9.9916	9.9915	9.9913	9.9912	9.9910	9.9909	9.9907	9.9906
	log tan	9.2887	9.2927	9.2967	9.3006	9.3046	9.3085	9.3123	9.3162	9.3200	9.3237
12	log sin	9.3179	9.3214	9.3250	9.3284	9.3319	9.3353	9.3387	9.3421	9.3455	9.3488
	log cos	9.9904	9.9902	9.9901	9.9899	9.9897	9.9896	9.9894	9.9892	9.9891	9.9889
	log tan	9.3275	9.3312	9.3349	9.3385	9.3422	9.3458	9.3493	9.3529	9.3564	9.3599
13	log sin	9.3521	9.3554	9.3586	9.3618	9.3650	9.3682	9.3713	9.3745	9.3775	9.3806
	log cos	9.9887	9.9885	9.9884	9.9882	9.9880	9.9878	9.9876	9.9875	9.9873	9.9871
	log tan	9.3634	9.3668	9.3702	9.3736	9.3770	9.3804	9.3837	9.3870	9.3903	9.3935
14	log sin	9.3837	9.3867	9.3897	9.3927	9.3957	9.3986	9.4015	9.4044	9.4073	9.4102
	log cos	9.9869	9.9867	9.9865	9.9863	9.9861	9.9859	9.9857	9.9855	9.9853	9.9851
	log tan	9.3968	9.4000	9.4032	9.4064	9.4095	9.4127	9.4158	9.4189	9.4220	9.4250
Degrees	Function	0′	6′	12′	18′	24′	30′	36′	42′	48′	54′

Table 19 (*continued*)   Common Logarithms of Sines, Cosines, and Tangents

Degrees	Function	0.0°	0.1°	0.2°	0.3°	0.4°	0.5°	0.6°	0.7°	0.8°	0.9°
15	log sin	9.4130	9.4158	9.4186	9.4214	9.4242	9.4269	9.4296	9.4323	9.4350	9.4377
	log cos	9.9849	9.9847	9.9845	9.9843	9.9841	9.9839	9.9837	9.9835	9.9833	9.9831
	log tan	9.4281	9.4311	9.4341	9.4371	9.4400	9.4430	9.4459	9.4488	9.4517	9.4546
16	log sin	9.4403	9.4430	9.4456	9.4482	9.4508	9.4533	9.4559	9.4584	9.4609	9.4634
	log cos	9.9828	9.9826	9.9824	9.9822	9.9820	9.9817	9.9815	9.9813	9.9811	9.9808
	log tan	9.4575	9.4603	9.4632	9.4660	9.4688	9.4716	9.4744	9.4771	9.4799	9.4826
17	log sin	9.4659	9.4684	9.4709	9.4733	9.4757	9.4781	8.4805	9.4829	9.4853	9.4876
	log cos	9.9806	9.9804	9.9801	9.9799	9.9797	9.9794	9.9792	9.9789	9.9787	9.9785
	log tan	9.4853	9.4880	9.4907	9.4934	9.4961	9.4987	9.5014	9.5040	9.5066	9.5092
18	log sin	9.4900	9.4923	9.4946	9.4969	9.4992	9.5015	9.5037	9.5060	9.5082	9.5104
	log cos	9.9782	9.9780	9.9777	9.9775	9.9772	9.9770	9.9767	9.9764	9.9762	9.9759
	log tan	9.5118	9.5143	9.5169	9.5195	9.5220	9.5245	9.5270	9.5295	9.5320	9.5345
19	log sin	9.5126	9.5148	9.5170	9.5192	9.5213	9.5235	9.5256	9.5278	9.5299	9.5320
	log cos	9.9757	9.9754	9.9751	9.9749	9.9746	9.9743	9.9741	9.9738	9.9735	9.9733
	log tan	9.5370	9.5394	9.5419	9.5443	9.5467	9.5491	9.5516	9.5539	9.5563	9.5587
20	log sin	9.5341	9.5361	9.5382	9.5402	9.5423	9.5443	9.5463	9.5484	9.5504	9.5523
	log cos	9.9730	9.9727	9.9724	9.9722	9.9719	9.9716	9.9713	9.9710	9.9707	9.9704
	log tan	9.5611	9.5634	9.5658	9.5681	9.5704	9.5727	9.5750	9.5773	9.5796	9.5819
21	log sin	9.5543	9.5563	9.5583	9.5602	9.5621	9.5641	9.5660	9.5679	9.5698	9.5717
	log cos	9.9702	9.9699	9.9696	9.9693	9.9690	9.9687	9.9684	9.9681	9.9678	9.9675
	log tan	9.5842	9.5864	9.5887	9.5909	9.5932	9.5954	9.5976	9.5998	9.6020	9.6042
22	log sin	9.5736	9.5754	9.5773	9.5792	9.5810	9.5828	9.5847	9.5865	9.5883	9.5901
	log cos	9.9672	9.9669	9.9666	9.9662	9.9659	9.9656	9.9653	9.9650	9.9647	9.9643
	log tan	9.6064	9.6086	9.6108	9.6129	9.6151	9.6172	9.6194	9.6215	9.6236	9.6257
23	log sin	9.5919	9.5937	9.5954	9.5972	9.5990	9.6007	9.6024	9.6042	9.6059	9.6076
	log cos	9.9640	9.9637	9.9634	9.9631	9.9627	9.9624	9.9621	9.9617	9.9614	9.9611
	log tan	9.6279	9.6300	9.6321	9.6341	9.6362	9.6383	9.6404	9.6424	9.6445	9.6465
24	log sin	9.6093	9.6110	9.6127	9.6144	9.6161	9.6177	9.6194	9.6210	9.6227	9.6243
	log cos	9.9607	9.9604	9.9601	9.9597	9.9594	9.9590	9.9587	9.9583	9.9580	9.9576
	log tan	9.6486	9.6506	9.6527	9.6547	9.6567	9.6587	9.6607	9.6627	9.6647	9.6667
25	log sin	9.6259	9.6276	9.6292	9.6308	9.6324	9.6340	9.6356	9.6371	9.6387	9.6403
	log cos	9.9573	9.9569	9.9566	9.9562	9.9558	9.9555	9.9551	9.9548	9.9544	9.9540
	log tan	9.6687	9.6706	9.6726	9.6746	9.6765	9.6785	9.6804	9.6824	9.6843	9.6863
26	log sin	9.6418	9.6434	9.6449	9.6465	9.6480	9.6495	9.6510	9.6526	9.6541	9.6556
	log cos	9.9537	9.9533	9.9529	9.9525	9.9522	9.9518	9.9514	9.9510	9.9506	9.9503
	log tan	9.6882	9.6901	9.6920	9.6939	9.6958	9.6977	9.6996	9.7015	9.7034	9.7053
27	log sin	9.6570	9.6585	9.6600	9.6615	9.6629	9.6644	9.6659	9.6673	9.6687	9.6702
	log cos	9.9499	9.9495	9.9491	9.9487	9.9483	9.9479	9.9475	9.9471	9.9467	9.9463
	log tan	9.7072	9.7090	9.7109	9.7128	9.7146	9.7165	9.7183	9.7202	9.7220	9.7238
28	log sin	9.6716	9.6730	9.6744	9.6759	9.6773	9.6787	9.6801	9.6814	9.6828	9.6842
	log cos	9.9459	9.9455	9.9451	9.9447	9.9443	9.9439	9.9435	9.9431	9.9427	9.9422
	log tan	9.7257	9.7275	9.7293	9.7311	9.7330	9.7348	9.7366	9.7384	9.7402	9.7420
29	log sin	9.6856	9.6869	9.6883	9.6896	9.6910	9.6923	9.6937	9.6950	9.6963	9.6977
	log cos	9.9418	9.9414	9.9410	9.9406	9.9401	9.9397	9.9393	9.9388	9.9384	9.9380
	log tan	9.7438	9.7455	9.7473	9.7491	9.7509	9.7526	9.7544	9.7562	9.7579	9.7597
Degrees	Function	0′	6′	12′	18′	24′	30′	36′	42′	48′	54′

Table 19 (*continued*)   Common Logarithms of Sines, Cosines, and Tangents

Degrees	Function	0.0°	0.1°	0.2°	0.3°	0.4°	0.5°	0.6°	0.7°	0.8°	0.9°
30	log sin	9.6990	9.7003	9.7016	9.7029	9.7042	9.7055	9.7068	9.7080	9.7093	9.7106
	log cos	9.9375	9.9371	9.9367	9.9362	9.9358	9.9353	9.9349	9.9344	9.9340	9.9335
	log tan	9.7614	9.7632	9.7649	9.7667	9.7684	9.7701	9.7719	9.7736	9.7753	9.7771
31	log sin	9.7118	9.7131	9.7144	9.7156	9.7168	9.7181	9.7193	9.7205	9.7218	9.7230
	log cos	9.9331	9.9326	9.9322	9.9317	9.9312	9.9308	9.9303	9.9298	9.9294	9.9289
	log tan	9.7788	9.7805	9.7822	9.7839	9.7856	9.7873	9.7890	9.7907	9.7924	9.7941
32	log sin	9.7242	9.7254	9.7266	9.7278	9.7290	9.7302	9.7314	9.7326	9.7338	9.7349
	log cos	9.9284	9.9279	9.9275	9.9270	9.9265	9.9260	9.9255	9.9251	9.9246	9.9241
	log tan	9.7958	9.7975	9.7992	9.8008	9.8025	9.8042	9.8059	9.8075	9.8092	9.8109
33	log sin	9.7361	9.7373	9.7384	9.7396	9.7407	9.7419	9.7430	9.7442	9.7453	9.7464
	log cos	9.9236	9.9231	9.9226	9.9221	9.9216	9.9211	9.9206	9.9201	9.9196	9.9191
	log tan	9.8125	9.8142	9.8158	9.8175	9.8191	9.8208	9.8224	9.8241	9.8257	9.8274
34	log sin	9.7476	9.7487	9.7498	9.7509	9.7520	9.7531	9.7542	9.7553	9.7564	9.7575
	log cos	9.9186	9.9181	9.9175	9.9170	9.9165	9.9160	9.9155	9.9149	9.9144	9.9139
	log tan	9.8290	9.8306	9.8323	9.8339	9.8355	9.8371	9.8388	9.8404	9.8420	9.8436
35	log sin	9.7586	9.7597	9.7607	9.7618	9.7629	9.7640	9.7650	9.7661	9.7671	9.7682
	log cos	9.9134	9.9128	9.9123	9.9118	9.9112	9.9107	9.9101	9.9096	9.9091	9.9085
	log tan	9.8452	9.8468	9.8484	9.8501	9.8517	9.8533	9.8549	9.8565	9.8581	9.8597
36	log sin	9.7692	9.7703	9.7713	9.7723	9.7734	9.7744	9.7754	9.7764	9.7774	9.7785
	log cos	9.9080	9.9074	9.9069	9.9063	9.9057	9.9052	9.9046	9.9041	9.9035	9.9029
	log tan	9.8613	9.8629	9.8644	9.8660	9.8676	9.8692	9.8708	9.8724	9.8740	9.8755
37	log sin	9.7795	9.7805	9.7815	9.7825	9.7835	9.7844	9.7854	9.7864	9.7874	9.7884
	log cos	9.9023	9.9018	9.9012	9.9006	9.9000	9.8995	9.8989	9.8983	9.8977	9.8971
	log tan	9.8771	9.8787	9.8803	9.8818	9.8834	9.8850	9.8865	9.8881	9.8897	9.8912
38	log sin	9.7893	9.7903	9.7913	9.7922	9.7932	9.7941	9.7951	9.7960	9.7970	9.7979
	log cos	9.8965	9.8959	9.8953	9.8947	9.8941	9.8935	9.8929	9.8923	9.8917	9.8911
	log tan	9.8928	9.8944	9.8959	9.8975	9.8990	9.9006	9.9022	9.9037	9.9053	9.9068
39	log sin	9.7989	9.7998	9.8007	9.8017	9.8026	9.8035	9.8044	9.8053	9.8063	9.8072
	log cos	9.8905	9.8899	9.8893	9.8887	9.8880	9.8874	9.8868	9.8862	9.8855	9.8849
	log tan	9.9084	9.9099	9.9115	9.9130	9.9146	9.9161	9.9176	9.9192	9.9207	9.9223
40	log sin	9.8081	9.8090	9.8099	9.8108	9.8117	9.8125	9.8134	9.8143	9.8152	9.8161
	log cos	9.8843	9.8836	9.8830	9.8823	9.8817	9.8810	9.8804	9.8797	9.8791	9.8784
	log tan	9.9238	9.9254	9.9269	9.9284	9.9300	9.9315	9.9330	9.9346	9.9361	9.9376
41	log sin	9.8169	9.8178	9.8187	9.8195	9.8204	9.8213	9.8221	9.8230	9.8238	9.8247
	log cos	9.8778	9.8771	9.8765	9.8758	9.8751	9.8745	9.8738	9.8731	9.8724	9.8718
	log tan	9.9392	9.9407	9.9422	9.9438	9.9453	9.9468	9.9483	9.9499	9.9514	9.9529
42	log sin	9.8255	9.8264	9.8272	9.8280	9.8289	9.8297	9.8305	9.8313	9.8322	9.8330
	log cos	9.8711	9.8704	9.8697	9.8690	9.8683	9.8676	9.8669	9.8662	9.8655	9.8648
	log tan	9.9544	9.9560	9.9575	9.9590	9.9605	9.9621	9.9636	9.9651	9.9666	9.9681
43	log sin	9.8338	9.8346	9.8354	9.8362	9.8370	9.8378	9.8386	9.8394	9.8402	9.8410
	log cos	9.8641	9.8634	9.8627	9.8620	9.8613	9.8606	9.8598	9.8591	9.8584	9.8577
	log tan	9.9697	9.9712	9.9727	9.9742	9.9757	9.9772	9.9788	9.9803	9.9818	9.9833
44	log sin	9.8418	9.8426	9.8433	9.8441	9.8449	9.8457	9.8464	9.8472	9.8480	9.8487
	log cos	9.8569	9.8562	9.8555	9.8547	9.8540	9.8532	9.8525	9.8517	9.8510	9.8502
	log tan	9.9848	9.9864	9.9879	9.9894	9.9909	9.9924	9.9939	9.9955	9.9970	9.9985
Degrees	Function	0′	6′	12′	18′	24′	30′	36′	42′	48′	54′

Table 19 (*continued*)   Common Logarithms of Sines, Cosines, and Tangents

Degrees	Function	0.0°	0.1°	0.2°	0.3°	0.4°	0.5°	0.6°	0.7°	0.8°	0.9°
45	log sin	9.8495	9.8502	9.8510	9.8517	9.8525	9.8532	9.8540	9.8547	9.8555	9.8562
	log cos	9.8495	9.8487	9.8480	9.8472	9.8464	9.8457	9.8449	9.8441	9.8433	9.8426
	log tan	0.0000	0.0015	0.0030	0.0045	0.0061	0.0076	0.0091	0.0106	0.0121	0.0136
46	log sin	9.8569	9.8577	9.8584	9.8591	9.8598	9.8606	9.8613	9.8620	9.8627	9.8634
	log cos	9.8418	9.8410	9.8402	9.8394	9.8386	9.8378	9.8370	9.8362	9.8354	9.8346
	log tan	0.0152	0.0167	0.0182	0.0197	0.0212	0.0228	0.0243	0.0258	0.0273	0.0288
47	log sin	9.8641	9.8648	9.8655	9.8662	9.8669	9.8676	9.8683	9.8690	9.8697	9.8704
	log cos	9.8338	9.8330	9.8322	9.8313	9.8305	9.8297	9.8289	9.8280	9.8272	9.8264
	log tan	0.0303	0.0319	0.0334	0.0349	0.0364	0.0379	0.0395	0.0410	0.0425	0.0440
48	log sin	9.8711	9.8718	9.8724	9.8731	9.8738	9.8745	9.8751	9.8758	9.8765	9.8771
	log cos	9.8255	9.8247	9.8238	9.8230	9.8221	9.8213	9.8204	9.8195	9.8187	9.8178
	log tan	0.0456	0.0471	0.0486	0.0501	0.0517	0.0532	0.0547	0.0562	0.0578	0.0593
49	log sin	9.8778	9.8784	9.8791	9.8797	9.8804	9.8810	9.8817	9.8823	9.8830	9.8836
	log cos	9.8169	9.8161	9.8152	9.8143	9.8134	9.8125	9.8117	9.8108	9.8099	9.8090
	log tan	0.0608	0.0624	0.0639	0.0654	0.0670	0.0685	0.0700	0.0716	0.0731	0.0746
50	log sin	9.8843	9.8849	9.8855	9.8862	9.8868	9.8874	9.8880	9.8887	9.8893	9.8899
	log cos	9.8081	9.8072	9.8063	9.8053	9.8044	9.8035	9.8026	9.8017	9.8007	9.7998
	log tan	0.0762	0.0777	0.0793	0.0808	0.0824	0.0839	0.0854	0.0870	0.0885	0.0901
51	log sin	9.8905	9.8911	9.8917	9.8923	9.8929	9.8935	9.8941	9.8947	9.8953	9.8959
	log cos	9.7989	9.7979	9.7970	9.7960	9.7951	9.7941	9.7932	9.7922	9.7913	9.7903
	log tan	0.0916	0.0932	0.0947	0.0963	0.0978	0.0994	0.1010	0.1025	0.1041	0.1056
52	log sin	9.8965	9.8971	9.8977	9.8983	9.8989	9.8995	9.9000	9.9006	9.9012	9.9018
	log cos	9.7893	9.7884	9.7874	9.7864	9.7854	9.7844	9.7835	9.7825	9.7815	9.7805
	log tan	0.1072	0.1088	0.1103	0.1119	0.1135	0.1150	0.1166	0.1182	0.1197	0.1213
53	log sin	9.9023	9.9029	9.9035	9.9041	9.9046	9.9052	9.9057	9.9063	9.9069	9.9074
	log cos	9.7795	9.7785	9.7774	9.7764	9.7754	9.7744	9.7734	9.7723	9.7713	9.7703
	log tan	0.1229	0.1245	0.1260	0.1276	0.1292	0.1308	0.1324	0.1340	0.1356	0.1371
54	log sin	9.9080	9.9085	9.9091	9.9096	9.9101	9.9107	9.9112	9.9118	9.9123	9.9128
	log cos	9.7692	9.7682	9.7671	9.7661	9.7650	9.7640	9.7629	9.7618	9.7607	9.7597
	log tan	0.1387	0.1403	0.1419	0.1435	0.1451	0.1467	0.1483	0.1499	0.1516	0.1532
55	log sin	9.9134	9.9139	9.9144	9.9149	9.9155	9.9160	9.9165	9.9170	9.9175	9.9181
	log cos	9.7586	9.7575	9.7564	9.7553	9.7542	9.7531	9.7520	9.7509	9.7498	9.7487
	log tan	0.1548	0.1564	0.1580	0.1596	0.1612	0.1629	0.1645	0.1661	0.1677	0.1694
56	log sin	9.9186	9.9191	9.9196	9.9201	9.9206	9.9211	9.9216	9.9221	9.9226	9.9231
	log cos	9.7476	9.7464	9.7453	9.7442	9.7430	9.7419	9.7407	9.7396	9.7384	9.7373
	log tan	0.1710	0.1726	0.1743	0.1759	0.1776	0.1792	0.1809	0.1825	0.1842	0.1858
57	log sin	9.9236	9.9241	9.9246	9.9251	9.9255	9.9260	9.9265	9.9270	9.9275	9.9279
	log cos	9.7361	9.7349	9.7338	9.7326	9.7314	9.7302	9.7290	9.7278	9.7266	9.7254
	log tan	0.1875	0.1891	0.1908	0.1925	0.1941	0.1958	0.1975	0.1992	0.2008	0.2025
58	log sin	9.9284	9.9289	9.9294	9.9298	9.9303	9.9308	9.9312	9.9317	9.9322	9.9326
	log cos	9.7242	9.7230	9.7218	9.7205	9.7193	9.7181	9.7168	9.7156	9.7144	9.7131
	log tan	0.2042	0.2059	0.2076	0.2093	0.2110	0.2127	0.2144	0.2161	0.2178	0.2195
59	log sin	9.9331	9.9335	9.9340	9.9344	9.9349	9.9353	9.9358	9.9362	9.9367	9.9371
	log cos	9.7118	9.7106	9.7093	9.7080	9.7068	9.7055	9.7042	9.7029	9.7016	9.7003
	log tan	0.2212	0.2229	0.2247	0.2264	0.2281	0.2299	0.2316	0.2333	0.2351	0.2368

Degrees	Function	0′	6′	12′	18′	24′	30′	36′	42′	48′	54′

Table 19 (*continued*)   Common Logarithms of Sines, Cosines, and Tangents

Degrees	Function	0.0°	0.1°	0.2°	0.3°	0.4°	0.5°	0.6°	0.7°	0.8°	0.9°
60	log sin	9.9375	9.9380	9.9384	9.9388	9.9393	9.9397	9.9401	9.9406	9.9410	9.9414
	log cos	9.6990	9.6977	9.6963	9.6950	9.6937	9.6923	9.6910	9.6896	9.6883	9.6869
	log tan	0.2386	0.2403	0.2421	0.2438	0.2456	0.2474	0.2491	0.2509	0.2527	0.2545
61	log sin	9.9418	9.9422	9.9427	9.9431	9.9435	9.9439	9.9443	9.9447	9.9451	9.9455
	log cos	9.6856	9.6842	9.6828	9.6814	9.6801	9.6787	9.6773	9.6759	9.6744	9.6730
	log tan	0.2562	0.2580	0.2598	0.2616	0.2634	0.2652	0.2670	0.2689	0.2707	0.2725
62	log sin	9.9459	9.9463	9.9467	9.9471	9.9475	9.9479	9.9483	9.9487	9.9491	9.9495
	log cos	9.6716	9.6702	9.6687	9.6673	9.6659	9.6644	9.6629	9.6615	9.6600	9.6585
	log tan	0.2743	0.2762	0.2780	0.2798	0.2817	0.2835	0.2854	0.2872	0.2891	0.2910
63	log sin	9.9499	9.9503	9.9506	9.9510	9.9514	9.9518	9.9522	9.9525	9.9529	9.9533
	log cos	9.6570	9.6556	9.6541	9.6526	9.6510	9.6495	9.6480	9.6465	9.6449	9.6434
	log tan	0.2928	0.2947	0.2966	0.2985	0.3004	0.3023	0.3042	0.3061	0.3080	0.3099
64	log sin	9.9537	9.9540	9.9544	9.9548	9.9551	9.9555	9.9558	9.9562	9.9566	9.9569
	log cos	9.6418	9.6403	9.6387	9.6371	9.6356	9.6340	9.6324	9.6308	9.6292	9.6276
	log tan	0.3118	0.3137	0.3157	0.3176	0.3196	0.3215	0.3235	0.3254	0.3274	0.3294
65	log sin	9.9573	9.9576	9.9580	9.9583	9.9587	9.9590	9.9594	9.9597	9.9601	9.9604
	log cos	9.6259	9.6243	9.6227	9.6210	9.6194	9.6177	9.6161	9.6144	9.6127	9.6110
	log tan	0.3313	0.3333	0.3353	0.3373	0.3393	0.3413	0.3433	0.3453	0.3473	0.3494
66	log sin	9.9607	9.9611	9.9614	9.9617	9.9621	9.9624	9.9627	9.9631	9.9634	9.9637
	log cos	9.6093	9.6076	9.6059	9.6042	9.6024	9.6007	9.5990	9.5972	9.5954	9.5937
	log tan	0.3514	0.3535	0.3555	0.3576	0.3596	0.3617	0.3638	0.3659	0.3679	0.3700
67	log sin	9.9640	9.9643	9.9647	9.9650	9.9653	9.9656	9.9659	9.9662	9.9666	9.9669
	log cos	9.5919	9.5901	9.5883	9.5865	9.5847	9.5828	9.5810	9.5792	9.5773	9.5754
	log tan	0.3721	0.3743	0.3764	0.3785	0.3806	0.3828	0.3849	0.3871	0.3892	0.3914
68	log sin	9.9672	9.9675	9.9678	9.9681	9.9684	9.9687	9.9690	9.9693	9.9696	9.9699
	log cos	9.5736	9.5717	9.5698	9.5679	9.5660	9.5641	9.5621	9.5602	9.5583	9.5563
	log tan	0.3936	0.3958	0.3980	0.4002	0.4024	0.4046	0.4068	0.4091	0.4113	0.4136
69	log sin	9.9702	9.9704	9.9707	9.9710	9.9713	9.9716	9.9719	9.9722	9.9724	9.9727
	log cos	9.5543	9.5523	9.5504	9.5484	9.5463	9.5443	9.5423	9.5402	9.5382	9.5361
	log tan	0.4158	0.4181	0.4204	0.4227	0.4250	0.4273	0.4296	0.4319	0.4342	0.4366
70	log sin	9.9730	9.9733	9.9735	9.9738	9.9741	9.9743	9.9746	9.9749	9.9751	9.9754
	log cos	9.5341	9.5320	9.5299	9.5278	9.5256	9.5235	9.5213	9.5192	9.5170	9.5148
	log tan	0.4389	0.4413	0.4437	0.4461	0.4484	0.4509	0.4533	0.4557	0.4581	0.4606
71	log sin	9.9757	9.9759	9.9762	9.9764	9.9767	9.9770	9.9772	9.9775	9.9777	9.9780
	log cos	9.5126	9.5104	9.5082	9.5060	9.5037	9.5015	9.4992	9.4969	9.4946	9.4923
	log tan	0.4630	0.4655	0.4680	0.4705	0.4730	0.4755	0.4780	0.4805	0.4831	0.4857
72	log sin	9.9782	9.9785	9.9787	9.9789	9.9792	9.9794	9.9797	9.9799	9.9801	9.9804
	log cos	9.4900	9.4876	9.4853	9.4829	9.4805	9.4781	9.4757	9.4733	9.4709	9.4684
	log tan	0.4882	0.4908	0.4934	0.4960	0.4986	0.5013	0.5039	0.5066	0.5093	0.5120
73	log sin	9.9806	9.9808	9.9811	9.9813	9.9815	9.9817	9.9820	9.9822	9.9824	9.9826
	log cos	9.4659	9.4634	9.4609	9.4584	9.4559	9.4533	9.4508	9.4482	9.4456	9.4430
	log tan	0.5147	0.5174	0.5201	0.5229	0.5256	0.5284	0.5312	0.5340	0.5368	0.5397
74	log sin	9.9828	9.9831	9.9833	9.9835	9.9837	9.9839	9.9841	9.9843	9.9845	9.9847
	log cos	9.4403	9.4377	9.4350	9.4323	9.4296	9.4269	9.4242	9.4214	9.4186	9.4158
	log tan	0.5425	0.5454	0.5483	0.5512	0.5541	0.5570	0.5600	0.5629	0.5659	0.5689
Degrees	Function	0′	6′	12′	18′	24′	30′	36′	42′	48′	54′

Table 19 (*continued*)   Common Logarithms of Sines, Cosines, and Tangents

Degrees	Function	0.0°	0.1°	0.2°	0.3°	0.4°	0.5°	0.6°	0.7°	0.8°	0.9°
75	log sin	9.9849	9.9851	9.9853	9.9855	9.9857	9.9859	9.9861	9.9863	9.9865	9.9867
	log cos	9.4130	9.4102	9.4073	9.4044	9.4015	9.3986	9.3957	9.3927	9.3897	9.3867
	log tan	0.5719	0.5750	0.5780	0.5811	0.5842	0.5873	0.5905	0.5936	0.5968	0.6000
76	log sin	9.9869	9.9871	9.9873	9.9875	9.9876	9.9878	9.9880	9.9882	9.9884	9.9885
	log cos	9.3837	9.3806	9.3775	9.3745	9.3713	9.3682	9.3650	9.3618	9.3586	9.3554
	log tan	0.6032	0.6065	0.6097	0.6130	0.6163	0.6196	0.6230	0.6264	0.6298	0.6332
77	log sin	9.9887	9.9889	9.9891	9.9892	9.9894	9.9896	9.9897	9.9899	9.9901	9.9902
	log cos	9.3521	9.3488	9.3455	9.3421	9.3387	9.3353	9.3319	9.3284	9.3250	9.3214
	log tan	0.6366	0.6401	0.6436	0.6471	0.6507	0.6542	0.6578	0.6615	0.6651	0.6688
78	log sin	9.9904	9.9906	9.9907	9.9909	9.9910	9.9912	9.9913	9.9915	9.9916	9.9918
	log cos	9.3179	9.3143	9.3107	9.3070	9.3034	9.2997	9.2959	9.2921	9.2883	9.2845
	log tan	0.6725	0.6763	0.6800	0.6838	0.6877	0.6915	0.6954	0.6994	0.7033	0.7073
79	log sin	9.9919	9.9921	9.9922	9.9924	9.9925	9.9927	9.9928	9.9929	9.9931	9.9932
	log cos	9.2806	9.2767	9.2727	9.2687	9.2647	9.2606	9.2565	9.2524	9.2482	9.2439
	log tan	0.7113	0.7154	0.7195	0.7236	0.7278	0.7320	0.7363	0.7406	0.7449	0.7493
80	log sin	9.9934	9.9935	9.9936	9.9937	9.9939	9.9940	9.9941	9.9943	9.9944	9.9945
	log cos	9.2397	9.2353	9.2310	9.2266	9.2221	9.2176	9.2131	9.2085	9.2038	9.1991
	log tan	0.7537	0.7581	0.7626	0.7672	0.7718	0.7764	0.7811	0.7858	0.7906	0.7954
81	log sin	9.9946	9.9947	9.9949	9.9950	9.9951	9.9952	9.9953	9.9954	9.9955	9.9956
	log cos	9.1943	9.1895	9.1847	9.1797	9.1747	9.1697	9.1646	9.1594	9.1542	9.1489
	log tan	0.8003	0.8052	0.8102	0.8152	0.8203	0.8255	0.8307	0.8360	0.8413	0.8407
82	log sin	9.9958	9.9959	9.9960	9.9961	9.9962	9.9963	9.9964	9.9965	9.9966	9.9967
	log cos	9.1436	9.1381	9.1326	9.1271	9.1214	9.1157	9.1099	9.1040	9.0981	9.0920
	log tan	0.8522	0.8577	0.8633	0.8690	0.8748	0.8806	0.8865	0.8924	0.8985	0.9046
83	log sin	9.9968	9.9968	9.9969	9.9970	9.9971	9.9972	9.9973	9.9974	9.9975	9.9975
	log cos	9.0859	9.0797	9.0734	9.0670	9.0605	9.0539	9.0472	9.0403	9.0334	9.0264
	log tan	0.9109	0.9172	0.9236	0.9301	0.9367	0.9433	0.9501	0.9570	0.9640	0.9711
84	log sin	9.9976	9.9977	9.9978	9.9978	9.9979	9.9980	9.9981	9.9981	9.9982	9.9983
	log cos	9.0192	9.0120	9.0046	8.9970	8.9894	8.9816	8.9736	8.9655	8.9573	8.9489
	log tan	0.9784	0.9857	0.9932	1.0008	1.0085	1.0164	1.0244	1.0326	1.0409	1.0494
85	log sin	9.9983	9.9984	9.9985	9.9985	9.9986	9.9987	9.9987	9.9988	9.9988	9.9989
	log cos	8.9403	8.9315	8.9226	8.9135	8.9042	8.8946	8.8849	8.8749	8.8647	8.8543
	log tan	1.0580	1.0669	1.0759	1.0850	1.0944	1.1040	1.1138	1.1238	1.1341	1.1446
86	log sin	9.9989	9.9990	9.9990	9.9991	9.9991	9.9992	9.9992	9.9993	9.9993	9.9994
	log cos	8.8436	8.8326	8.8213	8.8098	8.7979	8.7857	8.7731	8.7602	8.7468	8.7330
	log tan	1.1554	1.1664	1.1777	1.1893	1.2012	1.2135	1.2261	1.2391	1.2525	1.2663
87	log sin	9.9994	9.9994	9.9995	9.9995	9.9996	9.9996	9.9996	9.9996	9.9997	9.9997
	log cos	8.7188	8.7041	8.6889	8.6731	8.6567	8.6397	8.6220	8.6035	8.5842	8.5640
	log tan	1.2806	1.2954	1.3106	1.3264	1.3429	1.3599	1.3777	1.3962	1.4155	1.4357
88	log sin	9.9997	9.9998	9.9998	9.9998	9.9998	9.9999	9.9999	9.9999	9.9999	9.9999
	log cos	8.5428	8.5206	8.4971	8.4723	8.4459	8.4179	8.3880	8.3558	8.3210	8.2832
	log tan	1.4569	1.4792	1.5027	1.5275	1.5539	1.5819	1.6119	1.6441	1.6789	1.7167
89	log sin	9.9999	9.9999	0	0	0	0	0	0	0	0
	log cos	8.2419	8.1961	8.1450	8.0870	8.0200	7.9408	7.8439	7.7190	7.5429	7.2419
	log tan	1.7581	1.8038	1.8550	1.9130	1.9800	2.0591	2.1561	2.2810	2.4571	2.7581
Degrees	Function	0′	6′	12′	18′	24′	30′	36′	42′	48′	54′

CHAPTER 1
Exercise 1.1

1.	750	**11.**	3.3×10^{-3}	**19.**	1.234×10^{-2}	**27.**	5.84×10^2
3.	90,500	**13.**	1.23×10^{-2}	**21.**	8.86×10^6	**29.**	3.15×10^{-2}
5.	3.11	**15.**	7.5×10^{-5}	**23.**	7.86×10^4	**31.**	3.0527×10^{-2}
7.	0.0035	**17.**	1.75×10^8	**25.**	8.9×10	**33.**	4.69×10^5
9.	0.00107						

Exercise 1.2 A

1.	E, A	**9.**	E, E	**17.**	E	**25.**	E	**33.**	A, A, A
3.	E	**11.**	E, E, E	**19.**	E, A	**27.**	E	**35.**	A
5.	E, A	**13.**	A	**21.**	E, E	**29.**	A	**37.**	E, A
7.	E, A	**15.**	A	**23.**	E, E	**31.**	E, E	**39.**	A

Exercise 1.2B

	Approxi-mate Number	Signifi-cant Figures	Rounded off to Four Figures	Rounded off to Three Figures	Rounded off to Two Figures	Rounded off to One Figure
1.	0.44549	5	0.4455	0.445	0.45	0.4
3.	0.005180	4	0.005180	0.00518	0.0052	0.005
5.	4.49	3	—	4.49	4.5	4
7.	3,485,000	4	3,485,000	3,480,000	3,500,000	3,000,000
9.	78.0273	6	78.03	78.0	78	80
11.	0.003195	4	0.003195	0.00320	0.0032	0.003
13.	12,0$\tilde{0}$0	4	12,0$\tilde{0}$0	12,$\tilde{0}$00	12,000	10,000
15.	9.200×10^4	4	9.200×10^4	9.20×10^4	9.2×10^4	9×10^4

17.	25.2	**21.**	780	**25.**	633.08	**29.**	54.00	**33.**	$830 \, \text{cm}^3$
19.	206	**23.**	73.93	**27.**	0.832	**31.**	5.5 in.		

35. $0.05205 \, \text{g} < N < 0.05215 \, \text{g}$; N is between 0.05205 g and 0.05215 g

37. $5.55 \, \text{in.} < N < 5.65 \, \text{in.}$; N is between 5.55 in. and 5.65 in.

39. $526.15 \, \text{lb} < N < 526.25 \, \text{lb}$; N is between 526.15 lb and 526.25 lb

41. $6.0665 \, \text{in.} < N < 6.0675 \, \text{in.}$; N is between 6.0665 in. and 6.0675 in.

43. $0.1245 \, \text{mm} < N < 0.1255 \, \text{mm}$; N is between 0.1245 mm and 0.1255 mm

45.	16.8	**47.**	310,000	**49.**	YES	**51.**	3.142	**53.**	3.142

Exercise 1.3A

1.	50%	**11.**	62.5%	**21.**	$\frac{52}{100}$, 0.52	**31.**	$\frac{3}{10,000}$, 0.0003
3.	150%	**13.**	360%	**23.**	$\frac{1}{1}$, 1	**33.**	$\frac{1000}{100}$ = 10, 10
5.	60%	**15.**	0.2%	**25.**	$\frac{10}{100}$ or $\frac{1}{10}$, 0.10	**35.**	$\frac{225}{1000}$, 0.225
7.	90%	**17.**	4%	**27.**	$\frac{35}{10} = 3\frac{1}{2}$, 3.50	**37.**	$\frac{7}{800}$, 0.00875
9.	325%	**19.**	0.03%	**29.**	$\frac{5}{1000}$ or $\frac{1}{200}$, 0.005	**39.**	$\frac{5}{1200}$, 0.004 16$\frac{2}{3}$

Exercise 1.3B

1. 3.20 gal
3. $52 rounds off to $50
5. 0.0022 lb
7. 1883.75 mm rounds off to 1880 mm

9. 1 ohm   15. 576
11. 96.0   17. 1.200 g/cm²
13. $2.250   19. 0.090 acre-ft

Exercise 1.3C

1.	30%	**7.**	2.08%	**11.**	20%	**15.**	50%	**19.**	800%
3.	7.5%	**9.**	150%	**13.**	10%	**17.**	0.625%	**21.**	100%
5.	1.0%								

Exercise 1.3D

1.	680	**5.**	0.035 g	**9.**	895	**13.**	14	**17.**	384 miles
3.	1.03	**7.**	500 tons	**11.**	400	**15.**	560 sq ft	**19.**	96 cu yd

Exercise 1.3E

1. 138 miles
3. 20%
5. $2.90
7. 5.0%

9. (a) 0.09 ft
   (b) 30.09 ft
11. 122.4 lb
13. $275

15. 140% increase
17. 12$\frac{2}{9}$% increase
19. No. 100 + (10%)(100) = 110
    110 − (10%)(110) = 99

Exercise 1.3F

1.	$195.00	**5.**	$214.50	**9.**	$1954.15	**13.**	$41.24
3.	$5460.00	**7.**	$2103.60	**11.**	$46,118.41	**15.**	25% discount

Exercise 1.3G

1. $576, $2976
3. $843.75, $3343.75

5. $1.80, $541.80
7. $1838.98, $7038.98

9. $1573, $23,573

Exercise 1.3H

1. $3601.80, $1601.80
3. $1208.88, $408.88

5. $2226.75, $726.75
7. $3368.32, $1768.32

9. $63.60 more
11. $9195.75, $4695.75

Exercise 1.4A

1.	4 ft 9 in.	**7.**	3 lb 10 oz	**13.**	28 lb 5 oz	**19.**	4 hr 51 min 45 sec
3.	2 yd 3 in.	**9.**	25° 16′ 23″	**15.**	6 tons 1600 lb	**21.**	255° 21′ 58″
5.	12 gal 3 qt	**11.**	6 tons 700 lb	**17.**	31 hr 53 min 45 sec	**23.**	3 yr 10 mo 24 days

Exercise 1.4B

1. 27 lb   3. 25 gal 1 qt 1 pt   5. 53 yd 9 in.   7. 17 tons 125 lb   9. 336° 3′ 12″

Exercise 1.4C

1. 4 yd 2 ft 10 in.   3. 4 gal 3 qt 1 pt   5. 26° 18′ 41″   7. 5 tons 324 lb   9. 15 gal 2 qt 1 pt

Exercise 1.5

1.	Yard	**13.**	1 sq mile = 640 acres	**25.**	1 km = 1000 m
3.	Metric	**15.**	1 cu yd = 27 cu ft	**27.**	1 hkm = 10,000 cm
5.	1 yd = 36 in.	**17.**	1 chain = 66 ft = 22 yd	**29.**	1 kℓ = 1000 ℓ
7.	1 ft = 12 in.	**19.**	1 gal = 8 pt	**31.**	0.552 ℓ
9.	1 lb = 16 oz	**21.**	1 day = 24 hr	**33.**	1000 kg
11.	1 fathom = 2 yd	**23.**	1 min = 60 sec	**35.**	265 cm

Exercise 1.6

3. 1 cu ft = 7.48055 gal

$$\frac{1 \text{ cu ft}}{7.48055 \text{ gal}} = 1; \frac{7.48055 \text{ gal}}{1 \text{ cu ft}} = 1$$

5. 1 kg = 1000 g

$$\frac{1000 \text{ g}}{1 \text{ kg}} = 1; \frac{1 \text{ kg}}{1000 \text{ g}} = 1$$

7. 1 ℓ = 1.05668 qt

$$\frac{1 \ell}{1.05668 \text{ qt}} = 1; \frac{1.05668 \text{ qt}}{1 \ell} = 1$$

9. 1 mile = 5280 ft

$$\frac{1 \text{ mile}}{5280 \text{ ft}} = 1; \frac{5280 \text{ ft}}{1 \text{ mile}} = 1$$

11. 1 N.M. = 1.15155 S.M.

$$\frac{1 \text{ N.M.}}{1.15155 \text{ S.M.}} = 1; \frac{1.15155 \text{ S.M.}}{1 \text{ N.M.}} = 1$$

13. 1 cu ft = 28.3170 ℓ

$$\frac{1 \text{ cu ft}}{28.3170 \ell} = 1; \frac{28.3170 \ell}{1 \text{ cu ft}} = 1$$

15. 1 hkm = 100 m

$$\frac{1 \text{ hkm}}{100 \text{ m}} = 1; \frac{100 \text{ m}}{1 \text{ hkm}} = 1$$

17. 1 gal = 3.78543 ℓ

$$\frac{1 \text{ gal}}{3.78543 \ell} = 1; \frac{3.78543 \ell}{1 \text{ gal}} = 1$$

19. 1 bushel (bu) = 1.24446 cu ft

$$\frac{1 \text{ bu}}{1.24446 \text{ cu ft}} = 1; \frac{1.24446 \text{ cu ft}}{1 \text{ bu}} = 1$$

21. 1 g = 15.4324 gr

$$\frac{1 \text{ g}}{15.4324 \text{ gr}} = 1; \frac{15.4324 \text{ gr}}{1 \text{ g}} = 1$$

23. 1 S.M. = 80 ch

$$\frac{1 \text{ S.M.}}{80 \text{ ch}} = 1; \frac{80 \text{ ch}}{1 \text{ S.M.}} = 1$$

25. 1 km = 49.7096 ch

$$\frac{1 \text{ km}}{49.7096 \text{ ch}} = 1; \frac{49.7096 \text{ ch}}{1 \text{ km}} = 1$$

27. 1 cu yd = 201.974 gal

$$\frac{1 \text{ cu yd}}{201.974 \text{ gal}} = 1; \frac{201.974 \text{ gal}}{1 \text{ cu yd}} = 1$$

29. 1 ch = 4 rods

$$\frac{1 \text{ ch}}{4 \text{ rods}} = 1; \frac{4 \text{ rods}}{1 \text{ ch}} = 1$$

31. (a) 1000 units
(b) 0.000001 unit
(c) 10 units
(d) 0.000000001 unit

33. (a) A, α
(b) Π, π
(c) Φ, ϕ
(d) B, β

35. (a) 62.4 lb/cu ft
(b) 3950 miles
(c) 980 cm/sec^2 = 32.16 ft/sec^2
(d) 186,272 miles/sec $\approx 3 \times 10^{10}$ cm/sec

37. (a) 0.0809 lb
(b) 554 lb
(c) 55–57 lb
(d) 1203 lb
(e) 90–147 lb
(f) 65–88 lb
(g) 15–50 lb

39. (a) 0.00000900
(b) $-38°$F
(c) 14.43
(d) 19,000–48,000 lb/in.^{2}
(e) 1204.3 lb

41. (a) 0.078125
(b) 0.5625
(c) 0.875
(d) 0.453125
(e) 0.15625

43. (a) $0.166\frac{2}{3}$
(b) $62\frac{1}{2}\%$
(c) $\frac{13}{16}$
(d) 0.6875

Exercise 1.7A

1. $6\frac{2}{3}$ ft
3. 640 rods
5. 24 oz

7. 160 lb
9. 54 cu ft
11. 270,072 sq ft

13. 2880 min
15. 1540.99 gal
17. 260,000 cm

19. 814,632 gal

Exercise 1.7B

1. 65 abamperes
3. 9.869 atm
5. 228.346 in.
7. 1.267 sq cm
9. 935.1 gal

11. 1515 gal
13. 35.30 in. of mercury
15. 1.003 cu ft/sec
17. 12.35 lb

19. 459.36 in.
21. 198 cu ft
23. 39.46 circular mils
25. 2844 B.t.u./min

Exercise 1.7C

1. 0.48 Ω
3. 134 gal
5. 19,000

7. 49,000
9. 38,000 miles
11. 1.0 in.

13. 11.7 g
15. 632 lb
17. $3400

19. 2.3 mm
21. 19.3 kg
23. 207 tons

Exercise 1.8

1. 12.0
3. 17.0
5. 3.300
7. 12.01000

9. 5.12000
11. 4.4140
13. 51.00
15. 109.956

17. 22.2222
19. 98.00
21. 1734.94 sq units
23. $d = 38$ in.
$C = 119.381$ in.

25. 47.5
27. 0.0121951
29. $\frac{1}{29}$

Exercise 1.9

1. 63,000
3. 41.3

5. 7.8
7. 3.9

9. 5.5
11. 111,000

13. 33.7
15. 3100

17. 0.0803
19. 0.000350

CHAPTER 2

Exercise 2.1

1. Infinite length, i.e., length without limit
3. By naming any two distinct points on the line
5. By placing a letter or number near the point
7. Position
9. Any desired number
11. Between two specific points on the line
13. Perpendicular to each other
17. See Figure 2-7 with $AB = 5$ in. (Disregard $\overset{\frown}{AEB}$.)

Exercise 2.3

3. $P = 17\frac{1}{4}$ in.

Exercise 2.4

1. Four right angles (4 Rt ∢ 's)
3. Supplementary angles
5. Rt ∢
7. They are equal
9. Two angles having a common vertex and a common side
11. See Figure E-1.
15. Yes

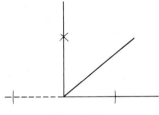

Figure E-1

Exercise 2.5A

1. A triangle
3. An equilateral triangle
5. A right triangle (Rt △)
7. A median
9. The base
11. One right angle and two complementary acute angles
13. Three unequal acute angles

15. $\dfrac{RO}{OC} = \dfrac{1}{2}, \qquad \dfrac{OC}{RO} = 2$

$\dfrac{RO}{RC} = \dfrac{1}{3}, \qquad \dfrac{RC}{RO} = 3$

$\dfrac{OC}{RC} = \dfrac{2}{3}, \qquad \dfrac{RC}{OC} = \dfrac{3}{2}$

17. Rt △
19. They are perpendicular
21. See Figure E-2

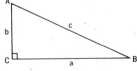

Figure E-2

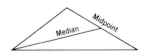

Figure E-3

23. $\sphericalangle A$ is opposite side BC (side a). Side AB (side c) is called the hypotenuse and is located opposite Rt $\sphericalangle C$

25. The line segment joining a vertex of a triangle and the midpoint of the side opposite the angle (Figure E-3)

Exercise 2.5B

1. A regular hexagon
3. The six triangles are all equilateral and congruent
9. See Figure 2-35 with $BC = \frac{1}{2}AB$
11. See Figure E-4. Draw along both edges of the ruler, turn it, and repeat
13. A square is a rhombus.

15. $W = \pi(10 \text{ in.})^2(4 \text{ in.})\left(\dfrac{555 \text{ lb}}{\text{cu ft}}\right)\left(\dfrac{1 \text{ cu ft}}{1728 \text{ in.}^3}\right)(0.827) = 334 \text{ lb}$

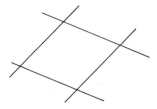

Figure E-4

Exercise 2.6

1. A chord
3. The radius
5. Concentric
7. See Figure E-5

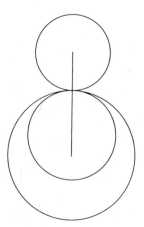

Figure E-5

9. Hexagon

11. See Figure E-6

13. See Figure E-7

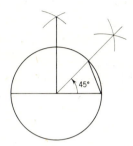

Figure E-6

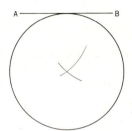

Figure E-7

15. $\sphericalangle AOB = \widehat{AB} = \dfrac{360°}{12} = 30°$

$\left.\begin{array}{l} \sphericalangle ACB = \frac{1}{2}\widehat{AB} = 15° \\[4pt] \sphericalangle ADB = \frac{1}{2}\widehat{AB} = 15° \end{array}\right\}$ [66]

17. Construct a circle with a diameter equal to the hypotenuse (Figure E-8). Construct an inscribed angle in one of the semicircles. [67]

19. See Figure E-9

21. $\dfrac{180°}{\pi} \approx 57.3°$

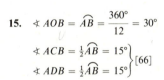

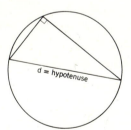

Figure E-8

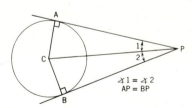

$\sphericalangle 1 = \sphericalangle 2$
$AP = BP$

Figure E-9

Exercise 2.7

1. 26 in.

3. 12 ft

5. 11.0 cm

Exercise 2.8

1. They are equal: $\sphericalangle 8 = \sphericalangle 7 = \sphericalangle 6 = \sphericalangle 5$.
3. $\sphericalangle 1 = \sphericalangle 2, \sphericalangle 5 = \sphericalangle 6, \sphericalangle 3 = \sphericalangle 4, \sphericalangle 7 = \sphericalangle 8$
5. They are parallel
7. $\sphericalangle 2$ and $\sphericalangle 3$, $\sphericalangle 6$ and $\sphericalangle 7$
9. $\sphericalangle 3, \sphericalangle 4, \sphericalangle 2$, and $\sphericalangle 1$
11. They are not equal
13. $\sphericalangle 2$
15. They would also be equal. Supplements of equal angles are equal
17. $\sphericalangle 7, \sphericalangle 6$, and $\sphericalangle 5$ (If you knew that $\sphericalangle 8 = 90°$, then all eight angles would be equal)
19. AB is not parallel to CD

21. $\dfrac{S_1}{T_1} = \dfrac{S_2}{T_2}$

23. $\dfrac{S_2}{S_1 + S_2} = \dfrac{d_2}{d_1 + d_2}$ or $\dfrac{T_2}{T_1 + T_2}$

25. $d_2 = 1\tilde{0}\,\text{ft}, T_1 = 1\tilde{0}\,\text{ft}$

Exercise 2.9

1. Congruent
3. Similar triangles are congruent only when corresponding sides are also equal
5. They are similar
7. They are proportional

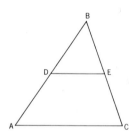

Figure E-10

11. See Figure E-10

    (a) $BD = DA$, $BE = EC$. DE cuts BC and BA into proportional segments and is therefore parallel to AC. Then $\sphericalangle A = \sphericalangle D, \sphericalangle C = \sphericalangle E$, and $\sphericalangle B = \sphericalangle B$. Therefore, the triangles are similar

    (b) $BD/BA = \frac{1}{2} = DE/AC$ (Corresponding sides of similar triangles are proportional)

13. 12 ft

Exercise 2.10

1. $\angle A = 90° - 27° \, 46' \, 31'' = 62° \, 13' \, 29''$
   $\angle B = 27° \, 46' \, 31''$
   $\angle C = 90° + \angle A = 152° \, 13' \, 29''$
   $\angle D = 90° + 27° \, 46' \, 31'' = 117° \, 46' \, 31''$
   $\angle E = 180° + \angle A = 242° \, 13' \, 29''$
3. $35°$
5. $\angle A = 142°, \angle B = 60°, \angle C = 48°$
7. $\angle B = 70.5°, \angle A = 109.5°$
9. Yes ($\angle C$ is a right angle), $d = 2.982$ mm
11. $F = 263$ lb
13. 142 lb
15. 1628 lb ≈ 1600 lb

17. 210 mm
19. $\angle 1 = 32°$
21. $x = 1.15$ in., $y = 1.85$ in., $z = 0.769$ in., $w = 0.481$ in.
23. $\angle 1 = \angle 6$ [53]
    $\angle 2 = \angle 4$ [11]
    side $FD \triangle 1 = $ side $FD \triangle 2$
    Congruent by [48]
25. $140°$
27. $d_2 = 0.216$ in.
29. 6 in.
31. $\angle 1 = 30°$, diameter $= 16$ units. $\overgroup{MN} = 30°$

CHAPTER 3

Exercise 3.0

1. (a) y yr
   (b) $2y$ yr
   (c) $\left(\dfrac{y}{2} \text{ or } \dfrac{1}{2}y\right)$ yr
   (d) $(y + 1)$ yr
   (e) $(y - 1)$ yr
3. (a) My age plus my sister's age
   (b) Twice my age
   (c) Three times my sister's age
   (d) Twice my sister's age plus three times my age
   (e) One-half my sister's age
   (f) One-third my age
   (g) My sister's age next year
   (h) My age 2 years ago
5. (a) Twice the length
   (b) Three times the width
   (c) Twice the length plus twice the width equals the perimeter
   (d) One-half the perimeter equals the length plus the width
   (e) The width equals one-half the length
   (f) The length equals twice the width
   (g) The length plus twice the width equals 64 in.
   (h) The width plus three times the thickness equals 32 in.
   (i) The volume equals the length times the width times the thickness
   (j) The length of the diagonal of the base equals the square root of the sum of the length squared plus the width squared
   (k) The square of the diagonal equals the square of the length plus the square of the width
7. (a) Twice the inside diameter plus the length
   (b) One-half the length plus the outside diameter
   (c) The difference in the diameters, or the outside diameter minus the inside diameter
   (d) The length minus the outside diameter
   (e) The inside diameter times the inside diameter, or the square of the inside diameter

(f) The inside radius equals one-half the inside diameter

(g) The circumference equals π times the outside diameter

(h) The wall thickness equals one-half the difference between the outside diameter and the inside diameter

(i) The outside diameter equals the inside diameter plus twice the wall thickness

Exercise 3.1 A

1. 12
3. -1593
5. $+3$
7. (a) $38°$
   (b) $45°$
   (c) $43°$
   (d) $53°$

(e) $37°$ (at 11:00 p.m.)
9. $5x$
11. $6R$
13. $8RT$
15. $\frac{7}{6}X$
17. $13W/10$
19. $-(1/30)Y$ or $-Y/30$

21. $-6ax + 6by$
23. $54ab + 90b$
25. $79.1Z + 35.6R$
27. Position
29. Change
31. Change
33. Position

Exercise 3.1B

1. $+105$
3. -20
5. -16
7. -4
9. $+2$
11. -3
13. -60
15. -54
17. -4
19. $+2$
21. $+\frac{1}{3}$
23. $-\frac{16}{35}$
25. -1

Exercise 3.1C

1. $+67$
3. $+24$
5. -18
7. $+518$
9. $+31s$
11. $+25d$
13. $-x$
15. $+116w$

Exercise 3.2 A

1. $1x - 4$
3. $1x + 1y$
5. $1(1x - 1y)$
7. $1\{1h + 1w\}$
9. $1(1x + 1y + 1z) - 1(1x + 1y - 1z)$

Exercise 3.2B

1. $2(1x - 3y) = 2x - 6y$
3. $1(1x - 4y) = 1x - 4y$
5. $-1(1x - 4) + 1(2x - 5) = 1x - 1$
7. $1[8x - 1y] - 1[3x - 1y] = 5x$
9. $-1(2x - 2) - 1(-1x + 2) - 1(3x - 2) = -4x + 2$
11. $2\{3x - 5\} - 1(1x - 8) = 5x - 2$
13. $2\{3x - 1[4x + 5(1x - 2) + 6] - 2\} = -12x + 4$

Exercise 3.3A

	Algebraic Expression	Numerical Coefficient	Literal Part	Base	Exponent
1.	$3y^5$	-3	y^5	y	5
3.	$-R^3$	-1	R^3	R	3
5.	$+x$	$+1$	x	x	1

7. $12x^3$
9. $24x^4y^3$
11. $32x^4y^3z^8$
13. $10y^3 + 15y^2 - 15y$
15. $-16x + 20z$
17. $88z^5 - 44z^4 + 66z^3 - 99z$
19. $-108n^4 + 96n^3 - 84n^2 - 72n$
21. $27m^3 - 45m^2$
23. $24xy^3 - 12xy^4 + 9x^2y^7 + 12x^3y^3$

Exercise 3.3B

1. $NRWZ^2$
3. $-3x^3 + 10x^2 + 3x + 2$
5. $-4x^4 + 3x - 8y - y^2 + y^3$
7. $-10a + 10b + 7c - 6$
9. $15x^2 - 10xy$
11. $72x^2 + 114xy - 10y^2$
13. $W^2 - 25$
15. $-2J^2 + 19Jk - 6Jn + 15kn - 35k^2$
17. $R^3 + R^2Z - 2RZ^3 - 2Z^4$
19. $z^4 - z^3 - 11z^2 + 5z + 30$

Exercise 3.4A

1. $9x^2$
3. $9d$
5. $27s^2t$
7. $12b^2f^4$
9. $16V^8$
11. $40K^{23}$
13. $11p^{25}$
15. $3s^2$
17. $6xyz$
19. 1

Exercise 3.4B

1. $6a + 7$
3. $12x^3 - 4y^2 + 6z^3$
5. $4r^2z - 3rz^2$
7. $(3x - 2)$
9. $k - 1$
11. $n - w$
13. $a + b$
15. $x^2 + xy + y^2$
17. $3a + 9$

Exercise 3.5

1. 225
3. 0
5. 121
7. -20
9. 16
11. $x^2 - 4$, 21
13. $2z^2 + 5z - 12$, -9
15. 15

Exercise 3.6

1. 1120 volts
3. 400 ft
5. (a) 376 cm/sec
   (b) 187.8 cm/sec rounds to 188 cm/sec
   (c) 112.7 cm/sec rounds to 113 cm/sec
   (d) Yes, decreasing depth results in lower pressure at the orifice.
7. 27,200 rounds to 27,000 lb/in.^{2}
9. 2.21 sec
11. 24.9 cm
13. 420 Ω
15. 6.04 km
17. 1.00×10^3 kg
19. 143.06 rounds to 140 horsepower
21. 95,000 cu ft
23. Doubling width, safe load is doubled. Doubling depth, safe load is increased 4 times.
25. (a) 1250 ft/sec, (b) 1010 ft/sec

CHAPTER 4

Exercise 4.1A

1. $x = 9$
3. $y = 8$
5. $n = -3$
7. $R = 7$
9. $P = 12$
11. $Z = 2.725$
13. $K = \frac{1}{12}$
15. $R = 16$
17. $P = 11.27$
19. $D = 18.90$
21. $x = 15$
23. $y = 7$
25. $u = 240$
27. $x = 12$
29. $y = 28$
31. $K = 3$
33. $x = 6$
35. $x = 20$
37. $x = 90$
39. $y = -8$
41. $x = 4$
43. $x = 20$
45. $R = 20$
47. $N = 15$
49. $K = 16$
51. $y = 10$
53. $y = 2$
55. $x = 7$
57. $y = 10$
59. $W = 3$

Exercise 4.1B

1. $y = \dfrac{15}{k}$ 9. $S = \dfrac{2T}{5}$ 17. $r = \dfrac{18 + P}{k}$ 23. $h = \dfrac{2A}{(b_1 + b_2)}$

3. $r = -\dfrac{4}{b}$ 11. $W = ar - 8$ 19. $E = IR$ 25. $P = \dfrac{2tS}{d}$

5. $x = \dfrac{3}{d}$ 13. $T = \dfrac{t + 3}{4}$ 21. $l = \dfrac{A}{w}$ 27. $R_3 = \dfrac{R_2 R_x}{R_1}$

7. $z = 4$ 15. $W = 6 - 2L$

Exercise 4.2A

Quantities Compared	Fraction Form	Colon Form	Decimal to 1 Form	Decimal Colon Form
1. 3 pt with 6 gal	$\frac{3}{48} = \frac{1}{16}$	1:16	0.0625 to 1	0.0625:1
3. 1 sq in. with 1 sq ft	$\frac{1}{144}$	1:144	0.00694 to 1	0.00694:1
5. 6 adults with 3 children	$\frac{6}{3} = \frac{2}{1}$	2:1	2 to 1	2:1
7. Perimeter of a 5-in. square with perimeter of an 18-in. square	$\frac{20}{72} = \frac{5}{18}$	5:18	0.278 to 1	0.278:1

9. $\dfrac{5}{12} = \dfrac{x}{120}$, $5:12 = x:120$, $x = 50$

11. $\dfrac{7}{1} = \dfrac{y}{3}$, $7:1 = y:3$, $y = 21$

13. $\dfrac{9}{K} = \dfrac{3}{1}$, $9:K = 3:1$, $K = 3$

15. 3 is to 5 as x is to 9

17. 11 is to J as 3 is to 42

19. $\dfrac{4}{X} = \dfrac{X}{16}$, $X = 8.00$

21. $\dfrac{1}{X} = \dfrac{X}{4}$, $X = 2.00$

23. $\dfrac{y}{X} = \dfrac{X}{4y}$, $X = 2y$

Exercise 4.2B

1. $\dfrac{r}{s} = \dfrac{3}{4}, \dfrac{r}{t} = \dfrac{3}{5}, \dfrac{s}{t} = \dfrac{4}{5}, \dfrac{r}{3} = \dfrac{s}{4} = \dfrac{t}{5}$

3. $\dfrac{5}{2} = \dfrac{K}{L}, \dfrac{5}{1} = \dfrac{K}{M}, \dfrac{2}{1} = \dfrac{L}{M}, \dfrac{5}{K} = \dfrac{2}{L} = \dfrac{1}{M}$

5. $a:d:g:j = b:e:h:k$

7. $a:b:c = d:e:f$

9. $d = 26\frac{10}{31}, e = 23\frac{7}{31}, g = 27\frac{1}{5}, i = 24\frac{4}{5}, k = 28\frac{4}{17}, l = 29\frac{3}{17}$

Exercise 4.3A

1. $R = kw$ 11. Direct, $P = kQ$ 21. Inverse, $S = k/t$

3. $P = kQ$ 13. Inverse, $P = k/T$ 23. Direct, $F = kI$

5. $Q = kN$ 15. Inverse, $P = k/V$ 25. Direct, $V = kd^2$

7. Direct, $Z = kW$ 17. Direct, $F = kA$ 27. Direct, $V = kd^2$

9. Direct, $R = kS$ 19. Inverse, $M = k/s$ 29. Direct, $A = kh^2$

Exercise 4.3B

1. Direct because $F/P = k$ (a constant)
3. Direct because $C/d = \pi$ (a constant)
5. Direct because $a/b = k$ (a constant)
7. Inverse because $Fd = W$ (a constant)

9. Direct because $W/F = d$ (a constant)
11. Direct because $P/D = d$ (a constant)
13. Inverse because $PA = F$ (a constant)
15. Inverse because $PV = k$ (a constant)

Exercise 4.3C

1. (a) P = perimeter, d = diagonal
   (b) $P = kd$
   (c) $k = \dfrac{P}{d} = \dfrac{420 \text{ mm}}{140 \text{ mm}} = 3$
   (d) $P = 3d$, (P and d measured in same units)

3. (a) A = average, S = sum
   (b) $A = kS$
   (c) $k = \dfrac{A}{S} = \dfrac{123 \text{ mm}}{369 \text{ mm}} = \dfrac{1}{3}$
   (d) $A = \dfrac{1}{3}S$, (S = sum of three numbers)

5. (a) W = weight of $8\frac{1}{2}$-in. × 11-in. paper, t = thickness of paper in millimeters
   (b) $W = kt$
   (c) $k = \dfrac{W}{t} = \dfrac{2.8 \text{ g}}{0.14 \text{ mm}} = 20 \text{ g/mm}$
   (d) $W = 20t$ g/mm, (t = thickness in millimeters, $8\frac{1}{2}$-in. × 11-in. size of paper remains constant)

7. (a) V = viscosity index, T = number of degrees absolute temperature
   (b) $V = \dfrac{k}{T}$
   (c) $k = VT = (20)(400) = 8000$
   (d) $V = \dfrac{8000}{T}$, (grade of oil is constant)

9. (a) R = resistance in ohms, L = length of wire in feet
   (b) $R = kL$
   (c) $k = \dfrac{R}{L} = \dfrac{99 \text{ ohms}}{550 \text{ ft}} = 0.18 \text{ ohm/ft}$
   (d) $R = 0.18 L$ ohm/ft, (L = length in feet, size of wire is constant)

11. (a) H = units of heat generated in B.t.u., t = time in seconds, I = current flow = constant
    (b) $H = kt$
    (c) $k = \dfrac{H}{t} = \dfrac{400 \text{ B.t.u.}}{250 \text{ sec}} = 1.6 \text{ B.t.u./sec}$
    (d) $H = (1.6)(t)$B.t.u./sec, (t = time in seconds, current remains constant)

Exercise 4.3D

1. R varies directly with s and inversely with t
3. A varies jointly with L and W.
5. S varies jointly with P and T and inversely as W^2

7. H varies jointly with R^2 and S and inversely as T

9. $R = kWT$

11. $C = \dfrac{kt}{P^2}$

13. $W = \dfrac{kST^2}{r^3}$

15. (a) $y = kx$, (b) $k = 30$, (c) $y = 30x$, (d) $y = 1200$

17. (a) $Z = \dfrac{kR}{Y}$, (b) $k = 189$, (c) $Z = \dfrac{189R}{Y}$, (d) $Z = 756$

19. $Z = kd^2$, (b) $k = 10$, (c) $Z = 10d^2$, (d) $Z = 302.5$

21. (a) $K = \dfrac{kst}{\sqrt{d}}$

(b) $k = \dfrac{K\sqrt{d}}{st} = \dfrac{(71.60)\sqrt{64.00}}{(4.000)(0.8950)} = 160.0$

(c) $K = \dfrac{160.0st}{\sqrt{d}}$

(d) $K = \dfrac{(160.0)(5.10)(1.02)}{\sqrt{2.25}} = 555$

23. (a) F = force, s = stretch of spring

(b) $F = ks$

(c) $k = \dfrac{F}{s} = \dfrac{80\,\text{lb}}{16\,\text{in.}} = 5\,\text{lb/in.}$

(d) $F = 5s$ lb/in. (With s measured in inches, the inch units cancel leaving force F in pounds—a proper force unit)

(e) $F = 5s$ (F is the force in pounds when s is the number of inches of spring stretch.)

(f) $F = 5(24\,\text{in.})\text{lb/in.} = 120\,\text{lb}$

25. (a) C = capacity of bin in tons, l = length of bin in feet, w = width of bin in feet, d = depth of coal in bin in feet

(b) $C = klwd$

(c) $k = \dfrac{C}{lwd} = \dfrac{320\,\text{tons}}{(40.0\,\text{ft})(20.0\,\text{ft})(12.0\,\text{ft})} = \dfrac{1}{30}\,\text{ton/ft}^3$

(d) $C = \dfrac{lwd}{30}\,\text{ton/ft}^3$ (with l, w, and d expressed in feet, the ft^3 units cancel leaving C in tons)

(e) $C = \dfrac{lwd}{30}$ (C = capacity in tons when l, w, and d are the numbers of feet of length, width, and depth, respectively.)

(f) $C = \dfrac{(82\,\text{ft})(46\,\text{ft})(18\,\text{ft})\,\text{tons}}{30\,\text{ft}^3} = 2263\,\text{tons} \approx 2300\,\text{tons}$

27. (a) N = number of trees/acre, d = distance between trees in feet

(b) $N = \dfrac{k}{d^2}$

(c) $k = Nd^2 = (43{,}560 \text{ trees})(1 \text{ ft})^2 = (43{,}560 \text{ trees})(\text{ft}^2)$

(d) $N = \dfrac{(43{,}560 \text{ trees})(\text{ft}^2)}{d^2}$ (With d expressed in feet, the ft^2 units cancel leaving N as the number of trees/acre.)

(e) $N = \dfrac{43{,}560}{d^2}$, ($N$ = number of trees/acre, d = number of feet between trees.)

(f) $N = \dfrac{(43{,}560 \text{ trees})(\text{ft}^2)}{5^2 \text{ ft}^2} = 1742 \text{ trees}$

Exercise 4.4A

1. A straight line
3. Direct
5. False
7. True
9. (a) Linear
   (g) Linear
   (h) Linear
11. $r = 2s + 6, s = \frac{1}{2}r - 3$
13. Solving $2x - 4y = 8$ for y, we obtain the second equation.
15. H—dependent on C. The amount of heat is physically dependent on the carbon content.
17. No. Their positions vary inversely with each other:

$$\text{Position}_1 = \frac{k}{\text{Position}_2}$$

Exercise 4.4B

1. Two straight lines intersecting at the point $x = 4, y = 2$
3. Independent, $x = 5, y = 3$
5. Inconsistent—no solution
7. Independent, $x = \frac{1}{4}, y = 2$
9. Independent, $x = 2.5, y = -1$
11. Independent, $x = 5, y = 4$
13. Independent, $W = 3, R = 4$
15. Independent, $R = -2, Z = 5$
17. Yes, because they all have the common solution $x = 8, y = 10$
19. 0.020 in., 0.050 in.

CHAPTER 5

Exercise 5.1A

1. $6a + 3ab$
3. $5a - 5$
5. $6x - 24x^2$
7. $12x^2z + 18xz^2 - 6xz$
9. $2NR - 2NK + 2N$
11–20. Answers are Problems 1–10

21. $6x(2x - y)$
23. $5a(b - c)$
25. $4y(2x - 4z + 1)$
27. $3R^2(6R^2 + 3N + R)$
29. $6Z^2Y(4Z^3Y^3 - 3Y + Z)$

Exercise 5.1B

1. $W = \dfrac{12}{2N - 3} = \dfrac{12}{7}$

3. $R = \dfrac{6 + 4W}{2W + 1} = \dfrac{14}{5}$

5. $L = \dfrac{-5}{K - 2} = -5$

7. $T = \dfrac{3 + S}{S^2 - S - 1} = 2$

9. $R = \dfrac{3}{J - 1} = -1$

11. $r^2 - s^2$
13. $x^2 - y^2$
15. $R^2 - K^2$
17. $25 - L^2$
19. $H^2 - 1$
21. $25L^2 - 4P^2$
23. $a^2 + 2ab + b^2$
25. $x^2 + 6x + 9$
27. $4x^2 + 20x + 25$
29. $T^2 - 2TW + W^2$
31. $S^2 - 2SN + N^2$
33. $y^2 - y - 2$
35. $R^2 - 3R - 4$
37. $Z^2 - 11Z + 24$
39. $W^2 + 10W + 9$
41. $3X^2 + 13X + 4$
43. $10x^2 + 14x - 12$
45. $24x^2 + 17x - 20$
47. $6R^2 + R - 40$
49. $33L^2 + 67L + 20$
53. $(m + w)r + (m + w)(-s) =$
    $mr + wr - ms - ws$
55. $(V + K)R + (V + K)(-T) =$
    $VR + KR - VT - KT$

57. $(3R - 4P)5N + (3R - 4P)2W =$
    $15RN - 20PN + 6RW - 8PW$
59. $(12N + 3K)Z + (12N + 3K)(-P) =$
    $12NZ + 3KZ - 12NP - 3KP$
61. $(x - y)(x + y)$
63. $(R^2 + 1)(R - 1)(R + 1)$
65. $(11 - 9W)(11 + 9W)$
67. $(Z - 1)(Z + 1)(Z^2 + 1)(Z^4 + 1)$
69. $(5x - 7y)(5x + 7y)$
71. $(x + y)^2$
73. $(r + s)^2$
75. $(3x + 2)^2$
77. $(b + 4)^2$
79. $(7x - 4)^2$
81. $2(3x - 4y)^2$
83. $(x + 3)(x + 4)$
85. $(N - 3)(N + 2)$
87. $(y + 7)(y + 5)$
89. $(m - 3)(m - 2)$
91. $2(x - 6)(x + 5)$
93. $(b + c)(x + y)$
95. $(2N + Z)(W - 3R)$
97. $(4H + 5R)(3Z - 2W)$
99. $(5L - 6T)(2W - 3Z + 2)$
101. $(K - 2R)(J - 5W)$
103. $7(R - 2W)$
105. $(3x + 2)(mb - pq)$
107. $(8S - 7T)(R - W)$
109. $(a + 1)(b + 1)$
111. $(x - 1)[(x - 1)^2 - 9(x - 1) + 8] =$
     $(x - 1)[\{(x - 1) - 8\}\{(x - 1) - 1\}] =$
     $(x - 1)(x - 9)(x - 2)$
113. $(x - \frac{1}{2})(x + \frac{1}{2})$
115. $(25Z^2 - T)(25Z^2 + T)$
117. $ax(3x - 2)(5x + 3)$

Exercise 5.2A

1. $x = 1, x = -1$
3. $s = 5, s = -2$
5. $P = 7, P = -4$
7. $x = -2, x = 1$
9. $x = 1, x = 1$ (called a double root)
11. $r = -\frac{3}{2}, r = 4$
13. $q = 5, q = 3$
15. $T = 2, T = -1$

Exercise 5.2B

1. $x^2 - 1$
3. $t^2 - 4t + 4$
5. $s^2 - 25$
7. $x = -2$ (double root)

9. $x_1 = \dfrac{3 + \sqrt{69}}{6}, x_2 = \dfrac{3 - \sqrt{69}}{6}$

11. $x_1 = \dfrac{-4 + \sqrt{-44}}{6}$ and $x_2 = \dfrac{-4 - \sqrt{-44}}{6}$ are imaginary roots.

13. $x_1 = \frac{3}{2}, x_2 = -\frac{1}{2}$ **15.** $t = 4$ (double root)

Exercise 5.3A

1. $-8a$

3. $5xy^2$

5. $a - x$

7. $-3x + 4y$

9. $\dfrac{-3}{x}, \dfrac{3}{-x}, \dfrac{-3}{-x}$

11. $\dfrac{-(x + 4)}{5}, -\dfrac{-(x + 4)}{-5}, -\dfrac{x + 4}{5}$

13. $-, +, -$

15. $-\frac{1}{2}$

17. $y - x$

19. $x - y$

21. $\dfrac{-1}{s + r}$

23. $3 - 2x$

25. $\dfrac{2}{x - y}$

27. $\dfrac{14y - 8k}{S + t}$

Exercise 5.3B

1. $\dfrac{6}{x - y}$

3. $\dfrac{-4a}{a + b}$

5. 3

7. $\dfrac{1}{x + y}$

9. $\dfrac{8}{x - y}$

11. $\dfrac{3a - 2b}{b}$

13. -2

Exercise 5.3C

1. $b + c$

3. -1

5. Prime

7. $\dfrac{x + 3}{x + 2}$

9. $\dfrac{a + 2b}{a + 3b}$

11. $\dfrac{8a - 5b}{36}$

13. $\dfrac{1 + x}{x}$

15. $\dfrac{a + b}{ab}$

17. $\dfrac{2x}{(x + 1)(x - 1)}$

19. $\dfrac{x + y + 3}{(x + y)(x - y)}$

21. $\dfrac{r_1 + r_2}{r_1 r_2}$

23. $\dfrac{6}{(2x - 3)(2x + 3)}$

25. $\dfrac{x + 2}{x(x + 1)(1 - x)}$

27. $\dfrac{yz + xz + xy}{xyz}$

Exercise 5.3D

1. $\dfrac{35}{x^3 + y^3}$

3. 7

5. 1

7. $\dfrac{4a - 4}{20a + 5}$

9. $\dfrac{a + b}{6(a + 2b)}$

11. $\frac{5}{2}$

Exercise 5.3E

1. $\frac{9}{4} = 2\frac{1}{4}$

3. $\dfrac{3x^2 + x}{15x - 3}$

5. $\dfrac{3x^2 - 1}{3 - 5x}$

7. $\dfrac{x^2 - 7x + 10}{x^2 + 4x + 3}$

Exercise 5.3F

1. $y = -2$
3. $s = \frac{3}{4}$
5. $V = 8$

7. $s = 12$
9. $y = 1$
11. $x = 2a$

13. $x = \dfrac{42c}{4c - 3}$
15. $s = 1$

Exercise 5.4A

1. $x = 6$
3. $x = \pm\frac{1}{2}$

5. $x = 5^3 = 125$
7. $x = 2z^2$

9. $x = -2^3 = -8$
11. $2^2 = 4$

13. $2^4 = 16$
15. ± 3

17. 9.27
19. ± 6.6

Exercise 5.4B

1. $x^{1/2}$
3. $3^{3/2}$

5. $x^{3/5}$
7. $x^{1/3}y^{2/3}$

9. $x^{4/2} = x^2$
11. $\sqrt{5}$

13. $\sqrt[3]{4^2y^2}$ or $\sqrt[3]{16y^2}$
15. $\sqrt[4]{28^3 z^3}$

Exercise 5.4C

1. $\sqrt{35}$
3. $\sqrt[5]{27}$
5. $\sqrt{36} = 6$
7. $\sqrt[3]{125} = 5$
9. $\sqrt{\dfrac{1}{6}} = \dfrac{\sqrt{6}}{6}$
11. $\dfrac{\sqrt{5}}{5}$

13. $\sqrt{4 \cdot 5} = 2\sqrt{5}$
15. 0.5
17. $\dfrac{\sqrt{6}}{3}$
19. $\dfrac{1}{3^3} = \dfrac{1}{27}$
21. 2
23. $\left(\frac{9}{4}\right)^{1/2} = \frac{3}{2}$

25. 8
27. $3^{3/3} = 3$
29. $\dfrac{1}{4y^2}$
31. $x^0 = 1$
33. $\dfrac{8}{x}$

Exercise 5.4D

1. $x = 148$
3. $x = 10$

5. $x = 5$
7. No solution

9. No solution
11. $x = 2$

13. $x = 2$

Exercise 5.5A

1. Five of 8 windows were not sold.
3. $2N = W$
5. The cylinder is three-fourths empty.
7. The lens admits 80% of the light.
9. The connecting rod weighs twice as much as a certain piston.
11. A salt solution is 80% water.
13. The tensile strength of aluminum wire is approximately one-half that of copper wire.
15. $D = \frac{1}{2}W$
17. $5R = 4T, \frac{5}{4}R = T$
19. The numerator of a fraction is 5 less than the denominator.
21. $q =$ number of quarters, $d =$ number of dimes, $q = d + 42, d = q - 42, q - d = 42$
23. $s =$ amount of silver in bar, $t =$ amount of tin in bar, $t = 3s, s = \frac{1}{3}t$
25. $r_1 =$ rate of flow for smaller pipe in gallons/minute, $r_2 =$ rate of flow for larger pipe in gallons/minute, $r_1 = \frac{4}{5}r_2, 5r_1 = 4r_2, r_2 = \frac{5}{4}r_1$, and $r_2 + \frac{4}{5}r_2 = 80$ gallons/minute.
27. See Figure E-11.

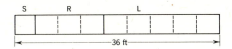

Figure E-11

Let S = length of shorter piece in feet, L = length of longer piece in feet, and R = length of remaining piece in feet. $L = 5S$ and $S = \frac{1}{3}R$. Also, $S + R + L = 48$ ft, $S = \frac{1}{5}L$, and $R = 3S$. From the figure, $9S = 36$ ft, $S = 4$ ft.

29. Let W = weight of fuel in pounds when tank is full, and T = weight of the tank in pounds: $\frac{1}{4}W = T, 4T = W, \frac{3}{4}W - 40$ lb $= T$

Exercise 5.5B

1. $A = (W)(W + 4) = 26\frac{1}{4}$ in.^{2}, $W = 3.5$ in. Length = 7.5 in.
3. $N(N + 11) = 242, N_1 = 11, N_2 = 22$
5. $N(N + 2) = 528, N_1 = 22, N_2 = 24$
7. $N^2 - 10 = 10 - N, N = 4$
9. $(d)(20 - 2d) = 50$ in.^{2}, $d = 5$ in., width = 10 in.
11. See Figure E-12.

$(S + 16)^2 = (S + 14)^2 + S^2, S = 10$ ft, $S + 14 = $ 24 ft, $S + 14 + 2 = 26$ ft

Figure E-12

13. 30 poles; X = number of poles/mile, $\dfrac{5280}{X}$ ft = distance between poles

$$\frac{5280 \text{ ft}}{X + 2} = \left(\frac{5280}{X} - 11\right) \text{ ft}, 11X^2 + 22X - 10560 = 0$$

$$X = 30$$

Exercise 5.5C

1. $\frac{40}{64}$
3. 50, 60
5. 13 in.
7. 20 in. by 30 in.
9. 4 in.
11. 12
13. 48 yr
15. 90
17. $2\tilde{0}$ in. by 36 in.
19. No. $\sqrt{2600} > \dfrac{200}{4}$
21. $2(12 + 12\sqrt{2})$ in.
23. 6.0 in.
25. 160 in.
27. (a) $4\frac{4}{9}$ hr, (b) $5\frac{5}{7}$ hr
29. 42 ft and 84 ft

Exercise 5.5D

1. $51\frac{3}{7}\% \approx 51.4\%$ acid
3. $1066\frac{2}{3}$ lb ≈ 1100 lb testing 0.010% and $213\frac{1}{3}$ lb ≈ 210 lb testing 0.040%
5. $72\tilde{0}$ ml
7. $6\frac{1}{4}\ell \approx 6.2\ell$ of 95% solution and $18\frac{3}{4}\ell \approx 19\ell$ of 15% solution
9. $6\tilde{0}00$ gal
11. 120 gal

CHAPTER 6

Exercise 6.1

1. (a) $P = 4s$
    (b) $P = 2a + 2h$
    (c) $P = 3s$
    (d) $P = 6s = 6R$
3. 280 rods, $4416 \approx 4400$ sq rods
5. 94.3 ft
7. 8$\hat{0}$0 sq in.
9. 1.50 sq in.

11. 6.93 sq in.
13. 35.4 cm
15. $A = 126$ sq in.
     $d = 9.14$ in.
17. 28.0 in.
19. $A\odot = 127\% A\square$
21. 10.3 sq in.
23. $r = 3.984$ in.

25. diameter of hole $= \sqrt{2}/2$
     (diameter of washer)
27. $h = 4.82$ in.
     $A = 117$ in.^{2}
29. 35,000 lb
31. 2,260,000 lb
33. 570 sq in.

Exercise 6.2

1. 1130 lb, $d = 44.1$ in.
3. 9410 cu ft, 761 lb
5. 166 in.
7. 28,000 cu yd
9. 42,000,000 lb
11. $V = 245$ cu ft; between 38,400 lb and 43,300 lb, depending on the weight of marble (157–177) lb/cu ft
13. 1872 tiles
15. 87,700,000 cu ft
17. 43.56 cu yd
19. 1350 bu
21. 1700 cu ft, 250,000 lb
23. 84,900 dice
25. With ice weighing 56 lb/cu ft, $W = 1.0 \times 10^7$ lb
27. 32$\tilde{0}$,000 cu yd
29. 44.2 cu ft, 331 gal
31. 4.90 ft
33. (a) 40,600 gal, (b) 3.9 gal
35. 415.1 cu yd becomes 420 cu yd (to two significant figures)
37. 904.8 cu in.
39. $d = 1\frac{19}{64}$ in.

Exercise 6.3

1. $A = 355$ sq ft (assuming given data are exact), $P = 83.4$ ft
3. $A = 96.75$ sq in., $P = 54.00$ in.
5. $V \approx 3.2$ cu in.
7. Approximately 22 acre-ft
9. $A = 27$ sq in.
11. 40 cu yd
13. $A \approx 540$ sq in., $A = 530$ sq in.
15. 4600 cu yd
17. 876 cu in.
19. #1. He could probably carry its weight (80 lb) without much difficulty. (#2 weighs 365 lb and #3 weighs 46 lb)
21. $h = 18$ in.
23. $V = 15,400$ cu ft, weight $= 751,000$ lb

CHAPTER 7

Exercise 7.1

7. See Figure E-13

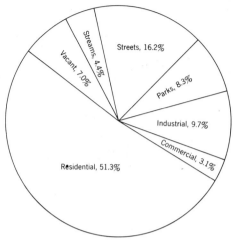

Land Use For Hannensville

Figure E-13

17. The graph for Problem 5. To magnify the differences requires a smaller scale ratio which allows for more accurate reading.
19. No. The rates of flow for the 8 days do not constitute 100% of the total flow for the 8 days.
21. (a) 29.76 in. of mercury
    (b) 29.58 in. of mercury
    (c) 5:00 p.m.

Exercise 7.2

3. The coordinates for each point on the line are equal; i.e., $y = x$
5. Each point has an x value $= 0$
7. A (8, 11), B (−8, 12), C (−11, −7.25), D (4, −7.5), E (−4, 7), F (13.8, 2.5), G (−8.6, −7.3), H (7.1, −3.7), I (−5.4, −3.5), J (3.8, 3.3), K (12.1, −4.8)
9. Each point on the line has an x value $= 3$
11. Each point on the line has an x value $= -8$
13. All points on the line have an abscissa value of 8
15. The coordinates for each point on the line are equal; i.e., $y = x$
17. See Figure E-14

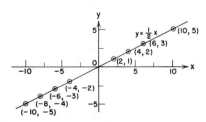

Figure E-14

19. Positive, I and IV; negative, II and III
21. Same in I and III; opposite in II and IV
23. Yes, the axis is a perpendicular bisector of the line segment joining the two points

Exercise 7.3

7. See Figure E-15

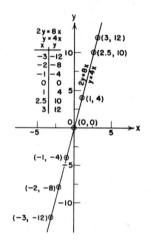

Figure E-15

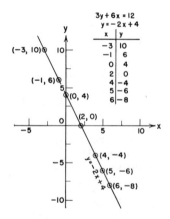

Figure E-16

11. See Figure E-16
13. The lines are parallel
15. The value of the constant term (*b*) in each equation (written in the form $y = mx + b$) equals the value of the ordinate at the point where the line crosses the *y* axis

19. See Figure E-17

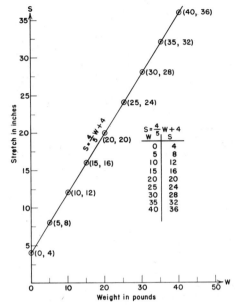

Figure E-17 Plot Showing Spring Stretch Against Weight Applied

Exercise 7.4

	Slope-Intercept Form	Slope	y-Intercept Point
1.	$y = 2x + 5$	$m = 2$	$(0, 5)$
3.	$y = 3x + 9$	$m = 3$	$(0, 9)$
5.	$y = \frac{4}{5}x - 4$	$m = \frac{4}{5}$	$(0, -4)$
7.	$y = 5x - 8$	$m = 5$	$(0, -8)$
9.	$y = -x + 1$	$m = -1$	$(0, 1)$

11. $y = 3x - 3$
13. $y = \frac{3}{4}x + 2$
15. $y = -\frac{7}{8}x - 1$
17. $y = 2x$ (constant term $= 0$)
19. $y = \frac{3}{4}x + 2$
21. $m = 0$
23. $m = 4$, intercept point is $(0, -5)$
25. See Figure E-18
27. $m = 3$, intercept point is $(0, -4)$
29. See Figure E-19
31. $m = 3$
33. $m = -\frac{5}{8}$
35. $m = \frac{1}{2}$
37. $m = -\frac{1}{3}$, $y = -\frac{1}{3}x + 2$
39. $m = \frac{1}{5}$, $y = \frac{1}{5}x + 1$

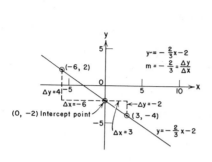

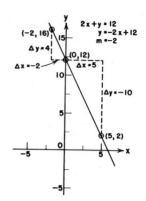

Figure E-18   Figure E-19

Exercise 7.5

1. $x = 3, y = 2$
3. $x = -2.5, y = 2.5$
5. $x = 5, y = 3$
7. By graphing, $r \approx 3.3$, $L \approx 570$. By algebraic solution, $r = 3\frac{1}{4}$, $L = 575$. The difference is due to approximations from graph
9. $x + y = 200$, $y - x = 40$, $y = 120$, $x = 80$
11. 190 ft

Exercise 7.6 A

1. See Figure E-20

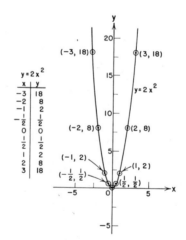

Figure E-20

3. x intercepts $(-2, 0)$ and $(3, 0)$, y intercept $(0, -6)$, vertex $(\frac{1}{2}, -6\frac{1}{4})$
5. x intercepts $(-5, 0)$ and $(5, 0)$, y intercept and vertex $(0, -25)$

7. *x* intercept, *y* intercept, and vertex are $(0, 0)$
9. *x* intercepts $\approx (\pm 2.9, 0)$, *y* intercept and vertex $(0, -16)$
11. *x* intercept, *y* intercept, and vertex are $(0, 0)$
13. See Figure E-21

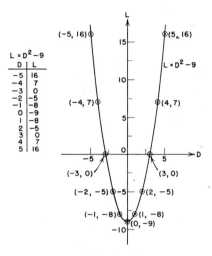

Figure E-21

Exercise 7.6B

1. See Figure E-22

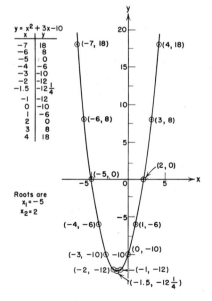

Figure E-22

3. Roots are -1.5 and 1

5. Roots are -1.5 and 4 (parabola opens down)

7. Roots are $\pm\sqrt{10} \approx \pm 3.2$

9. See Figure E-23

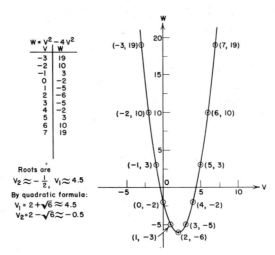

Figure E-23

Exercise 7.6C

1. Opens up; roots are $x_1 = -5$, $x_2 = 5$, $V(0, -25)$. See Figure E-24

3. Opens up; roots are $x_1 = -5$, $x_2 = 4$, $V(-\frac{1}{2}, -20\frac{1}{4})$. See Figure E-25

5. Opens down; roots are $x_1 = -\frac{2}{3}$, $x_2 = \frac{3}{2}$, $V(\frac{5}{12}, 7\frac{1}{24})$. See Figure E-26

7. Opens up, roots are $x_1 = -2$, $x_2 = 4$, $V(1, -9)$. See Figure E-27

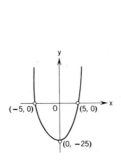

Figure E-24

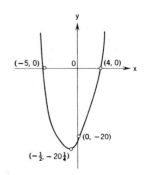

Figure E-25

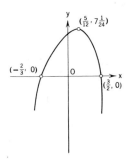

Figure E-26

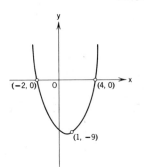

Figure E-27

9. $V(5, 416)$ maximum point
11. $V(5, 6)$ maximum point
13. $V(-2, -4)$ minimum point
15. Let x = depth in inches. Then width = $(20 - 2x)$. Area = $-2x^2 + 20x$. Maximum area with $x = 5$ in. is 50 sq in. See Figure E-28

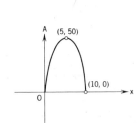

Figure E-28

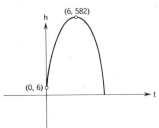

Figure E-29

17. $t_{max} = 6$ sec, $h_{max} = 582$ ft, time in air ≈ 12.03 sec. See Figure E-29
19. Let x = first number and $(60 - x)$ = second number. Then, $P = x(60 - x) = -x^2 + 60x$. $P_{max} = 900$ when both numbers equal 30.

Exercise 7.7

1. $s = 3t + 2$ 3. $w = 2r^2$ 5. $h = \dfrac{12}{v}$ 7. $P = \dfrac{6}{t^2}$ 9. $y = \dfrac{48}{\sqrt{x}}$

Exercise 7.8

1. See Figure E-30

3. See Figure E-31

5. See Figure E-32

7. See Figure E-33

9. See Figure E-34

11. See Figure E-35

13. $y > -x + 4$

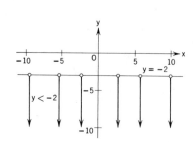

Figure E-30

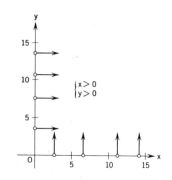

Figure E-31

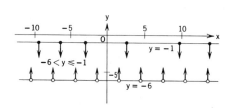

Figure E-32

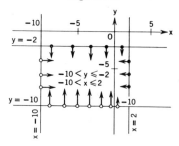

Figure E-33

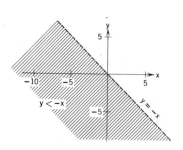

Figure E-34

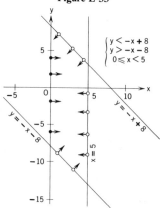

Figure E-35

CHAPTER 8

Exercise 8.1 A

1. $120°\ 54'\ 21''$
3. (a) $54°\ 47'\ 18''$
   (b) $70°\ 31'\ 19''$
   (c) $46°\ 54'\ 51''$

5. (a) $1\tilde{0}$
(b) 15
(c) 22.0
(d) 87.5
(e) $\frac{1}{8}$
(f) $\frac{1}{4}$
(g) $\frac{1}{2}$
(h) $\frac{3}{4}$

7. 37° and 53°
9. 5.0 ft
11. (a) 0.66π in. $\approx$ 2.07 in.
(b) 0.33π in. $\approx$ 1.04 in.
13. (a) 24
(b) 3.4 in.
15. 23 in.
17. 71°

Exercise 8.1B

1. (a) 360°
(b) 45°
(c) 270°
(d) 150°
(e) $\pi/4$
(f) $\pi/3$
(g) $\pi/2$
(h) π
3. 2.0π ft $\approx$ 6.3 ft
5. 16.00π miles $\approx$ 50.27 miles

7. $\frac{1}{3}$ radian
9. 5.000 radians
11. $V = 150$ in./sec
13. $\omega = 950/8$ radians/sec $\approx$ 120 radians/sec
15. 235 radians
17. $\frac{1}{4}$ radian
19. 4800 rpm
21. 8.0 radians/sec
23. 40 and 55
25. 20.0 ft

Exercise 8.2

1. $\sin R = \frac{3}{5}$ $\csc R = \frac{5}{3}$
$\cos R = \frac{4}{5}$ $\sec R = \frac{5}{4}$
$\tan R = \frac{3}{4}$ $\cot R = \frac{4}{3}$

3. $PH = \sqrt{85} \approx 9.22$ ft

$\sin H = \dfrac{7}{\sqrt{85}} \approx 0.759$

$\cos H = \dfrac{6}{\sqrt{85}} \approx 0.651$

$\tan H = \frac{7}{6} \approx 1.17$
$\cot H = \frac{6}{7} \approx 0.857$

$\sec H = \dfrac{\sqrt{85}}{6} \approx 1.54$

$\csc H = \dfrac{\sqrt{85}}{7} \approx 1.32$

5. $\cos 72° = 0.30902$
$\cot 49° = 0.86929$
$\csc 28° = 2.1301$

7. sine and cosine ratios $= \dfrac{\text{leg}}{\text{hypotenuse}}$ which is less than 1 because leg < hypotenuse
9. $\angle B \approx 35°$
11. $\angle R \approx 49°$

Exercise 8.3

1. See Figure E-36

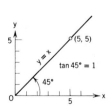

Figure E-36

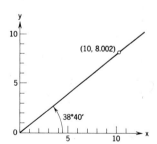

Figure E-37

3. See Figure E-37

6. Angle of Problem 1—sin 45° = 0.707 = $\dfrac{7.07}{10}$ (Figure E-38)

Angle of Problem 3—sin 38° 40′ = 0.62479 ≈ $\dfrac{6.25}{10}$ (Figure E-39)

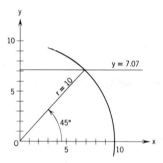

Figure E-38

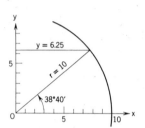

Figure E-39

Exercise 8.4

1. See Figure E-40

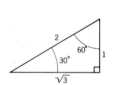

Figure E-40

$\sin 30° = \cos 60° = \tfrac{1}{2}$

$\cos 30° = \sin 60° = \dfrac{\sqrt{3}}{2}$

$\tan 30° = \cot 60° = \dfrac{1}{\sqrt{3}} = \dfrac{\sqrt{3}}{3}$

$\csc 30° = \sec 60° = 2$

$\sec 30° = \csc 60° = \dfrac{2}{\sqrt{3}} = \dfrac{2\sqrt{3}}{3}$

3. (a) $AO = \dfrac{10}{\sqrt{3}} = \dfrac{10\sqrt{3}}{3}$ ft

(b) $AB = 10.0$ mm

(c) $OB = \dfrac{15\sqrt{3}}{2}$ cm

5. $AOR = 93.0$ ft

7. $h = 39$ ft

9. (a) $h = 40.0$ ft
   (b) $V = 2480$ cu yd

11. 462 ft

13. (a) $FO = 465$ ft
    (b) $PF = 658$ ft
    (c) $TO = 268$ ft
    (d) $PT = 537$ ft
    (e) $FT = 197$ ft

Exercise 8.5

1. 0.4147
3. 0.7790
5. 0.1363
7. 0.7408

9. $\theta = 29°\ 35'$
11. $\theta = 23°\ 8'$
13. $s = 4.59$ ft
15. 0.832 ft

17. (a) 17.1 ft
      (b) 20.1 ft
      (c) 5.11 ft
19. 7090 ft

Exercise 8.6

1. $\sin 0.620° = 0.0108$
3. $\sin 62° = 0.883$
5. $\sin 90° = 1.00$
7. $\tan 7.40° = 0.1299$
9. $\tan 45° = 1.000$
11. $\tan 84.25° = 9.93$
13. $\cos 25.5° = \sin 64.5° = 0.903$
15. $\cos 62°\ 12' = 0.466$

17. $\csc 0.620° = 92.4*$
       (by interpolation 94.0)
19. $\csc 62.0° = 1.133$
21. $\theta = 7.4°$
23. $\theta = 19.4°$
25. $\theta = 2.95°$
27. $\theta = 0.95°$
29. $\theta = 19.9°$

Exercise 8.7

1. 41.1 to two figures = 41
3. 16.47 to three figures = 16.5
5. 0.394
7. 23.2
9. 0.462

11. 0.117
13. 19.3
15. 1.20×10^3
17. 534
19. 2.70×10^3

21. 50.7
23. 92.1
25. 45.4
27. 97.5
29. 3310

Exercise 8.8

1. $A = 61.0°,\ B = 29.0°$
3. $A = 37.2°,\ B = 52.8°$
5. $A = 15.7°,\ B = 74.3°$
7. $c = 51.5$ ft
9. $\angle A = 49.8°,\ c = 74.7$ ft
11. $a = 12.7$ in.
13. $\theta = 38.3°,\ Z = 39.7$ units
15. $X_L = 46.0$ ohms, $Z = 69.9$ ohms
17. $\theta = 51.0°,\ E = 67.3$ units
19. $E_L = 51$ volts, $E_R = 110$ volts
21. $OB = 1310$ ft, $PB = 841$ ft
23. latitude = $22\tilde{0}$ ft, departure = $41\tilde{0}$ ft

25. (a) $\sin \theta = \dfrac{\text{latitude}}{\text{course distance}}$, latitude = C.D. $\sin \theta$

   (b) $\cos \theta = \dfrac{\text{departure}}{\text{course distance}}$, departure = C.D. $\cos \theta$

   (c) $\tan \theta = \dfrac{\text{latitude}}{\text{departure}}$

27. latitude = 631 ft, departure = 1520 ft

29. $\theta = 37.3°$, C.D. = 259 ft

   * When the difference between consecutive entries in Table 16 becomes relatively large, linear interpolation becomes less accurate than using the slide rule.

Exercise 8.9

1. (a) Negative   (d) Negative   (g) Negative
    (b) Negative   (e) Negative
    (c) Negative   (f) Positive

3. $\sin 150° = \sin 30° = 0.500$

$$\tan 150° = -\tan 30° = \frac{-\sqrt{3}}{3} \approx -0.5774$$

5. $\sin 315° = -\sin 45° = \dfrac{-\sqrt{2}}{2} \approx -0.7071$

$\sec 315° = \sec 45° = \sqrt{2} \approx 1.414$

7. $\sin \theta = \dfrac{6}{\sqrt{61}}$ $\tan \theta = \dfrac{6}{-5}$ $\sec \theta = \dfrac{\sqrt{61}}{-5}$

$\cos \theta = \dfrac{-5}{\sqrt{61}}$ $\cot \theta = \dfrac{-5}{6}$ $\csc \theta = \dfrac{\sqrt{61}}{6}$

9. (a) 45°   (d) 60°   (g) 30°
    (b) 60°   (e) 30°   (h) 80°
    (c) 30°   (f) 60°

11. $\sin 180° = \dfrac{y}{r} = \dfrac{0}{r} = 0$

$\cos 180° = \dfrac{x}{r} = \dfrac{x}{-x} = -1, \quad (r = -x)$

$\tan 180° = \dfrac{y}{x} = \dfrac{0}{x} = 0$

$\cot 180° = \dfrac{x}{y} = \dfrac{x}{0}, \quad \text{not defined}$

$\sec 180° = \dfrac{r}{x} = -1, \quad (r = -x)$

$\csc 180° = \dfrac{r}{y} = \dfrac{r}{0}, \quad \text{not defined}$

13. $\sin 360° = \dfrac{y}{r} = \dfrac{0}{r} = 0$

$\cos 360° = \dfrac{x}{r} = 1, \quad (x = r)$

$\tan 360° = \dfrac{y}{x} = \dfrac{0}{x} = 0$

$\cot 360° = \dfrac{x}{y} = \dfrac{x}{0}, \quad \text{not defined}$

$$\sec 360° = \frac{r}{x} = 1, \quad (r = x)$$

$$\csc 360° = \frac{r}{y} = \frac{r}{0} , \quad \text{not defined}$$

15. 210°
17. 144° 20′
19. 114° 20′
21. θ in quadrant III

$\sin \theta = \dfrac{y}{r}$, negative,   ($y$ negative)

$\cos \theta = \dfrac{x}{r}$, negative,   ($x$ negative)

$\tan \theta = \dfrac{y}{x}$, positive,   (both x and y negative)

$\cot \theta = \dfrac{x}{y}$, positive,   (both x and y negative)

$\sec \theta = \dfrac{r}{x}$, negative,   ($x$ negative)

$\csc \theta = \dfrac{r}{y}$, negative,   ($y$ negative)

Exercise 8.10 A

 1. $\measuredangle B = 31.7°$ or $148.3°$; no, two triangles are possible.
 3. $\measuredangle B = 124.9°$ or $55.1°$; no, two triangles are possible.
 5. $c = 15.6$; yes
 7. 102.9°
 9. $c = 11.7$ units
11. (a) 347 ft, (b) 35.0°

Exercise 8.10 B

 1. 9.7144 −10   **7.** 153.6°
 3. 9.3529 −10   **9.** 59.20 ft
 5. 16.3°

Exercise 8.11 A

 1. Vector (weight is a force acting vertically downward)
 3. Scalar (path indicates direction only)
 5. Scalar (no direction involved)
 7. Scalar (no direction specified)
 9. Scalar (no direction specified)
11. See Figure E-41
13. See Figure E-42
15. See Figure E-43
17. See Figure E-44
19. The magnitude of the resultant is doubled.

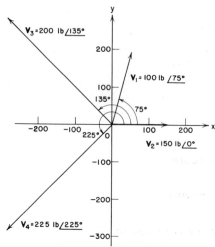

(Vector V₂ lies on top of x axis)

Figure E-41

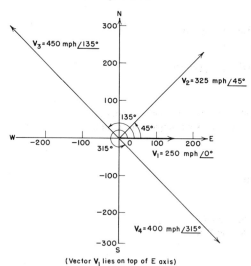

(Vector V₁ lies on top of E axis)

Figure E-42

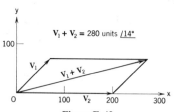

Figure E-43

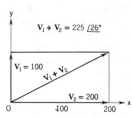

Figure E-44

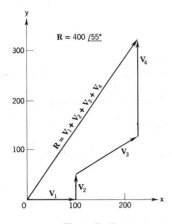

Figure E-45

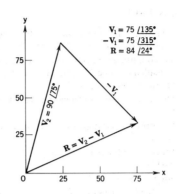

Figure E-46

21. See Figure E-45
23. See Figure E-46

Exercise 8.11B

1. $70.5 \big/ 60.0°$
3. x component $= +118.2 \approx +120$, y component $= +92.4 \approx +92$
5. x component $= -277$ lb, y component $= +480$ lb
7. x component $= +94.6$, y component $= +119$
9. $C = 13.86 \approx 14$ lb

Exercise 8.11C

1. $|N| = 51.1$ tons, $|F_1| = 13.3$ tons
3. 8.47 sq yd
5. latitude $= 1260$ ft, departure $= 1180$ ft
7. Course distance for $AB = (512')(\cos 22.3°) = 474$ ft

9.

Station	Latitude	Departure
B	220 ft	170 ft
C	390 ft	−215 ft
D	−233 ft	−211 ft
E	−292 ft	309 ft

11. $|F_1| = 31.2$ lb
13. Will move left, decreased by 103 lb
15. (a) 396 sq ft   (b) 522 cm^2
17. (a) 160 sq in.   (b) 17 in.
19. (a) 205 sq in.   (b) 14.0 in. by 14.6 in.
21. 5.0×10^2 lb
23. $E_L = 77.0$ volts, $E_R = 195$ volts
25. $E_{line} = 119$ volts, $E_R = 109$ volts
27. $\sphericalangle\,\theta = 52.1°$, $I_{line} = 20.6$ amp
29. $I_R = 0.0394$ amp, $I_{line} = 0.0413$ amp
31. (a) 268 lb, (b) 615 lb

APPENDIX A

Inventory Test I

1. Addition. The two numbers are to be added; 12 is to be added to 7.
2. Subtraction. Subtract 7 from 16. (16 − 7 also means the difference between 16 and 7)
3. Two quantities are equal (or equivalent). For example, $12 = (3 \times 4)$, 12 in. = 1 ft, and 30 min = $\frac{1}{2}$ hr
4. × (called a "times" sign); 18×41. You also could show this multiplication as (18)(41)
5. Division. Divide 42 by 7
6. Division symbol. It means that 42 is to be divided by 7
7. $\frac{42}{7}$
8. Parentheses. To group two or more terms together or to show multiplication of two or more factors
9. Brackets and braces: [ ] and { }
10. (8)(12)
11. Square root symbol
12. $7 < 10$ (The symbol "<" means less than; $7 < 10$ is read "seven is less than ten.")
13. The symbol "$\approx$" means approximately equal. For example, $\frac{7}{9} \approx \frac{7}{10}$ is read "seven-ninths is approximately equal to seven-tenths"
14. 0, 1, 2, 3, 4, 5, 6, 7, 8, and 9. Ten (10 is not a digit. It is a combination of two digits, the "one" and the "zero."
15. ⑦ ② ⓪ ⑤ · ⑥ ⓪ ③
16. (b) $\frac{5}{5}$ is equal to 1 and therefore represents a unit
    (c) 1 yd is a unit of length
    (e) 1 hr is a unit of time
    (f) 1 ton is a unit of weight
    Answers (a) and (d) are fractional parts of a unit
17. (b) 126   (d) $\frac{125}{25} = 5$ (an integer)   (f) 2,365,487
18. A whole number. Examples are 5, 48, 207, and 3475
19. $n, n + 1, n + 2, n + 3$. When n is any whole number. For example, if $n = 20$, the four consecutive integers would be 20, 21, 22, and 23
20. $n, n + 2, n + 4, n + 6$. When n is an even integer (one that can be divided by 2). For example, if $n = 20$, four consecutive even integers would be 20, 22, 24, and 26
21. $n, n + 2, n + 4, n + 6$. Where n is an odd integer (a whole number that cannot be divided by 2). For example, if $n = 19$, the four consecutive odd integers would be 19, 21, 23, and 25
22. A prime number is an integer (other than the integer 1) whose only divisors are 1 and itself. For example, 13 is a prime number because the only numbers that divide 13 are 13 and 1. In contrast, the number 12 is not prime because it can be divided by 2, 3, 4, and 6
23. 2, 3, 5, 7, 11, 13, 17, 19, 23, 29, 31, 37, 41, 43, 47, 53. (As an interesting project, look up "Eratosthenes—Sieve of" in an encyclopedia)
24. A composite number contains factors other than 1 and itself. For example, 12 is a composite number which can be written in factored form as 3×4, 2×6, or $2 \times 2 \times 3$
25. 7, 14, 21 (A multiple of a number is any product of the number and an integer)
26. 3, 9, 12
27. 21 (21 is the smallest multiple of 7 which is also a multiple of 3)
28. 2. An even number is any number whose last digit is a 0, 2, 4, 6, or 8. For example, the number 3156 can be divided by 2 because the last digit, 6, is even
29. 3. The number 26,922 can be divided by 3 because the sum of its digits, $2 + 6 + 9 + 2 + 2 = 21$, can be divided by 3
30. Zero or a 5

31. The sum of its digits can be divided evenly by 9. $7 + 9 + 2 + 3 +$ the last digit, must be divisible by 9. $7 + 9 + 2 + 3 = 21$. Because you know that 27 is divisible by 9, add a 6 for the fifth digit, obtaining 27 for the sum of the digits and 79,236 for the required number

32. 17 is less than 19; 17 is not equal to 19.

33. Addends

34. Add 21, 23, and 25 together. The answer to addition is frequently called the sum

35. A fraction

36. The numerator and the denominator are called the *terms* of a fraction

37. The numerator. The denominator. Terms of the fraction

38. (a) Prime fraction. The fraction is in lowest terms because the 7 and 10 have no common factors except the factor 1

(b) Common fraction. The term common fraction refers to the form in which the fraction is written, i.e., two integers one above the other. Other examples of common fractions are $\frac{3}{4}$, $\frac{5}{9}$, and $\frac{8}{15}$. Consider $\frac{3.5}{7} = \frac{1}{2}$; $\frac{3.5}{7}$ is not a common fraction whereas $\frac{1}{2}$ is. Also, $0.5 = \frac{1}{2}$ but 0.5 is not a common fraction

(c) Decimal fraction is skipped because the decimal fraction equivalent for $\frac{7}{10}$ is written 0.7

(d) Proper fraction; $\frac{7}{10}$ is proper because the numerator is smaller than the denominator.

(e) Mixed number. Skipped because a mixed number is the sum of a whole number and a fraction. In this case, there is no whole number indicated

(f) Improper fraction. Skipped because an improper fraction has a numerator equal to or greater than the denominator. With $7 < 10$, the fraction is proper

(g) Integer. Skipped because an integer is a whole number whereas $\frac{7}{10}$ is a fractional part

(h) Ratio. A ratio between two numbers can be written in fractional form; $\frac{7}{10}$ expresses the ratio of 7 compared with 10.

(i) Digit. Skipped because $\frac{7}{10}$ is not one of the defined digits: 0, 1, 2, 3, 4, 5, 6, 7, 8, or 9

39. Cancellation is the term applied to dividing the numerator and the denominator by a common factor. For example,

$$\frac{24}{30} = \frac{4 \times 6}{5 \times 6}$$

Because 6 is a common factor of 24 and 30, divide each term by 6, obtaining $\frac{4}{5}$ for the reduced equivalent fraction. The 6's are "cancelled," usually shown as

$$\frac{\overset{4}{\cancel{24}}}{\underset{5}{\cancel{30}}} \text{ or } \frac{4 \times \cancel{6}}{5 \times \cancel{6}} = \frac{4}{5}$$

40. Yes; $\frac{3}{4} = \frac{6}{8}$. If the terms 3 and 4 are both multiplied by the same nonzero number, an equivalent fraction is obtained. Reducing may result in higher as well as lower terms

41. Yes. Changing the terms of a fraction by multiplying or dividing both terms by the same nonzero number results in a fraction equal in value to the original fraction. Fractions equal in value are called equivalent fractions

42. $\frac{3}{4}$, $\frac{6}{8}$, $\frac{30}{40}$

43. A mixed number is the sum of a whole number and a common fraction; $5\frac{3}{4}$ is a mixed number representing the sum of 5 and $\frac{3}{4}$. The plus sign is omitted but is implied in the way the number is read: "five *and* three fourths." Examples are $7\frac{2}{3}$ and $12\frac{4}{9}$

44. (a) $\frac{1}{7}$ (b) $\frac{1}{2/5}$ or $\frac{5}{2}$ (Note that the reciprocal of a common fraction can be obtained by inverting the fraction) (c) $\frac{1}{0.25}$ or $\frac{1}{25/100} = \frac{100}{25} = 4$, (d) $\frac{1}{(6+8)} = \frac{1}{14}$

45. $0.723\frac{1}{4}$. A mixed decimal consists of a decimal fraction and a common fraction. $0.723\frac{1}{4}$ does not mean $0.723 + \frac{1}{4}$ but rather $0.723 + \frac{1}{4}$ of the place value of the 3, i.e., $\frac{1}{4}$ of 0.001:

$$0.723\tfrac{1}{4} = 0.723 + (\tfrac{1}{4} \times 0.001) = 0.723 + 0.00025 = 0.72325$$

Numbers should not be written in the form of $0.723\frac{1}{4}$. If the needed accuracy demands that the $\frac{1}{4}$ be included, then 0.72325 is a much easier and less confusing form. Otherwise the number should be rounded to 0.723

46. Yes. If you multiply 12 by the reciprocal of 4, i.e., $12 \times \frac{1}{4} = 3$, you obtain the same result as when you divide 12 by the number 4. This is the reason why you multiply by the reciprocal of a fraction when division by the fraction is indicated

47. Added. Subtracted

48. $\frac{6}{8}, \frac{1}{8}$ (Similar fractions are fractions having the same denominator)

49. $\frac{5}{24}, \frac{7}{24}, \frac{13}{24}$

50. By the place value of the last digit in the decimal. For example, 0.2935 has an understood denominator of 10,000 because the 5 is in the ten-thousandths position. The fraction is read: "two thousand nine hundred thirty-five ten-thousandths"

51. 1000. The digit 2 is in the thousand*ths* position. The fraction is read: "three hundred twelve one-thousandths." (Frequently the *one-thousandths* is shortened to just *thousandths*)

Inventory Test II

1. (a) 41,000.040. If you wrote the decimal part as .041, you overlooked the hyphen that ties the one with the thousandths; .041 is read "forty-one thousandths"
   (b) Sixty-eight thousand one hundred thirty-five hundred-thousandths. Except for the denominator part of the decimal the .68135 is read the same as the whole number 68,135, i.e., "sixty-eight thousand one hundred thirty-five." The denominator is determined by the location of the 5 with respect to the decimal point. Because 5 is in the hundred-thousandths position, add hundred-thousandths to the reading

Hundred trillions	Ten trillions	Trillions	Hundred billions	Ten billions	Billions	Hundred millions	Ten millions	Millions	Hundred thousands	Ten thousands	Thousands	Hundreds	Tens	Units		Tenths	Hundredths	Thousandths	Ten-thousandths	Hundred-thousandths	Millionths	Ten-millionths	Hundred-millionths	Billionths
2	9	5,	3	8	6,	8	7	4,	9	3	7,	8	5	2	.	3	8	5	4	1	2	9	7	3

Trillion   Billion   Million   Thousand   (Ones)

2. 76,650. *Solution*:

      12 33*
    66,815
     2,864
       385
        97
     6,489
    —————
    76,650

* Carry digits shown in small type.

3. 57,816,178. *Solution*:

60,819,032	minuend
3,002,854	subtrahend
57,816,178	difference

To check, add the difference and the subtrahend to obtain the minuend:

57,816,178
3,002,854
60,819,032

4. 334,230,112 or 334,200,000 rounded to four significant figures. Carry digits need not be written down but can be added mentally to the partial products.

```
  2111
  6432
  111
  7543
 89654
  3728
------
717232
179308
627578
268962
------
334230112
```

5. 8697 or 87̃00 rounded to three significant figures.* *Solution*:

```
           8697
      589)5122533
          4712
          ----
           4105
           3534
           ----
            5713
            5301
            ----
             4123
             4123
             ----
                0
```

To check division, multiply quotient (before rounding off) by the divisor and add the remainder

6. 5705.0026 becomes 5705.003 when rounded to thousandths. Addition may be checked by adding digits in reverse order, i.e., from bottom to top or top to bottom
Solution:

```
   222 31
  889.030
   46.977
    3.0851
    0.0205
  1
 4765.890
 --------
 5705.0026
```

* The tilde ∼ is used to show the last significant zero digit.

7. 336.8125 becomes 336.81 when rounded off to hundredths. *Solution*:

$$\begin{array}{ll} \text{minuend} & 423.5025 \\ \text{subtrahend} & \underline{86.69} \\ \text{difference} & \overline{336.8125} \end{array}$$

To check, add the difference (before rounding off) and the subtrahend

8. 1683.3348 becomes 1683 when rounded to four significant figures. *Solution*:

$$\begin{array}{r} 7\,5\,2 \\ 6\,5\,2 \\ 5\,4\,1 \\ 8\,9\,7.3 \\ 1.8\,7\,6 \\ \hline 5\,3\,8\,3\,8 \\ 6\,2\,8\,1\,1 \\ 7\,1\,7\,8\,4 \\ 8\,9\,7\,3 \\ \hline 1\,6\,8\,3.3\,3\,4\,8 \end{array}$$

To check, go back over each step of the work or reverse positions of the two numbers and multiply again

9. 6.90. *Solution*: First, to make the divisor a whole number, multiply both divisor and dividend by 100. Then, locate the decimal point in the quotient directly above its new position in the dividend:

$$\begin{array}{r} 6.9 \\ 4.8\,7.)\overline{3\,3.6\,0.3} \\ 2\,9\,2\,2 \\ \hline 4\,3\,8\,3 \\ 4\,3\,8\,3 \\ \hline 0 \end{array}$$

To check, multiply the divisor by the quotient (before rounding off) and add the remainder

10. (a) $\frac{11}{17}$. *Solution*: Because both terms of $\frac{132}{204}$ are even numbers, you can divide each term by 2, obtaining $\frac{66}{102}$; dividing each term by 2 again, you have $\frac{33}{51}$. Now divide each term by 3, obtaining $\frac{11}{17}$. (This is in lowest terms since both the 11 and 17 have no factor in common (except the factor 1)

(b) $\frac{35}{84}$. *Solution*: The problem could be stated $\frac{5}{12} = \frac{?}{84}$. If you multiply the 12 by 7, you obtain the 84, so you only need to multiply the 5 by 7 also. This gives

$$\frac{5 \times 7}{12 \times 7} = \frac{35}{84}$$

for the required fraction

(c) $7\frac{20}{36} = 6\frac{56}{36}$. *Solution*: $7\frac{5}{9} = 7\frac{20}{36} = 6\frac{56}{36}$. First, multiply the 5 and 9 by 4 which gives $7\frac{20}{36}$. Then, for the second part of the problem, i.e., $7\frac{20}{36} = 6\frac{?}{36}$, you can see that the 7 has been decreased by 1. To account for the 1, express the 1 as $\frac{36}{36}$ and add it to the $\frac{20}{36}$. This gives you $\frac{56}{36}$. The final result becomes $6\frac{56}{36}$

11. (a) $12\frac{5}{7}$. *Solution*: To simplify the improper fraction $\frac{89}{7}$, divide the numerator 89 by the denominator 7. This gives 12 with a remainder (5) which is written as the fraction $\frac{5}{7}$; $\frac{89}{7} = 12\frac{5}{7}$

  (b) $\frac{158}{15}$. *Solution*: $10\frac{8}{15} = \frac{158}{15}$. To make the change, change the whole number (10) to fifteenths. Because $1 = \frac{15}{15}$, 10 equals $\frac{150}{15}$. This is added to the fraction $\frac{8}{15}$, making a total of $\frac{158}{15}$

12. 120. *Solution*: The simplest and easiest way to obtain the LCD is to factor each denominator to prime factors:

$$\frac{2}{5\cdot 1}, \frac{5}{2\cdot 3}, \frac{7}{2\cdot 2\cdot 2}, \frac{1}{3\cdot 1}, \frac{5}{2\cdot 2\cdot 3}$$

Then select one of each different valued factor, using any repeated factor the highest number of times it appears in any single denominator. By selecting factors of 1, 2, 3, and 5 and using the 2 as a factor three times, the LCD is equal to $1\cdot 2\cdot 2\cdot 2\cdot 3\cdot 5$, or 120

13. (a) $185\frac{103}{120}$. *Solution*:

$$12\frac{2}{5} = 12\frac{48}{120}$$
$$67\frac{5}{6} = 67\frac{100}{120}$$
$$38\frac{7}{8} = 38\frac{105}{120}$$
$$49\frac{1}{3} = 49\frac{40}{120}$$
$$\frac{17\frac{5}{12} = 17\frac{50}{120}}{183\frac{343}{120}}$$

  which simplifies to $185\frac{103}{120}$.

  (b) $29\frac{13}{72}$. *Solution*:

$$46\frac{5}{9} = 46\frac{40}{72}$$

$$\text{LCD} = \overbrace{3\cdot 3}^{9}\cdot \overbrace{2\cdot 2\cdot 2}^{8} = 72$$

Subtracting

$$\frac{17\frac{3}{8} = 17\frac{27}{72}}{29\frac{13}{72}}$$

14. $\frac{21}{64}$. *Solution*:

$$\frac{9}{16} \times \frac{7}{12} = \frac{9 \times 7}{16 \times 12} = \frac{63}{192} = \frac{21}{64}$$

or

$$\frac{\overset{3}{\cancel{9}}}{16} \times \frac{7}{\underset{4}{\cancel{12}}} = \frac{3 \times 7}{16 \times 4} = \frac{21}{64}$$

15. $2\frac{4}{5}$. *Solution*: $\frac{7}{8} \div \frac{5}{16}$ becomes

$$\frac{7}{8} \times \frac{16}{5} = \frac{7 \times 16}{8 \times 5} = \frac{112}{40} = 2\frac{32}{40} = 2\frac{4}{5}$$

or

$$\frac{7}{\cancel{8}} \times \frac{\overset{2}{\cancel{16}}}{5} = \frac{14}{5} = 2\frac{4}{5}$$

(Remember that the divisor is inverted and the operation changed to multiplication.)

16. 247. *Solution*: First, change the mixed numbers to improper fractions, obtaining $\frac{15}{4} \times \frac{38}{5} \times \frac{26}{3}$. This then will simplify to

$$\frac{\overset{1}{\cancel{15}}}{\underset{\underset{1}{\cancel{2}}}{\cancel{4}}} \times \frac{\overset{19}{\cancel{38}}}{\underset{1}{\cancel{5}}} \times \frac{\overset{13}{\cancel{26}}}{\underset{1}{\cancel{3}}} = \frac{1 \times 19 \times 13}{1 \times 1 \times 1} = \frac{247}{1} = 247$$

17. $\frac{205}{342}$. *Solution*: First, change to improper fractions, obtaining $\frac{41}{9} \div \frac{38}{5}$. Then invert the divisor and multiply:

$$\frac{41}{9} \times \frac{5}{38} = \frac{41 \times 5}{9 \times 38} = \frac{205}{342}$$

18. (a) 630,050. *Solution*: To multiply by 10,000, move the decimal point four places to the right (63,0050) obtaining 630,050

(b) 0.0605. *Solution*: To divide by 100, move the decimal point two places to the left (.06.05) obtaining 0.0605

19. (a) 0.685. *Solution*: Divide the numerator (137) by the denominator (200):

$$\begin{array}{r} 0.6\,8\,5 \\ 200\overline{)1\,3\,7.0\,0\,0} \\ \underline{1\,2\,0\,0} \\ 1\,7\,0\,0 \\ \underline{1\,6\,0\,0} \\ 1\,0\,0\,0 \\ \underline{1\,0\,0\,0} \\ 0 \end{array}$$ (Annex zeros as needed)

(b) $\frac{11}{400}$. *Solution*: If you can read a decimal fraction, you can write it as a common fraction. This fraction is read "two hundred seventy-five ten-thousandths," so write 275 over the denominator 10,000. This fraction reduces to $\frac{11}{400}$

20. 419. *Solution*: To average, add the 7 numbers together, obtaining a sum of 2933. Then divide the sum by 7 (the number of addends) 2933 ÷ 7 = 419, for the average.

Practice Problems Set 1

1. One hundred thirty-eight thousand, five hundred forty-seven; 138,547
3. One hundred seventy thousand, sixteen; 170,016
5. One thousand twenty-nine and five-tenths; 1,029.5

7. One thousand two hundred-five hundred-thousandths
9. Forty-five and forty-five thousandths
11. Forty-seven thousandths (also forty-seven one-thousandths)
13. Fifty-two hundred-thousandths
15. Eighty-one and eighty-one hundredths
17. 40.007
19. 0.20
21. 0.0016
23. 3.21
25. 0.410
27. 0.00054
29. 4,000,004

Practice Problems Set 2

1. 8082   3. 21,043   5. 448,816   7. 104,497   9. 308,462

Practice Problems Set 3

1. 4104   3. 5750   5. 4922   7. 85,214,858   9. 11,891

Practice Problems Set 4

1. 63,684   3. 39,984   5. 1,550,050   7. 29,960   9. 37,471

Practice Problems Set 5

1. $393\frac{1}{8}$ 3. $3308\frac{1}{9}$ 5. $243\frac{2}{5}$ 7. $252\frac{27}{326}$ 9. $188\frac{183}{391}$

Practice Problems Set 6

1. 200.8786   3. 974.7055   5. 242.6877   7. 814.332   9. 270.026

Practice Problems Set 7

1. 326.793   3. 1478.3199   5. 5380.9868   7. 0.08999   9. 1.00987

Practice Problems Set 8

1. 553.4369   3. 787.4176   5. 269.6484   7. 54,400.65039   9. 4937.2840

Practice Problems Set 9

1. 12.1   3. 3.2889   5. 0.23   7. 29   9. $8\tilde{0}0$

Practice Problems Set 10

1. $\frac{21}{28}$ 3. $\frac{4}{5}$ 5. $\frac{60}{100}$ 7. $3\frac{10}{25} = 2\frac{35}{25}$ 9. $\frac{7}{12}$

Practice Problems Set 11

1. $5\frac{3}{5}$ 3. $14\frac{2}{7}$ 5. $1\frac{3}{5}$ 7. $\frac{64}{5}$ 9. $\frac{177}{4}$ 11. $\frac{62}{8}$

Practice Problems Set 12

1. $LCD = 2 \cdot 2 \cdot 3 = 12$
3. $LCD = 2 \cdot 2 \cdot 3 \cdot 5 = 60$
5. $LCD = 2 \cdot 2 \cdot 2 \cdot 7 = 112$
7. $LCD = 2 \cdot 2 \cdot 2 \cdot 2 \cdot 3 \cdot 5 = 240$
9. $LCD = 2 \cdot 2 \cdot 2 \cdot 2 \cdot 3 \cdot 3 = 144$

Practice Problems Set 13

1. $2\frac{1}{4}$ 3. $2\frac{31}{60}$ 5. $\frac{107}{112}$ 7. $3\frac{38}{45}$ 9. $11\frac{15}{16}$

Practice Problems Set 14

1. $\frac{1}{2}$ **3.** $\frac{2}{63}$ **5.** $\frac{7}{12}$ **7.** $\frac{3}{5}$ **9.** 3

Practice Problems Set 15

1. $\frac{15}{16}$ **3.** $\frac{35}{8} = 4\frac{3}{8}$ **5.** $\frac{27}{56}$ **7.** 12 **9.** $9\frac{1}{3}$

Practice Problems Set 16

1. $24\frac{11}{12}$ **3.** $83\frac{3}{5}$ **5.** 76 **7.** 40 **9.** 280

Practice Problems Set 17

1. $\frac{28}{39}$ **3.** $1\frac{7}{188}$ **5.** $\frac{11}{18}$ **7.** $6\frac{14}{19}$ **9.** $1\frac{1}{5}$

Practice Problems Set 18

1. 861.5 **3.** 0.6814164 **5.** 0.06868 **7.** 0.00005 **9.** 45,000

Practice Problems Set 19

1. 0.8750 **3.** 0.5556 **5.** 0.3125 **7.** 0.0833 **9.** $\frac{125}{1000} = \frac{1}{8}$ **11.** $\frac{55}{100} = \frac{11}{20}$

Practice Problems Set 20

1. 39 **3.** $0.2892 \approx 0.289$ **5.** 15.78 in. **7.** 18,228 lb **9.** 3.501 in.

Inventory Test III

1. (a) Eighty-three million, six hundred eighty-five thousand, four hundred thirty-five and sixty-eight thousand, one hundred thirty-five hundred-thousandths
   (b) 6, 831, 542, 967

2. 1265. *Solution*:

$$
\begin{array}{r}
23^* \\
48 \\
725 \\
396 \\
87 \\
9 \\
\hline
1265
\end{array}
$$

3. 2177. *Solution*:

$$
\begin{array}{r}
3925 \\
1748 \\
\hline
2177
\end{array}
$$

4. 22,011 or 22,000 when rounded off to two significant figures. *Solution*:

$$
\begin{array}{r}
42^* \\
32 \\
253 \\
87 \\
\hline
1771 \\
2024 \\
\hline
22011
\end{array}
$$

* Carry digits are shown in small type. They may be carried mentally.

5. 385 or 400 when rounded to one significant figure. *Solution:*

$$
\begin{array}{r}
385 \\
7)\overline{2695} \\
21 \\
\overline{59} \\
56 \\
\overline{35} \\
35 \\
\overline{0}
\end{array}
$$

6. 418.810 becomes 418.8 when rounded off to nearest tenth. *Solution:*

$$
\begin{array}{r}
1\ 11 \\
_1 3.86 \\
29.145 \\
_1\ 0.705 \\
385.1 \\
\hline
418.810
\end{array}
$$

7. 15.444 becomes 15.4 when rounded off to tenths. *Solution:*

83.900   (Zeros may be annexed if they make the
68.456   subtraction easier)
15.444

8. 9.156 becomes 9.2 when rounded off to two significant figures. *Solution:*

$$
\begin{array}{r}
1 \\
2\ 5 \\
3.2\ 7 \\
2.8 \\
\hline
2\ 6\ 1\ 6 \\
6\ 5\ 4 \\
\hline
9.1\ 5\ 6
\end{array}
$$

You obtain thousandths in the product because you multiply hundredths times tenths

9. 110.2 becomes 110 when rounded off to two significant figures. *Solution:* Multiply both divisor and dividend by 10 to make the divisor a whole number

$$
\begin{array}{r}
1\ 1\ 0.2 \\
6.7)\overline{7\ 3\ 8\ 4.5} \\
6\ 7 \\
\hline
6\ 8 \\
6\ 7 \\
\hline
1\ 4\ 5 \\
1\ 3\ 4 \\
\hline
1\ 1\quad \text{remainder}
\end{array}
$$

10. (a) $\frac{3}{13}$. *Solution*:

$$\frac{18}{78} = \frac{3 \cdot \cancel{6}}{13 \cdot \cancel{6}} = \frac{3}{13}$$

(b) $\frac{20}{35}$. *Solution*:

$$\frac{4 \cdot 5}{7 \cdot 5} = \frac{20}{35}$$

(c) $6\frac{6}{10} = 5\frac{32}{20}$, *Solution*: $6\frac{3}{5} = 6\frac{6}{10} = 5\frac{32}{20}$. First, $\frac{3}{5} = \frac{6}{10}$. Second, $\frac{6}{10} = \frac{12}{20}$. Third, the 1 borrowed from the 6 is changed to $\frac{20}{20}$ and added to the $\frac{12}{20}$, giving $\frac{32}{20}$.

11. (a) $\frac{33}{5}$. *Solution*: The whole number 6 is equal to $\frac{30}{5}$ which, added to the $\frac{3}{5}$, gives $\frac{33}{5}$. Mechanically, you multiply the whole number (6) by the number in the denominator (5) and add the number in the numerator (3). This sum is then written over the original denominator:

$$\frac{6 \times 5 + 3}{5} = \frac{33}{5}$$

(b) $3\frac{1}{3}$. *Solution*: Divide the numerator (30) by the denominator (9) and simplify the remaining fraction: $30 \div 9 = 3\frac{3}{9} = 3\frac{1}{3}$.

12. 60. *Solution*: Factor the denominators to prime factors:

$$\frac{2}{3}, \frac{4}{5}, \frac{3}{2 \cdot 2}, \frac{7}{2 \cdot 5}.$$

The LCD $= 2 \cdot 2 \cdot 3 \cdot 5 = 60$

13. (a) $84\frac{19}{48}$. *Solution*: The LCD $= 2 \cdot 2 \cdot 2 \cdot 2 \cdot 3 = 48$.

$$42\frac{3}{4} = 42\frac{36}{48}$$
$$26\frac{1}{3} = 26\frac{16}{48}$$
$$15\frac{5}{16} = 15\frac{15}{48}$$
$$\overline{83\frac{67}{48}} = 84\frac{19}{48}$$

(b) $11\frac{31}{40}$. *Solution*: The LCD $= 2 \cdot 2 \cdot 2 \cdot 5 = 40$.

$$27\frac{3}{8} = 27\frac{15}{40} = 26\frac{55}{40}$$
$$15\frac{3}{5} = 15\frac{24}{40} \quad 15\frac{24}{40}$$
$$\overline{11\frac{31}{40}}$$

14. $\frac{12}{35}$. *Solution*: $\frac{3}{5} \times \frac{4}{7} = \frac{12}{35}$.

15. $\frac{6}{7}$. *Solution*:

$$\frac{3}{4} \div \frac{7}{8} = \frac{3}{4} \times \frac{8}{7} = \frac{24}{28} = \frac{6}{7} \text{ or } \frac{3}{\cancel{4}} \times \frac{\cancel{8}^{2}}{7} = \frac{6}{7}$$

16. 27. *Solution*: $3\frac{3}{4} \times 7\frac{1}{5} = \frac{\cancel{15}^{3}}{\cancel{4}_1} \times \frac{\cancel{36}^{9}}{\cancel{5}_1} = \frac{27}{1} = 27$

17. $\frac{207}{64}$ or $3\frac{15}{64}$. *Solution*:

$$8\tfrac{5}{8} \div 2\tfrac{2}{3} = \tfrac{69}{8} \div \tfrac{8}{3}$$

$$\tfrac{69}{8} \times \tfrac{3}{8} = \tfrac{207}{64} = 3\tfrac{15}{64}$$

18. (a) 605.4. *Solution*: $6.054 \times 100 = 605.4$. Move decimal point two places to the right

   (b) 0.0006035. *Solution*: $6.035 \div 10,000 = 0.0006035$. Move the decimal point four places to the left

19. (a) 0.625. *Solution*: Divide the numerator by the denominator: $5 \div 8 = 0.625$. Check: $0.625 = \frac{625}{1000} = \frac{5}{8}$

   (b) $\frac{427}{2000}$. *Solution*:

$$0.2135 = \frac{2135}{10,000} = \frac{427}{2000}$$

20. $77\frac{3}{7}$. *Solution*: Determine the sum of the 7 scores and divide by 7: $542 \div 7 = 77\frac{3}{7}$

APPENDIX C

Exercise C.1

1. 128
3. 14
5. 32,768
7. $6 + 7 = 13$
9. $\sqrt{4096} = 2^{12/2} = 2^6 = 64$
11. (a) $5^2 = 25$ (b) $10^3 = 1000$ (c) $2^6 = 64$ (d) $5^1 = 5$ (e) $4^0 = 1$ (f) $10^2 = 100$ (g) $4^3 = 64$
13. (a) $(32)(256) = (2^5)(2^8) = 2^{13} = 8192$ (b) $\dfrac{16,384}{1024} = \dfrac{2^{14}}{2^{10}} = 2^4 = 16$

   (c) $64^2 = (2^6)^2 = 2^{12} = 4096$ (d) $\sqrt[5]{32,768} = 2^{15/5} = 2^3 = 8$

Exercise C.4

1. 1	**5.** -6	**9.** 7	**13.** 381,250	**17.** 38,125,000
3. 1	**7.** 5	**11.** 0.038125	**15.** 3812.5	**19.** 333.3

Exercise C.5

1. 1.5563	**5.** 4.5843	**9.** 2.9899	**13.** 3.36	**17.** 0.0742
3. $8.4082 - 10$	**7.** 1.5391	**11.** 141	**15.** 53,400	**19.** 7.943×10^{-7}

Exercise C.6

1. 0.5863	**5.** $8.0913 - 10$	**9.** 5.6594	**13.** 47.49	**17.** 0.2242
3. 2.9944	**7.** 4.5849	**11.** 736.5	**15.** 0.02849	**19.** 1.415

Exercise C.7

1. 913,000	**2.** 67,800 cu ft	**5.** 21.73	**7.** 139	**9.** 2.32

Exercise C.8

1. 8.26	**3.** 5.213	**5.** 5.52×10^{-11}	**7.** 9420	**9.** 209

Index

EQUIVALENTS

Note: in the following tables $\frac{3}{0}$, $\frac{4}{0}$, $\frac{5}{0}$, etc., indicate the number of zeros which follow the decimal point. Thus 1 ft = $0.\frac{3}{0}1894$ S.M. = 0.0001894 S.M.

Equivalent Lengths

	Inches (in.)	Feet (ft)	Yards (yd)	Rods (rd)	Chains (ch)	Statute miles	Nautical miles	Meters (m)	Kilometers (km)
1 in.	1	0.08333	0.02778	$0.\frac{2}{0}5051$	$0.\frac{2}{0}1263$	$0.\frac{4}{0}1578$	$0.\frac{4}{0}1371$	0.02540	$0.\frac{4}{0}254$
1 ft	12	1	0.33333	0.06061	0.01515	$0.\frac{3}{0}1894$	$0.\frac{3}{0}1645$	0.30480	$0.\frac{3}{0}304$
1 yd	36	3	1	0.18182	0.04545	$0.\frac{3}{0}5682$	$0.\frac{3}{0}4934$	0.91440	$0.\frac{3}{0}914$
1 rd	198	16.5	5.5	1	0.25	$0.\frac{2}{0}3125$	$0.\frac{2}{0}2714$	5.02921	$0.\frac{2}{0}502$
1 ch	792	66	22	4	1	0.01250	0.01085	20.1168	0.0201
1 S.M.	63360	5280	1760	320	80	1	0.86839	1609.35	1.6093
1 N.M.	72962.5	6080.20	2026.73	368.497	92.1243	1.15155	1	1853.25	1.8532
1 m	39.37	3.28083	1.09361	0.19884	0.04971	$0.\frac{3}{0}6214$	$0.\frac{3}{0}5396$	1	0.001
1 km	39370	3280.83	1093.61	198.838	49.7096	0.62137	0.53959	1000	1

Equivalent Areas

	Square inches	Square feet	Square yards	Square rods	Acres	Square miles	Square centimeters	Square meters	Square kilometers
1 sq in.	1	$0.\frac{2}{0}6944$	$0.\frac{3}{0}7716$	$0.\frac{4}{0}2551$	$0.\frac{6}{0}1594$	$0.\frac{9}{0}2491$	6.452	$0.\frac{3}{0}6452$	$0.\frac{9}{0}6452$
1 sq ft	144	1	0.11111	$0.\frac{2}{0}3673$	$0.\frac{4}{0}2296$	$0.\frac{7}{0}3587$	929.0	0.09290	$0.\frac{7}{0}929$
1 sq yd	1296	9	1	0.03306	$0.\frac{3}{0}2066$	$0.\frac{6}{0}3228$	8361.3	0.83613	$0.\frac{6}{0}8361$
1 sq rd	39204	272.25	30.25	1	0.00625	$0.\frac{5}{0}9766$	252930	25.2930	$0.\frac{4}{0}2529$
1 acre	6272640	43560	4840	160	1	$0.\frac{2}{0}1563$	40468700	4046.87	$0.\frac{2}{0}404$
1 sq mi	—	27878400	3097600	102400	640	1	—	2589999	2.5900
1 sq cm	0.1550	0.001076	$0.\frac{3}{0}1196$	$0.\frac{5}{0}3954$	—	—	1	$0.\frac{3}{0}1$	$0.\frac{9}{0}1$
1 sq m	1550.00	10.7639	1.19599	0.03954	$0.\frac{3}{0}2471$	$0.\frac{6}{0}3861$	10000	1	$0.\frac{5}{0}1$
1 sq km	—	10763867	1195985	39536.6	247.104	0.38610	—	1000000	1

Equivalent Volumes

	Cubic inches	Cubic feet	Cubic yards	Liquid quarts	Liquid gallons	Bushels	Cubic decimeters (liters)
1 cu in.	1	$0.\frac{3}{0}5787$	$0.\frac{4}{0}2143$	0.01732	$0.\frac{2}{0}4329$	$0.\frac{3}{0}4650$	0.0163
1 cu ft	1728	1	0.03704	29.9221	7.48055	0.80356	28.317
1 cu yd	46656	27	1	807.896	201.974	21.6962	764.55
1 liq qt	57.75	0.03342	$0.\frac{2}{0}1238$	1	0.25	0.02686	0.9463
1 liq gal	231	0.13368	$0.\frac{2}{0}4951$	4	1	0.10742	3.7854
1 bu	2150.42	1.24446	0.04609	37.2368	9.3092	1	35.239
1 ℓ	61.0234	0.03531	$0.\frac{2}{0}1308$	1.05668	0.26417	0.02838	1

Liquid quarts and gallons × 0.86 = dry quarts and gallons.

Equivalent Velocities and Accelerations

	Feet/second	Miles/hour	Knots	Meters/second	Kilometers/hour
1 ft/sec	1	0.68182	0.59209	0.30480	1.09728
1 mi/hr	1.46667	1	0.86839	0.44704	1.6093
1 knot	1.68894	1.15155	1	0.51479	1.8532
1 m/sec	3.28083	2.23693	1.94254	1	3.6
1 km/hr	0.91134	0.62137	0.53959	0.27778	1

Equivalent Forces, Weights, and Masses

(45° latitude at sea level)

	Grains	Ounces (avoir)	Pounds (avoir)	Short tons (2000 lb)	Dynes	Newtons	Grams	Kilograms
1 grain	1	$0.\frac{2}{0}2286$	$0.\frac{3}{0}1429$	$0.\frac{7}{0}7143$	63.546	$0.\frac{3}{0}6355$	0.0648	$0.\frac{4}{0}6480$
1 ounce	437.5	1	0.06250	$0.\frac{4}{0}3125$	28013.8	0.28014	28.35	0.02835
1 pound	7000	16	1	0.00050	444822	4.44822	453.59	0.45359
1 ton	14000000	32000	2000	1	—	8896.44	907185	907.185
1 dyne	0.015737	$0.\frac{4}{0}3534$	$0.\frac{5}{0}22481$	—	1	$0.\frac{4}{0}1000$	$0.\frac{2}{0}10197$	$0.\frac{5}{0}10197$
1 newton	1573.7	3.53397	0.22481	$0.\frac{3}{0}11241$	100000	1	101.97	0.10197
1 gram	15.4324	0.035274	$0.\frac{2}{0}22046$	$0.\frac{5}{0}1102$	980.67	$0.\frac{2}{0}98067$	1	0.00100
1 kg	15432.4	35.2740	2.20462	$0.\frac{2}{0}1102$	980665	9.80665	1000	1